The Princeton Review®

AP® BIOLOGY
PREMIUM PREP

2021 Edition

The Staff of The Princeton Review

PrincetonReview.com

Penguin
Random
House

The Princeton Review
110 East 42nd St., 7th Floor
New York, NY 10017
Email: editorialsupport@review.com

Published in the United States by Penguin Random House LLC,
New York, and in Canada by Random House of Canada,
a division of Penguin Random House Ltd., Toronto.

ISBN: 978-0-525-56942-8
eBook ISBN: 978-0-525-56980-0
ISSN: 2688-1810

AP is a trademark registered and owned by the College Board,
which is not affiliated with, and does not endorse, this product.

The Princeton Review is not affiliated with Princeton University.

Editor: Aaron Riccio
Production Editors: Sarah Litt and Emma Parker
Production Artist: Kris Ogilvie
Content Contributors: Katie Chamberlain, Gabrielle Budzon

Printed in the United States of America.

10 9 8 7 6 5 4 3 2 1

2021 Edition

Editorial

Rob Franek, Editor-in-Chief
David Soto, Director of Content Development
Stephen Koch, Survey Manager
Deborah Weber, Director of Production
Gabriel Berlin, Production Design Manager
Selena Coppock, Managing Editor
Aaron Riccio, Senior Editor
Meave Shelton, Senior Editor
Chris Chimera, Editor
Eleanor Green, Editor
Orion McBean, Editor
Brian Saladino, Editor
Patricia Murphy, Editorial Assistant

Penguin Random House Publishing Team

Tom Russell, VP, Publisher
Alison Stoltzfus, Publishing Director
Amanda Yee, Associate Managing Editor
Ellen Reed, Production Manager
Suzanne Lee, Designer

Acknowledgments

The Princeton Review would like to thank Katie Chamberlain,
Gabrielle Budzon, Deborah Weber, Kris Ogilvie, Sarah Litt, and
Emma Parker for their hard work on revisions to this edition.

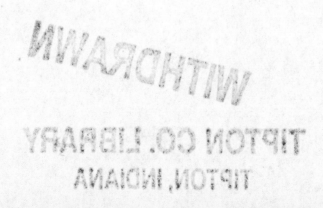

Contents

Get More (Free) Content
at **PrincetonReview.com/prep**

As easy as **1·2·3**

1 Go to PrincetonReview.com/prep and enter the following ISBN for your book:
9780525569428

2 Answer a few simple questions to set up an exclusive Princeton Review account. *(If you already have one, you can just log in.)*

3 Enjoy access to your **FREE** content!

Once you've registered, you can...

- Get our take on any late-breaking developments to the AP Biology Exam

- Get valuable advice about the college application process, including tips for writing a great essay and where to apply for financial aid

- If you're still choosing between colleges, use our searchable rankings of *The Best 386 Colleges* to find out more information about your dream school.

- Access a fifth and sixth practice test, as well as comprehensive study guides and a variety of printable resources, including bubble sheets and the AP Biology equation and formula tables

- Check to see if there have been any corrections or updates to this edition

Need to report a potential **content** issue?

Contact **EditorialSupport@review.com** and include:

- full title of the book
- ISBN
- page number

Need to report a **technical** issue?

Contact **TPRStudentTech@review.com** and provide:

- your full name
- email address used to register the book
- full book title and ISBN
- Operating system (Mac/PC) and browser (Firefox, Safari, etc.)

Look For These Icons Throughout The Book

 PREMIUM PORTAL

 ONLINE PRACTICE TESTS

 PROVEN TECHNIQUES

 APPLIED STRATEGIES

 OTHER REFERENCES

Part I
Using This Book
to Improve Your
AP Score

- Preview: Your Knowledge, Your Expectations
- Your Guide to Using This Book
- How to Begin

PREVIEW: YOUR KNOWLEDGE, YOUR EXPECTATIONS

Welcome to your *AP Biology Premium Prep, 2021 Edition*. Your route to a high score on the AP Biology Exam depends a lot on how you plan to use this book. To help you determine your approach, respond to the following questions.

1. Rate your level of confidence about your knowledge of the content tested by the AP Biology Exam.

    A. Very confident—I know it all
    B. I'm pretty confident, but there are topics for which I could use help
    C. Not confident—I need quite a bit of support
    D. I'm not sure

2. If you have a goal score in mind, circle your goal score for the AP Biology Exam.

    5   4   3   2   1   I'm not sure yet

3. What do you expect to learn from this book? Circle all that apply to you.

    A. A general overview of the test and what to expect
    B. Strategies for how to approach the test
    C. The content tested by this exam
    D. I'm not sure yet

YOUR GUIDE TO USING THIS BOOK

AP Biology Premium Prep, 2021 Edition is organized to provide as much—or as little—support as you need, so you can use this book in whatever way will be most helpful to improving your score on the AP Biology Exam.

*   The remainder of **Part I** will provide guidance on how to use this book and help you determine your strengths and weaknesses.

*   **Part II** of this book contains Practice Test 1 along with the answers and explanations. (A bubble sheet is provided before Test 1. Additional sheets can be found at the back of the book, and you can also print more out from your free online student tools.) We recommend that you take this test before going any further in order to realistically determine:
    o   your starting point right now
    o   which question types you're ready for and which you might need to practice
    o   which content topics you are familiar with and which you will want to carefully review

    Once you have nailed down your strengths and weaknesses with regard to this exam, you can focus your test preparation, build a study plan, and use your time efficiently.

- **Part III** of this book will:
  - provide information about the structure, scoring, and content of the AP Biology Exam
  - help you to make a study plan
  - point you toward additional resources

- **Part IV** of this book will explore:
  - how to attack multiple-choice questions
  - how to write high-scoring free-response answers
  - how to manage your time to maximize the number of points available to you

- **Part V** of this book covers the content you need for your exam.

- **Part VI** of this book contains Practice Tests 2, 3, and 4, and their answers and explanations. Compare your progress between these tests and with Practice Test 1. If you answer a certain type of question wrong several times, you probably need to review it. If you answer it wrong only once, you may have run out of time or been distracted. In either case, comparing your progress will allow you to focus on the factors that caused the discrepancy in scores and to be as prepared as possible on the day of the test.

You may choose to prioritize some parts of this book over others, or you may work through the entire book. Your approach will depend on your needs and how much time you have. Let's now examine how to make this determination.

HOW TO BEGIN

1. **Take Practice Test 1**

Before you can decide how to use this book, you need to take a practice test. Doing so will give you insight into your strengths and weaknesses, and the test will also help you make an effective study plan. If you're feeling test-phobic, remind yourself that a practice test is just a tool for diagnosing yourself—it's not how well you do that matters. As long as you try your best, you can glean invaluable information from your performance to guide your preparation.

So, before you read further, take Practice Test 1 starting at page 13 of this book. Be sure to finish in one sitting, following the instructions that appear before the test.

Note that the originally planned administration of the new AP Biology test was scheduled for May 11, 2020, but delayed because of the coronavirus. All of the content in this book accurately describes what the College Board plans for its exam, but in the event there are any late-breaking changes, please register your book online (it's free!) to see if there are any updates.

Go online to access two additional AP Biology practice exams, A and B.

2. **Check Your Answers**

Using the answer key on page 34, count the number of multiple-choice questions you answered correctly and the number you missed. Don't worry about the explanations for now and don't worry about why you missed questions. We'll get to that soon.

3. **Reflect on the Test**

After you take your first test, respond to the following questions:

- How much time did you spend on the multiple-choice questions?
- How much time did you spend on each free-response question?
- How many multiple-choice questions did you miss?
- Do you feel you had the knowledge to address the subject matter of the free-response questions?
- Do you feel you wrote well-organized, thoughtful responses to the free-response questions?

4. **Read Part III of this Book and Complete the Self-Evaluation**

As discussed in the Guide section above, Part III will provide information on how the test is structured and scored. It will also set out areas of content that are tested.

As you read Part III, re-evaluate your answers to the questions above. At the end of Part III, you will revisit and refine those questions. You will then be able to make a study plan, based on your needs and time available, that will allow you to use this book most effectively.

5. **Engage with Parts IV and V as Needed**

Notice the word *engage*. You'll get more out of this book if you use it intentionally than if you read it passively, hoping for an improved score through osmosis.

The strategy chapters in Part IV will help you think about your approach to the question types on this exam. Part IV will open with a reminder to think about how you approach questions now and then close with a reflection section asking you to think about how or whether you will change your approach in the future.

The content chapters in Part V are designed to provide a review of the content tested on the AP Biology Exam, including the level of detail you need to know and how the content is tested. You will have the opportunity to assess your mastery of the content of each chapter through test-appropriate questions.

I know we just said this on the previous page, but it's worth repeating! The strategy, content, and practice test chapters in this 2020 Edition were up-to-date, as of print, with the newly redesigned AP Biology test (May 11, 2020). Some scoring information was not yet available (as the test has not yet been administered), but if you register your book online, you'll have free access to any further updates.

6.   Take Another Test and Assess Your Performance

Once you feel you have developed the strategies you need and gained the knowledge you lacked, you should take Practice Test 2, which starts on page 329 of this book. You should finish in one sitting, following the instructions at the beginning of the test.

When you are done, check your answers to the multiple-choice sections. See if a teacher will read your essays and provide feedback.

Once you have taken the test, reflect on what areas you still need to work on and revisit the chapters in this book that address those deficiencies. Once you feel confident, take Practice Test 3, and repeat the process. Take Practice Test 4, and repeat the process again. Through this type of reflection and engagement, you will continue to improve.

7.   Keep Working

After you have revisited certain chapters in this book, continue the process of testing, reflection, and engaging with the content in this book and online. Consider what additional work you need to do and how you will change your strategic approach to different parts of the test.

As we will discuss in Part III, there are other resources available to you, including a wealth of information at AP Students, the official site of the AP Exams. You can continue to explore areas that can stand improvement and engage in those areas right up to the day of the test.

Part II
Practice Test 1

Practice Test 1

Completely darken bubbles with a No. 2 pencil. If you make a mistake, be sure to erase mark completely. Erase all stray marks.

1. YOUR NAME:
(Print)

Last   First   M.I.

SIGNATURE: _____ DATE: _____ / ___ /

HOME ADDRESS:
(Print)

Number and Street

City   State   Zip Code

PHONE NO. :
(Print)

IMPORTANT: Please fill in these boxes exactly as shown on the back cover of your test book.

2. TEST FORM

3. TEST CODE

4. REGISTRATION NUMBER

5. YOUR NAME

First 4 letters of last name | FIRST INIT | MID INIT

6. DATE OF BIRTH

Month	Day	Year
◯ JAN		
◯ FEB		
◯ MAR		
◯ APR		
◯ MAY		
◯ JUN		
◯ JUL		
◯ AUG		
◯ SEP		
◯ OCT		
◯ NOV		
◯ DEC		

7. SEX
◯ MALE
◯ FEMALE

The Princeton Review®

1 Ⓐ Ⓑ Ⓒ Ⓓ
2 Ⓐ Ⓑ Ⓒ Ⓓ
3 Ⓐ Ⓑ Ⓒ Ⓓ
4 Ⓐ Ⓑ Ⓒ Ⓓ
5 Ⓐ Ⓑ Ⓒ Ⓓ
6 Ⓐ Ⓑ Ⓒ Ⓓ
7 Ⓐ Ⓑ Ⓒ Ⓓ
8 Ⓐ Ⓑ Ⓒ Ⓓ
9 Ⓐ Ⓑ Ⓒ Ⓓ
10 Ⓐ Ⓑ Ⓒ Ⓓ
11 Ⓐ Ⓑ Ⓒ Ⓓ
12 Ⓐ Ⓑ Ⓒ Ⓓ
13 Ⓐ Ⓑ Ⓒ Ⓓ
14 Ⓐ Ⓑ Ⓒ Ⓓ
15 Ⓐ Ⓑ Ⓒ Ⓓ
16 Ⓐ Ⓑ Ⓒ Ⓓ

17 Ⓐ Ⓑ Ⓒ Ⓓ
18 Ⓐ Ⓑ Ⓒ Ⓓ
19 Ⓐ Ⓑ Ⓒ Ⓓ
20 Ⓐ Ⓑ Ⓒ Ⓓ
21 Ⓐ Ⓑ Ⓒ Ⓓ
22 Ⓐ Ⓑ Ⓒ Ⓓ
23 Ⓐ Ⓑ Ⓒ Ⓓ
24 Ⓐ Ⓑ Ⓒ Ⓓ
25 Ⓐ Ⓑ Ⓒ Ⓓ
26 Ⓐ Ⓑ Ⓒ Ⓓ
27 Ⓐ Ⓑ Ⓒ Ⓓ
28 Ⓐ Ⓑ Ⓒ Ⓓ
29 Ⓐ Ⓑ Ⓒ Ⓓ
30 Ⓐ Ⓑ Ⓒ Ⓓ
31 Ⓐ Ⓑ Ⓒ Ⓓ
32 Ⓐ Ⓑ Ⓒ Ⓓ

33 Ⓐ Ⓑ Ⓒ Ⓓ
34 Ⓐ Ⓑ Ⓒ Ⓓ
35 Ⓐ Ⓑ Ⓒ Ⓓ
36 Ⓐ Ⓑ Ⓒ Ⓓ
37 Ⓐ Ⓑ Ⓒ Ⓓ
38 Ⓐ Ⓑ Ⓒ Ⓓ
39 Ⓐ Ⓑ Ⓒ Ⓓ
40 Ⓐ Ⓑ Ⓒ Ⓓ
41 Ⓐ Ⓑ Ⓒ Ⓓ
42 Ⓐ Ⓑ Ⓒ Ⓓ
43 Ⓐ Ⓑ Ⓒ Ⓓ
44 Ⓐ Ⓑ Ⓒ Ⓓ
45 Ⓐ Ⓑ Ⓒ Ⓓ
46 Ⓐ Ⓑ Ⓒ Ⓓ
47 Ⓐ Ⓑ Ⓒ Ⓓ
48 Ⓐ Ⓑ Ⓒ Ⓓ

49 Ⓐ Ⓑ Ⓒ Ⓓ
50 Ⓐ Ⓑ Ⓒ Ⓓ
51 Ⓐ Ⓑ Ⓒ Ⓓ
52 Ⓐ Ⓑ Ⓒ Ⓓ
53 Ⓐ Ⓑ Ⓒ Ⓓ
54 Ⓐ Ⓑ Ⓒ Ⓓ
55 Ⓐ Ⓑ Ⓒ Ⓓ
56 Ⓐ Ⓑ Ⓒ Ⓓ
57 Ⓐ Ⓑ Ⓒ Ⓓ
58 Ⓐ Ⓑ Ⓒ Ⓓ
59 Ⓐ Ⓑ Ⓒ Ⓓ
60 Ⓐ Ⓑ Ⓒ Ⓓ

AP® Biology Exam

SECTION I: Multiple-Choice Questions

DO NOT OPEN THIS BOOKLET UNTIL YOU ARE TOLD TO DO SO.

At a Glance

Total Time
1 hour and 30 minutes
Number of Questions
60
Percent of Total Score
50%
Writing Instrument
Pencil required

Instructions

Section I of this examination contains 60 multiple-choice questions.

Indicate all of your answers to the multiple-choice questions on the answer sheet. No credit will be given for anything written in this exam booklet, but you may use the booklet for notes or scratch work. After you have decided which of the suggested answers is best, completely fill in the corresponding oval on the answer sheet. Give only one answer to each question. If you change an answer, be sure that the previous mark is erased completely. Here is a sample question and answer.

Sample Question   Sample Answer

Chicago is a

(A)   state
(B)   city
(C)   country
(D)   continent

Use your time effectively, working as quickly as you can without losing accuracy. Do not spend too much time on any one question. Go on to other questions and come back to the ones you have not answered if you have time. It is not expected that everyone will know the answers to all the multiple-choice questions.

About Guessing

Many candidates wonder whether or not to guess the answers to questions about which they are not certain. Multiple-choice scores are based on the number of questions answered correctly. Points are not deducted for incorrect answers, and no points are awarded for unanswered questions. Because points are not deducted for incorrect answers, you are encouraged to answer all multiple-choice questions. On any questions you do not know the answer to, you should eliminate as many choices as you can, and then select the best answer among the remaining choices.

BIOLOGY
SECTION I
60 Questions
Time—90 minutes

Directions: Each of the questions or incomplete statements below is followed by four suggested answers or completions. Select the one that is best in each case and then fill in the corresponding oval on the answer sheet.

Questions 1–5 refer to the following passage.

The following table lists the intracellular osmolality concentrations of four osmoconforming organisms that are known to mimic the osmotic conditions of their surroundings.

Table 1. Osmolality concentrations of osmoconforming organisms

Organism 1	100 mOsm kg^{-1}
Organism 2	400 mOsm kg^{-1}
Organism 3	500 mOsm kg^{-1}
Organism 4	150 mOsm kg^{-1}
Organism 5	350 mOsm kg^{-1}

1. If a large amount of salt was added to the sealed tank where organism 4 was being kept, what would be the effect on the intracellular osmolality of organism 4?

    (A) Increase initially and then return to the initial state
    (B) Increase and maintain the increased state
    (C) Decrease and then return to the initial state
    (D) Decrease and maintain the decreased state

2. Which of the following would likely have the largest direct effect on data in the table?

    (A) A mutation in a cytoskeleton protein
    (B) A mutation in a protein in the sodium-potassium pump complex
    (C) A mutation in a lipid in the outer chloroplast membrane
    (D) A mutation in a nuclear pore channel protein

3. Which could decrease the osmolalities of the organisms?

    (A) Sodium is added to the sealed tank.
    (B) Chloride is injected into the organism.
    (C) Glucose is ingested by the organism.
    (D) Water is added to the sealed tank.

4. Which of the following statements best describes the insides of these organisms compared to the environment in the tank when they reach osmoconformity?

    (A) Due to the influx of water into the organisms' cells, the organisms' insides are hypertonic when compared to the tank.
    (B) Compared to the tank, the loss of water out of the organisms' cells makes the organisms' insides hypotonic.
    (C) The organisms' insides are isotonic since the intracellular osmolality concentrations match the osmolality concentrations of the tank.
    (D) The flow of water into and out of the organisms' insides causes the cells to be retrotonic.

5. A population of organisms that are not osmoconformers is dropped into the tank with Organism 1. The intracellular osmolality of Organism 1 begins to slightly decrease. Which of the following could explain this?

    (A) The new organisms were hypotonic compared to the tank and they lost water via osmosis.
    (B) The new organisms expelled waste products that increased the tank's osmolality.
    (C) The new organism competed with organism 1 for food and caused it to become active.
    (D) The new organism ingested large amounts of fluid and the tank fluid level decreased.

GO ON TO THE NEXT PAGE.

Questions 6–9 refer to the following passage.

The active site of the enzyme Ritzolinine (RZN45) contains three positively charged lysine residues. When ascorbic acid is present, binding of JB-76, the substrate of RZN45, decreases. The reaction rate is affected by the presence of ascorbic acid as shown in the figure below. It is thought that a daily supplement of Vitamin C might aid those suffering from Ritzolierre's Disease, which is caused by elevated levels of RZN45.

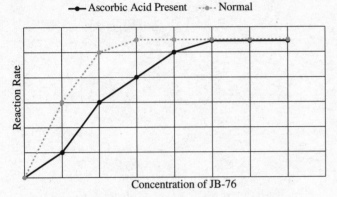

Figure 1. Reaction rate based on ascorbic acid

6. Which of the following is likely true?

   (A) RZN45 and JB-76 have similarly charged amino acids at their active sites.
   (B) Ritzolinine is stabilized in the presence of Vitamin C.
   (C) Ascorbic acid and JB-76 each have a pocket of negatively charged amino acids.
   (D) JB-76 and ascorbic acid have an identical number of amino acids.

7. Which is the best explanation for the differing effects on reaction rate caused by ascorbic acid at low and high concentrations of JB-76?

   (A) At high concentrations of JB-76, there are fewer free active sites for the ascorbic acid to bind to.
   (B) At low concentrations of JB-76, there are fewer free allosteric sites for the RZN45 to bind to.
   (C) At high concentrations of JB-76, there are more free allosteric sites for the ascorbic acid to bind to.
   (D) At low concentrations of JB-76, there are more free active sites for the RZN45 to bind to.

8. A patient with Ritzolierre's Disease would likely benefit from which of the following:

   I.   Injections of RZN45
   II. Injections of JB-76
   III. Injections of ascorbic acid

   (A) I only
   (B) II only
   (C) III only
   (D) II and III

9. If Figure 1 were to include additional data for higher concentrations of JB-76, how would the reaction rate change?

   (A) The reaction rate would be lower than any reaction rate shown in Figure 1.
   (B) The reaction rate would be higher than any reaction rate shown in Figure 1.
   (C) The reaction rate would be equal to the highest reaction rate shown in Figure 1.
   (D) The reaction rate would increase gradually for each concentration of JB-76 added.

10. The Galapagos Islands contain about 15 species of finches that vary in terms of beak size and strength. The differences are believed to have occurred as a result of allopatric speciation acting on the finch populations. Which of the following best explains how the finch populations developed different beaks?

   (A) As finches migrated between islands, alleles were transferred between populations through the process of gene flow.
   (B) Genetic drift occurred due to a hurricane near the islands causing a bottleneck effect.
   (C) Through convergent evolution, the finches developed comparable features due to being exposed to comparable selective pressures.
   (D) Geographic isolation between the different islands kept the populations from breeding with each other, and each population evolved separately.

GO ON TO THE NEXT PAGE.

11. Figure 1 shows a cell at the onset of meiosis (just prior to initiation of prophase I) and the resulting 4 gametes achieved at the end of meiosis. Select the statement that is likely true.

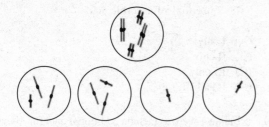

Figure 1. Resulting gametes from meiosis

(A) Meiosis occurred as it should.
(B) There was a nondisjunction event in Meiosis I.
(C) There was a nondisjunction event in Meiosis II.
(D) Two Barr bodies were produced instead of four gametes.

Questions 12–16 refer to the following passage.

A gene responsible for production of hair pigment in dogs is called Fursilla (frsl). A map of the Fursilla locus is shown in Figure 1, below. When expressed, it results in darkly pigmented dog fur. When unexpressed, the hair is devoid of pigmentation and appears pure white. Expression of Fursilla (frsl) depends on the binding/lack of binding of several proteins: Nefur (NEFR), Lesfur (LSFR), and Dirkfur (DRKFR). Figure 2 shows the relative levels of frsl transcript as measured via RT-qPCR when each protein is overexpressed or unexpressed in the cell.

Start site

| 1220 | 1295 | | 1617 |

Figure 1. Map of the Fursilla region of chromosome 8

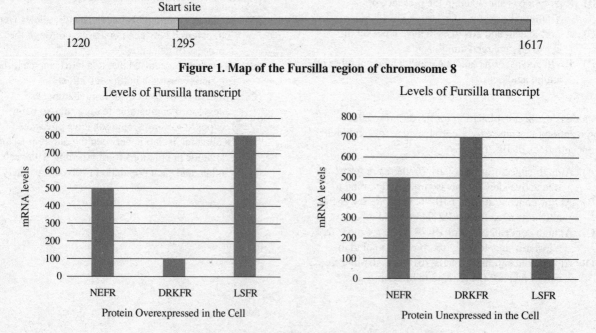

Figure 2: Relative amounts of Fursilla transcript

GO ON TO THE NEXT PAGE.

12. Scientists identified a mutant allele of the Fursilla gene caused by a deletion of bases 1231-1295 on chromosome 8. What is the likely effect of this mutation?

    (A) No transcription and no translation of Fursilla will occur.
    (B) Transcription will not occur, but translation will be unaffected.
    (C) Transcription of Fursilla will occur, but no translation will occur.
    (D) Both transcription and translation of Fursilla will be unaffected.

13. Which of the following would likely produce a dog with the darkest fur?

    (A) High levels of LSFR and high levels of DRKFR
    (B) High levels of LSFR and low levels of DRKFR
    (C) Low levels of LSFR and high levels of DRKFR
    (D) Low levels of LSFR and low levels of DRKFR

14. Based on the data, which of the following most likely describes how the binding of proteins affects the production of Fursilla transcript?

    (A) Because Fursilla transcript levels change when certain proteins are overexpressed or unexpressed in the cell, Fursilla must interfere with the other proteins.
    (B) Because Fursilla transcript levels stay the same whether NEFR is overexpressed or unexpressed, NEFR must bind inhibit the production of Fursilla transcript.
    (C) Because Fursilla transcript levels decrease when LSFR is unexpressed in the cell, LSFR must compete for the binding site or RNA polymerase.
    (D) Because Fursilla transcript levels increase when DRKFR is unexpressed in the cell, DRKFR must compete for the binding site of RNA polymerase.

15. An additional protein, Baldidog (BLD), was identified and found to interact with the DRKFR protein. When they bind together, it has been shown that they fit together because of:

    (A) Complementary nucleotides sequences
    (B) Amino acid pockets with complementary conformations
    (C) Opposing regions rich in cytosines and thymines
    (D) Hydrophobic lipids and hydrophobic molecules

16. How would the overexpression of BLD affect the transcript levels of Fursilla?

    (A) Fursilla transcript levels would increase.
    (B) Fursilla transcript levels would decrease.
    (C) Fursilla transcript levels would be unaffected.
    (D) Fursilla transcript would completely disappear.

17. Water molecules experience intermolecular forces when interacting with each other. Which of the following would most likely occur if water became a nonpolar molecule?

    (A) Polar solutes would dissolve in water forming aqueous solutions of various concentrations.
    (B) The amount of liquid water on the planet would increase, and the amount of water vapor in the air would decrease.
    (C) The melting and boiling points of water would increase.
    (D) Solid water would be more dense than liquid water, so ice would no longer float.

Questions 18–21 refer to the following passage.

A chain of three small islands was found to be the home to a small species of mouse. The angle of jaw opening was found to vary significantly. The average size angle of the maximal jaw opening found in mice at 10 locations on the three small islands is shown in Figure 1, below.

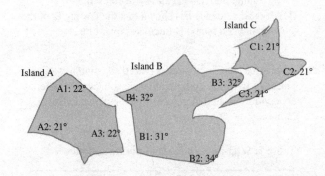

Figure 1. Angles of maximal jaw opening (in degrees) for mice found at various island positions

18. Which of the following might account for the variations in maximal jaw angle?

    (A) Different wind speeds on each island
    (B) Different heights of trees on each island
    (C) Different sizes of seeds on each island
    (D) Differences in altitude on each island

19. If the mice from Island A and the mice from Island B were placed together, what would likely happen?

    (A) The mice would mate, but it is impossible to predict jaw angle.
    (B) The mice would mate, and the jaw angle would be approximately 27°.
    (C) The mice would mate, and the jaw angle would be either 21° or 32°.
    (D) The mice would not be capable of mating.

GO ON TO THE NEXT PAGE.

20. Skeletons have been found that indicate that mice with jaw angles less than 30 degrees were once found on Island B. Which explanation best accounts for this.

(A) A small jaw angle hinders the survival of mice on Island B.
(B) A large jaw angle hinders the survival of mice on Island A.
(C) A small jaw angle encourages the survival of mice on Island C.
(D) A large jaw angle encourages the survival of mice on Island A.

21. Which of the following statements best supports that the jaw angle on Island B changed via punctuated equilibrium?

(A) Fossil evidence has shown that over time the jaw angle on Island B slowly increased.
(B) Fossil evidence has shown that the jaw angle of the mice on Island B increased very quickly.
(C) Fossil evidence has shown that over time the jaw angle on Island B slowly decreased.
(D) Fossil evidence has shown that the jaw angle of the mice on Island B fluctuates over time.

22. A scientist studying osmosis placed six similarly sized pieces of potato in sodium chloride solutions of different concentrations and then measured the percent change in mass.

Table 1. % change in mass of potato samples

Sodium chloride concentration	% change in mass
0.1 M	+17.5
0.3 M	+5.0
0.5 M	-7.5
0.7 M	-16.2
0.9 M	-26.1
1.1 M	-25.1

The percent change in mass is closest to 0% at which of the following approximate concentrations of the solutions?

(A) 0.01 M
(B) 0.25 M
(C) 0.3 M
(D) 0.45 M

Questions 23–26 refer to the following passage.

Proteins often need the equivalent of a shipping label so that they can be sent intracellularly or extracellularly to the correct location. These labels are typically found in three forms: signal sequences, target signals, or localization signals. A signal sequence is a stretch of hydrophobic amino acid residues that is only located within polypeptides destined to fold into EITHER extracellular proteins, intermembrane proteins that will anchor to the cell surface, or proteins that will be members of the secretory pathway that packs and ships things towards the cell surface. A targeting signal is a sequence that identifies the proteins of the secretory pathway. A localization signal is a sequence that labels proteins that are destined to go to specific organelles (that are not part of the secretory pathway). The table below shows seven eukaryotic proteins, and the sequences have been identified in each.

Table 1. Sequences identified within eukaryotic proteins

Protein	Signal Sequence	Targeting Signal	Localization Sequence
HRIET1	X	X	
HAZL2			
NUH8			X
K8TE	X		
TELEE	X	X	
TMSDG	X	X	
LNACT			X

23. Which of following could be a description for HRIET1?

(A) A protein found in the blood that plays a role in the immune system
(B) A protein found in the Golgi apparatus that helps tether vesicles
(C) A protein found in the nucleolus aiding in ribosome assembly
(D) A protein found in the cytosol serving as a cytoskeletal anchor

24. One of the proteins was found to be a receptor for a large protein that is produced in the brain and travels through the blood until it can dock at the target cells that it stimulates. Which of the proteins is likely this receptor?

(A) TMSDG
(B) NUH8
(C) HAZL2
(D) K8TE

GO ON TO THE NEXT PAGE.

25. Two proteins were identified in the electron transport chain. One was found to be active at the beginning of the chain, and the other was found to be active at the final stages of the chain. Which two proteins were these?

    (A) HRIET1 and HAZL2
    (B) NUH8 and LNACT
    (C) TELEE and K8TE
    (D) TMSDG and TELEE

26. If a scientist was hoping to isolate and purify HAZL2, which of the following techniques would they use?

    (A) Collect only the nuclear fraction
    (B) Collect only the cytosolic fraction
    (C) Collect only the extracellular fraction
    (D) Collect only the lipid membrane fraction

Questions 27–31 refer to the following passage.

In a heavily populated suburb, two cougars were once spotted roaming in a small field. Local wildlife experts, though not surprised, warned the public to be aware of their surroundings and to keep small pets protected. A group of local junior high school students were curious about the population of cougars in the area since no one they asked had ever seen one in the area. With the help of local wildlife enthusiasts and carefully placed motion-activated wildlife cameras, the group of students recorded sightings of animals in a local forest preserve for their entire four years of high school. The results are shown below.

All Animals

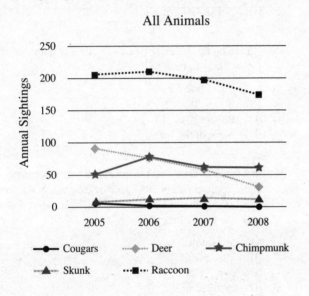

Cougars

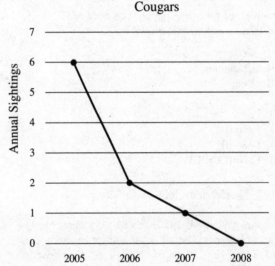

27. Raccoons eat a diet rich in berries. Which statement is the most likely to describe the berry population?

    (A) A hard frost eliminated nearly all of the berries in 2007, but 2008 was a milder winter.
    (B) A virus was introduced in 2006 that eliminated a fly using the berry bushes as a niche habitat.
    (C) Cougars destroyed much of the berry habitat in 2005.
    (D) The berry population remained steady, with a slight decrease due to lower rainfall in 2008.

28. Although the data is limited, which of the following populations is the most likely to be a primary food source for the cougars?

    (A) Skunk
    (B) Deer
    (C) Raccoon
    (D) Chipmunk

29. Which change will cause fewer skunks to be sighted in the future?

    (A) An increase in the number of cougars
    (B) An increase in the number of raccoons
    (C) An increase in the number of deer
    (D) It is impossible to make this determination

GO ON TO THE NEXT PAGE.

30. It has been shown that the population size of chipmunks is directly tied to the number of acorns dropped. This can be scientifically summarized by which of the following statements?

    (A) Acorn number is a density-dependent population factor and affects the carrying capacity of the chipmunk population.
    (B) Acorn number is a density-dependent population factor and does not affect the carrying capacity of the chipmunk population.
    (C) Acorn number is a density-independent population factor and affects the carrying capacity of the chipmunk population.
    (D) Acorn number is a density-independent population factor and does not affect the carrying capacity of the chipmunk population.

31. Which of the following populations does NOT demonstrate exponential growth?

    I. Cougar
    II. Chipmunk
    III. Deer

    (A) I only
    (B) I and II
    (C) I and III
    (D) I and II and III

Question 32–35 refer to the following passage.

A certain gene has been identified on chromosome 2 in a species of butterfly. The mRNA transcript has been found to be 1578 base pairs in length. A portion of the transcript is shown below. The area shown in grey is part of a known ribosome binding site.

5' UUACGUGCAUGCGUAGCUAGCUAGCUAUGCUAUCGCUUUAAGAGCAAAAAA 3'
15   30   45

**Figure 1. Portion of the mRNA transcript
of a species of butterfly**

32. If bases 0–15 were deleted, what would be the likely consequence?

    (A) No transcription would be allowed to occur.
    (B) No translation would be allowed to occur.
    (C) Neither transcription nor translation would occur.
    (D) No consequence would occur as these bases are non-coding.

33. A transcript was identified in the nucleus that is similar to this transcript. How is it likely different from the mRNA transcript?

    (A) The nuclear transcript likely contained introns.
    (B) The nuclear transcript likely contained exons.
    (C) The nuclear transcript likely contained a poly-A tail.
    (D) The nuclear transcript likely contained a 3' GTP-Cap.

34. RNA polymerase transcribes DNA into mRNA by matching base pairs of nucleotides. Which of the following would be the segment of DNA on the coding strand corresponding to bases 0–6 in Figure 1?

    (A) 5' AAUGCA 3'
    (B) 5' TTACGT 3'
    (C) 3' AATGCA 5'
    (D) 3' TTUGCT 5'

35. Where on the strand shown in Figure 1 is a transcription factor likely to bind?

    (A) Bases 0 to 15
    (B) Bases 20 to 25
    (C) Bases -20 to -15
    (D) None of the above

36. In a small ecosystem, it is estimated that producers equal approximately 2987 g/m^2 of biomass. What would be a good guess of how much biomass is found within secondary consumers?

    (A) 5974 g/m^2
    (B) 2987 g/m^2
    (C) 298.7 g/m^2
    (D) 29.87 g/m^2

GO ON TO THE NEXT PAGE.

Question 37–40 refer to the following passage.

The enzymatic catalysis of a reaction essential in the production of dog saliva is mediated by the protein ARKKKK60491. ARKKKK60491 is a homodimer, with each subunit being 337 amino acids. Figure 1, below, indicates a modular structure of the homodimer. Three positions are indicated on the model. Position A is the active site. Position B is a region known to have a large number of nonpolar residues. Position C is known to have a large number of charged residues.

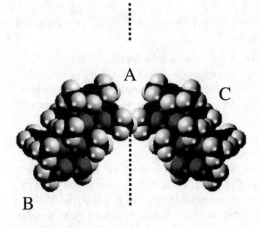

Figure 1. Modular structure of the homodimer

37. A mutation is discovered in the gene for ARKKKK60491 that converts positively charged lysine residue into negatively charged glutamic acid residue. This single change directly impacts the location where the substrate to ARKKKK60491 attaches during catalysis. Which position is likely affected by this change?

(A) Position A
(B) Position B
(C) Position C
(D) None of the positions will be affected

38. Which of the following is Position B most likely to assist the protein with?

(A) Stabilization of the substrate into the transition state during catalysis
(B) Transportation of catalyzed cargo from the nucleus to the mitochondria
(C) Creation of a homodimer through catalysis of homodimerization
(D) Attachment to the membrane by embedding into the phospholipid bilayer

39. Which of the following statements best predicts the effect on the protein's structure if the subunits of the homodimer fail to attach?

(A) The primary structure will change because the order of amino acids will differ.
(B) The secondary structure will differ because hydrogen bonds will occur in different locations and the protein will fold into a different shape.
(C) The tertiary structure will be destroyed because the side chain interactions will not occur and the protein will not fold into a 3-dimensional shape.
(D) The quaternary structure will be destroyed because the different amino acid chains will not join together.

40. According to Figure 1, within the protein ARKKKK60491, which of the following locations contain polypeptide N-termini?

I. Position A
II. Position B
III. Position C

(A) I only
(B) I and II
(C) II and III
(D) I and III

GO ON TO THE NEXT PAGE.

Questions 41–43 refer to the following passage.

There are four major types of macromolecules, and each type has specific functions and characteristics. Proteins are made of amino acids and function as enzymes, receptors, hormones, signals, and receptors. Lipids are made of fatty acids and glycerol and function as membrane components and energy storage. Carbohydrates are made of monosaccharides and function as energy storage. Nucleic acids are made of nucleotides and provide genetic information for organisms.

41. Fats are a type of lipid that are generally soluble in organic solvents and generally insoluble in water. Which of the following questions will best guide the researchers to determine if a substance is a fat?

(A) At what temperature does the substance boil?
(B) What type of mixture does the substance form when mixed with water?
(C) Does treating the substance with acid affect its appearance?
(D) Have chemical reactions been recorded between the substance and other chemicals?

42. Sugars are a type of carbohydrate typically made up of hydrogen, carbon, and oxygen. The hydrogen:oxygen ratio is typically 2:1. Which of the following statements explains how carbohydrates function as a stored form of energy?

(A) Excess glucose is stored as glycogen in muscles and the liver.
(B) Glucose is broken down during photosynthesis to act as an energy source.
(C) Glycerol is released into the bloodstream when stored fat is used as an energy source.
(D) Glucagon is a hormone that stores excess energy in the pancreas.

43. All four types of macromolecules are organic compounds. Which of the following statements describes a difference between organic and inorganic compounds?

(A) Only organic compounds contain oxygen.
(B) Only organic compounds contain covalently bonded carbon atoms.
(C) Only organic compounds contain hydrogen bonds.
(D) Only organic compounds contain ionizing chemical groups.

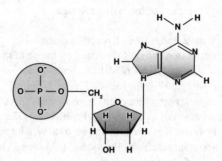

Figure 1. Structure of adenine

44. Which of the following could NOT be represented by Figure 1?

(A) A monomer used to build a protein
(B) A nucleotide
(C) A reactant used to synthesize nucleic acids
(D) A chemical used to relay genetic information

Questions 45–49 refer to the following passage.

The Dandranian Nettle is a plant native to the Eemagee nation. The nettle produces tiny flowers that come in two colors: white and pink. The petals of the flowers come in two varieties: ruffled or slotted. After years of studying them, it has been widely acknowledged that both traits follow classic Mendelian inheritance patterns of dominance. The table below shows floral phenotypes in a small garden in an isolated greenhouse, where the dominant phenotypes are the most plentiful and researchers assume the population is in Hardy Weinberg equilibrium.

Table 1. Population of various floral phenotypes

Phenotype	#
White	4900
Pink	400
Ruffled	700
Slotted	4600

45. A true-breeding white nettle is crossed with a true-breeding pink nettle. What will be the ratio of the offspring?

(A) 100% White
(B) 3 White: 1 Pink
(C) 1 White: 1 Pink
(D) 1 White: 3 Pink

GO ON TO THE NEXT PAGE.

46. The F1 generation shown in Table 1 was crossed with a true-breeding pink nettle. What percentage of offspring would have pink flowers?

    (A) 0%
    (B) 25%
    (C) 50%
    (D) 100%

47. A cross is performed between one nettle that is known to be heterozygous for both traits and a nettle that is pink and ruffled. Which of the following would lead researchers to believe the traits are linked?

    (A) Several white flowers are identified with ruffled petals.
    (B) Nearly all white flowers also have slotted petals and nearly all pink flowers have ruffled petals.
    (C) Petals are discovered that are both ruffled and slotted.
    (D) Flowers are discovered that have two colors of flowers and two shapes of petals.

48. What frequency of the alleles in the greenhouse garden are recessive?

    (A) 0.14
    (B) 0.27
    (C) 0.38
    (D) 0.56

49. What is the frequency of plants in the greenhouse garden that are heterozygous for the flower color trait?

    (A) 0.27
    (B) 0.39
    (C) 0.5
    (D) 0.74

GO ON TO THE NEXT PAGE.

Question 50–53 refer to the following passage.

G-protein coupled receptors (GPCRs) are a common type of protein receptor containing seven transmembrane segments and bearing intracellular and extracellular portions. The structure of a typical GPCR and its associated partners is shown below. GPCRs are coupled to G-proteins consisting of three subunits (α, β and γ) that bind the nucleotides GTP or GDP on the intracellular side of the GPCR. The presence or absence of a ligand on the extracellular side of the GPCR determines whether GDP or GTP will bind to the intracellular G-protein. Extracellular ligand binding initiates GDP being exchanged for GTP and the separation of the G-protein. One subunit will travel to join with a secondary partner and will initiate a cascade of signaling effects within the cell. In the example shown below, the secondary partner is adenylyl cyclase and the signaling is cAMP upregulation.

Figure 1. G-protein coupled receptor joining with adenylyl cyclase

50. A setup like a GPCR is unnecessary in which of the following situations?

(A) The signaling event is very specific
(B) The ligand is a hormone traveling in the bloodstream
(C) The effector molecules are located inside the cytoplasm
(D) The ligand is a small nonpolar molecule

51. Based on Figure 1, which of the following is the first step in the activation of cAMP-dPK?

(A) The exchange of GTP for GDP
(B) The binding of epinephrine
(C) The separation of α-GTP
(D) Phosphorylation of intracellular enzymes

52. Which of the following situations is most similar to extracellular binding to a GPCR?

(A) A police officer pulling over a speeding car on the expressway
(B) A celebrity chef visiting the kitchen of another restaurant
(C) A driver giving their order to a person through a drive-up window
(D) A taxi driver picking up a group of passengers

53. What does a GPCR's capability of making a ligand-induced conformational change allow?

(A) Binding of the ligand to the phospholipid head groups within the membrane
(B) Communication between the outside of the cell and the inside of the cell
(C) A solid attachment to the intermembrane region of the cell membrane
(D) Intracellular binding of the GDP-associated G-protein complex

Question 54–58 refer to the following passage.

A small family farm has documented Tonduly carrot growth for a hundred years in a Midwest microclimate. The average lengths and masses of the carrots are shown in the table below. The same ecosystem was home to a species of vegetable weevil that attacks the carrots underground. Other than carrots, the weevils also preyed on radishes found in the ecosystem. This particular species of weevil spends its whole life beneath the soil, and the average depth at which these weevils were found each year is shown in the table below.

Table 1. Average lengths and masses of carrots and depth at which weevils are found (by decade)

Year	1920	1930	1940	1950	1960	1970	1980	1990	2000	2010	2020
Average carrot mass (g)	61	62	61	65	58	61	60	62	65	61	59
Average carrot length (cm)	21.5	21.2	20.3	19.8	19.0	18.7	18.1	17.6	17.5	17.9	17.7
Average weevil depth (cm)	19.2	20.2	20.0	19.7	19.3	20.3	19.4	20.2	20.1	19.8	19.9

54. Which carrot phenotype was selected for over time?

(A) Thinner carrots
(B) Longer carrots
(C) Shorter carrots
(D) Heavier carrots

55. If a weevil were introduced to the ecosystem that lived at a depth of 17 centimeters, how would the carrot length change in the coming years?

(A) It would likely return to its original long size.
(B) It would likely get longer than it has been in the past.
(C) It would likely get shorter than it has been in the past.
(D) It would not likely change at all with a weevil at 17 centimeters.

56. The average carrot consumer prefers long carrots. What should be done to promote the growth of longer carrots?

I. Remove the weevil population
II. Cross the longest carrot of each generation
III. Introduce a different carrot predator that lives at 10 centimeters

(A) I only
(B) I and II
(C) II and III
(D) I and II and III

57. If true, which of the following would account for the average weevil depth changing to 17 centimeters in 2030?

(A) The average carrot length changed to 15 centimeters.
(B) The radishes completely disappeared from the garden.
(C) Goats arrived to eat the tops of the carrot plants above the surface.
(D) A small percentage of the weevil population died from a virus.

58. According to Table 1, which of the following traits is most linked to the carrot's fitness?

(A) Weight
(B) Length
(C) Color
(D) Flavor

59. A dance by the Lenoxian bird is one of the most highly complex mating rituals. Precision and flamboyancy are key to wooing a mate. The natural predator of the Lenoxian bird also enjoys the spirited dance and often strikes during the courtship ritual. Which of the following statements is likely true?

(A) Sexual selection of the most exciting dancers is the driving selective pressure.
(B) Selective pressure against the exciting dancers is the driving selective force.
(C) The Lenoxian bird will soon become extinct if the predatory action does not cease.
(D) There can exist a careful balance between multiple opposing selective pressures.

60. Antibiotic resistance occurs when drugs or chemicals used to treat an illness become less effective. Some varieties of *Neisseria gonorrhoeae* have become resistant to the antibiotic penicillin. Which of the following best explains how the bacteria became resistant?

(A) The bacteria experienced hybrid vigor through gaining improved traits from their parents.
(B) Through natural selection, bacteria with a random mutation that provided resistance survived and multiplied to create more bacteria.
(C) The bacteria were influenced by other species in their environment through coevolution.
(D) The bacteria evolved to fill different niches from other species in the environment through adaptive radiation.

STOP

END OF SECTION I

IF YOU FINISH BEFORE TIME IS CALLED, YOU MAY CHECK YOUR WORK ON THIS SECTION. DO NOT GO ON TO SECTION II UNTIL YOU ARE TOLD TO DO SO.

BIOLOGY
SECTION II
6 Questions
Writing Time—90 minutes

<u>Directions:</u> Questions 1 and 2 are long free-response questions that should require about 25 minutes each to answer and are worth 8–10 points each. Questions 3 through 6 are short free-response questions that should require about 10 minutes each to answer and are worth 4 points each.

Read each question carefully and completely. Write your response in the space provided following each question. Only material written in the space provided will be scored. Answers must be written out in paragraph form. Outlines, bulleted lists, or diagrams alone are not acceptable unless specifically requested.

1. The microbiome is a collection of microbes, such as bacteria and viruses, that live in/on the human body. Sites of colonization include the skin, nasal passages, mouth, and gut. Links between the microbiome and many different conditions, such as allergies, obesity, depression, and many others have been suggested. During antibiotic use, the body can often suffer repercussions due to the destruction of the natural microbiome. The bacterial community composition varies greatly between body sites. Even among the skin microbiome, there is a great variety in the amounts and types of bacteria found in different locations.

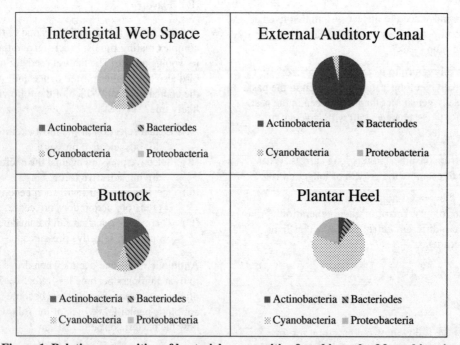

Figure 1. Relative composition of bacterial communities found in each of four skin microbiomes

An experiment was performed to assess the results of skin bacterial transplants to alternative sites of the body. A 2-inch square area of interdigital skin was swabbed with a sterile swab to collect bacteria. An additional swab was opened but not touched to the skin. Each swab was then rubbed on either an alternate skin site or the same type of skin site. After 7 days, the skin of the site of deposition was swabbed and plated on a dish to assess bacterial growth. The types of bacteria present were then cataloged.

GO ON TO THE NEXT PAGE.

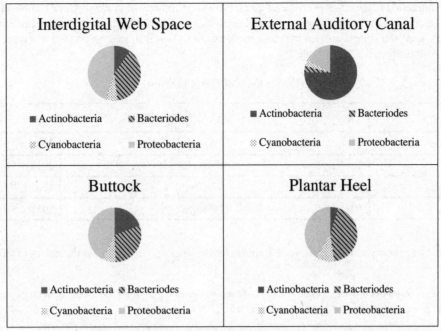

Figure 2. The results of rubbing a swab of interdigital skin on four skin sites

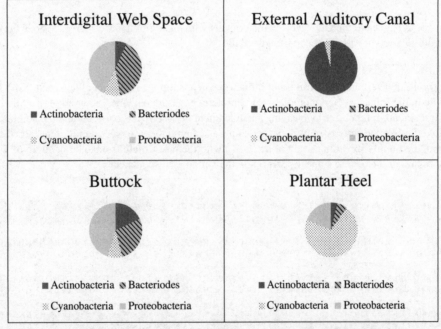

Figure 3. The results of rubbing a clean swab on four skin sites

(a) Bacterial communities containing different species of bacteria are no different than other communities with multiple species. **Explain** what factors contribute to a carrying capacity.

(b) **Identify** the two important controls that are used in this experiment. In addition, the experiment includes a 7-day waiting period between when the bacterial transplant takes place. **Justify** why this is necessary.

(c) **Interpret** the effect of the transplants between interdigital web space and the buttock, external auditory canal, and plantar heel areas.

(d) If each deposition site is pretreated with known antibiotics, **explain** how this would likely affect the results.

GO ON TO THE NEXT PAGE.

2. Researchers were interested in the effect of temperature on the respiration rate of crickets. Three chambers were prepared and kept at three different temperatures. A cricket was placed in each chamber. The CO_2 levels in the chamber were recorded at 0 minutes, 5 minutes, and 10 minutes to assess the changing levels of CO_2 over time. This process was repeated 50 times at each temperature. The results are indicated in Table 1.

Table 1. Levels of CO_2 over time.

	CO_2 levels(ppm $\pm$ 2 SE_x)		
Temperature (°C)	0 min	5 min	10 min
20	1034 ± 26	1074 ± 251	2012 ± 71
35	1560 ± 55	1670 ± 120	1785 ± 100
50	2021 ± 35	2040 ± 55	2051 ± 5

(a) **Describe** how CO_2 is a measure of respiration. **Explain** why temperature might affect the rate of respiration.

(b) On a set of axes, **construct** an appropriately labeled graph to demonstrate the CO_2 levels with changing temperature. Please include all temperatures, timepoints, and error bars.

(c) **Identify** which temperature led to having the highest respiration rate in the crickets.

(d) **Predict** what would happen to the CO_2 levels in the crickets with an additional 5°C increase above the highest temperature. **Justify** your prediction with evidence from the data.

3. A researcher was trying to determine why a strain of influenza spreads more easily than other strains. She suspected that the answer might lie in the Influenza Neuraminidase (NA) protein. NA is an enzyme known to play a role in releasing viral particles during viral spread. It functions by cleaving an attachment between viral particles and the cell membrane. The researcher evaluated the enzyme activity of each of the strains and found them to be quite different. She then compared the NA protein from the rapidly spreading strain of flu (NA_H) with the NA protein from another strain of flu (NA_A). The sequences were similar, but there were some differences in the amino acid sequence.

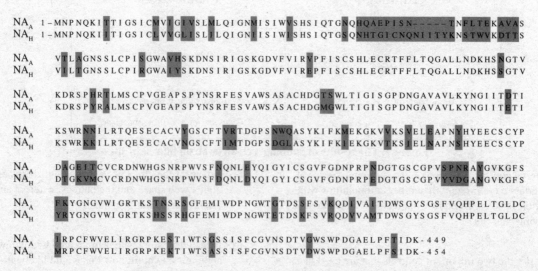

Figure 1. Amino acid sequence alignment between NA_H and NA_A

GO ON TO THE NEXT PAGE.

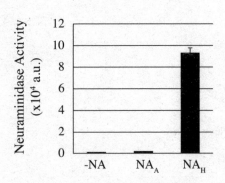

Figure 2. Neuraminidase enzyme activity

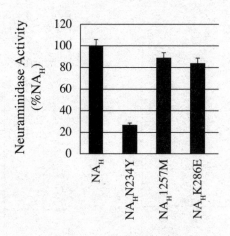

Figure 3. Neuraminidase activity of NA_H and NA_H with mutations

Point mutations were introduced in the nucleic acid sequence to modify the sequence of NA_H to make it more like NA_A. The enzyme activity was assessed.

(a) **Explain** how changing a single nucleotide can change the sequence of the protein.

(b) Given what is known about the neuraminidase enzyme, **describe** what is measured to provide the results in Figure 2 and Figure 3.

(c) Will an NA_H virus with an N234Y substitution be a rapid spreading or slow spreading virus?

(d) **Justify** your prediction.

4. The school is growing tomatoes using a hydroponics system. Hydroponics is a way of growing plants in a controlled system without traditional soil containing nutrients. Instead, nutrients that would be supplied by soils are added to water tanks at specific concentrations before the solution is pumped onto the plant root systems. This type of system allows for increased control over nutrient levels and allows for reduced water usage compared to outdoor crops because of minimal evaporation.

An out-of-control thermostat in the science wing of the school resulted in a temperature that is known to increase thylakoid permeability to H^+ ions.

(a) **Identify** which cellular process thylakoids are associated with.

(b) **Explain** why ions in the hydroponic tank cannot pass into the roots by simple diffusion.

(c) **Predict** the immediate effect this would have on the cell's ability to produce reactants necessary for carbon fixation.

(d) **Justify** your prediction.

GO ON TO THE NEXT PAGE.

5.

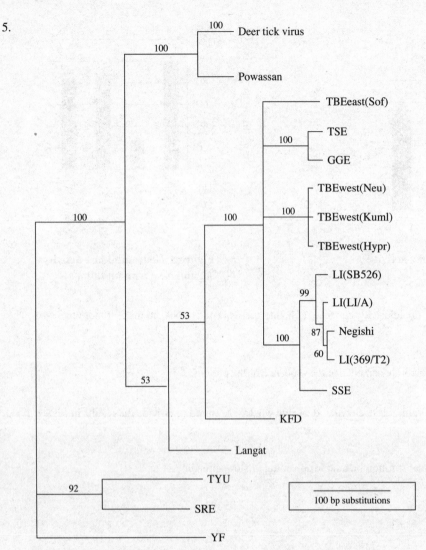

Figure 1. Phylogenetic tree illustrating the relationship of deer tick virus to other tick-borne encephalitis group viruses

In Figure 1, the values above branches indicate bootstrapped confidence values. Branch lengths are proportional to percent similarity in viral envelope gene sequence. TBE: tick borne encephalitis; TSE: Turkish sheep encephalitis; GGE: Greek goat encephalitis; LI: louping ill virus; SSE: Spanish sheep encephalitis; KFD: Kyasanur Forest disease virus; TYU: Tyuleniy virus; SRE: Saumarez Reef virus.

(a) **Describe** what is represented by a node in a phylogenetic tree.

(b) **Identify** two of the viruses that are LEAST similar to GGE.

(c) In 50 years, new data have identified that TBEwest(Hypr) is no longer in circulation, but it seems to have evolved in two directions: TBEwest1 and TBEwest2. Using the empty tree below, correctly **indicate** TBEwest(neu), TBEwest(Kuml), TBEwest1m, and TBEwest2.

(d) This phylogenetic tree was created using nucleotide sequences. If protein sequences were used instead, **explain** why this might increase the similarity between the viruses?

GO ON TO THE NEXT PAGE.

6. In a city in Australia there is a population of moths that live in the grass, rarely taking flight except for short journeys. There are two phenotypes for wing color: green and beige. The largest predator of moths are birds and bats that prey upon the moths as they rest upon blades of grass. Green moths fare better in wet conditions when the grass takes on a lush green color and beige moths fare better in dry conditions when the grass turns a dry brown color.

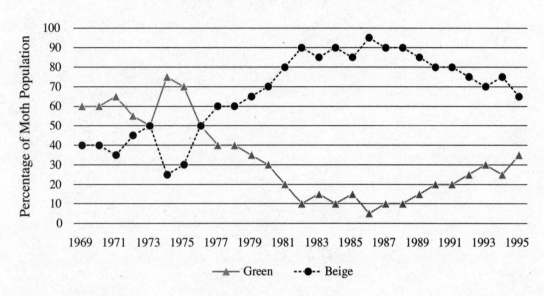

Percentage of Moths of Each Color

(a) **Identify** which color moth was the most plentiful between 1969 and 1995?

(b) **Explain** why the lines of the graph never exceed 100?

(c) **Predict** during which years there was a drought?

(d) **Explain** how natural selection led to the rise of the beige moth in the late 1970s?

STOP

END OF EXAM

Practice Test 1:
Answers and
Explanations

PRACTICE TEST 1 ANSWER KEY

Question Number	Question Answer	Correct? ✎	See Chapter #
1	B		5
2	B		5
3	D		5
4	C		5
5	A		5
6	C		6
7	A		6
8	C		6
9	C		6
10	D		7
11	B		7
12	A		9
13	B		9
14	D		9
15	B		9
16	A		9
17	D		4
18	C		10
19	A		10
20	A		10
21	B		10
22	D		5
23	B		7
24	D		7
25	B		7
26	B		7
27	D		11
28	B		11
29	D		11
30	A		11

Question Number	Question Answer	Correct? ✎	See Chapter #
31	D		11
32	B		9
33	A		9
34	B		9
35	D		9
36	D		11
37	A		6
38	D		6
39	D		6
40	C		6
41	B		6
42	A		6
43	B		6
44	A		6
45	A		8
46	C		8
47	B		8
48	B		8
49	B		8
50	D		7
51	B		7
52	C		7
53	B		7
54	C		8
55	C	✎	11
56	B		11
57	B		11
58	B		10
59	D		10
60	B		7

PRACTICE TEST 1 EXPLANATIONS

Section I: Multiple Choice

1. **B** Adding salt would increase the osmolality (a similar measure to osmolarity) and since Organism 4 is an osmoconformer it would change its intracellular osmolality/osmolarity to mimic the surroundings. This means that it would increase its osmolality and maintain it unless further changes are made to the system.

2. **B** The only choice that has direct ties to maintaining osmotic balance is the sodium-potassium pump.

3. **D** Since the organisms mimic their surroundings, the only choice that would decrease their osmolalities would be one that decreases the osmolality of their surrounding tank. Choice (A) increases the tank osmolality. Choices (B) and (C) directly increase the organism osmolality. Choice (D) decreases the tank's osmolality.

4. **C** The goal of an osomconformer is to mimic the osmolarity of its surroundings. An isotonic solution has the same concentration as its surroundings. Therefore, an osmoconformer would become isotonic to its tank.

5. **A** If Organism 1 decreases osmolality, it means that the tank must have decreased in osmolality. In other words, it gained water or lost solute. Choice (B) increases the tank solute. Choice (D) decreases the water. Choice (C) does not directly relate to the osmolality. Only (A) would cause the tank to gain water as water is removed from the new organism by osmosis.

6. **C** RZN45 is the enzyme and it binds to JB-76. Ascorbic acid (vitamin C) is an inhibitor of this reaction and binding to JB-76 is reduced, which suggests the inhibitor binds to the same place as the substrate. The graph confirms that the inhibition is competitive since it shows a decreased reaction rate when ascorbic acid is present. Choice (A) is not true because similarly charged amino acids would repel each other and RZN45 and JB-76 bind to each other. There is nothing to suggest that choice (B) is correct. Choice (C) would make sense since both ascorbic acid and JB-76 bind to the same place so they probably have similar structures. Choice (D) could be true, but it is unlikely. Even things that are the same shape do not necessarily have the same number of amino acids. Proteins are often hundreds of amino acids long, so having the exact same number is like a needle in a haystack.

7. **A** The graph shows that at low/medium levels of substrate the reaction rate is low in the presence of ascorbic acid and then at high levels of substrate the reaction rate is high, even with ascorbic acid present. This is because with lots of substrate, the ascorbic acid does not stand a chance at finding a free active site to bind to because the superfluous amount of substrate is likely to be bound at most of the sites.

8. **C** Ritzolierre's Disease is caused by elevated enzyme levels. Only option III would decrease the enzyme activity levels. Options I and II would raise the levels and make the disease worse.

9.   **C**   As shown in the graph, the amount of substrate has reached a saturation point and adding additional substrate will not increase the rate further. The reaction rate will remain at the same high level.

10.   **D**   The islands are separated by ocean. Allopatric speciation occurs when populations are isolated from each other. The finch populations are far away from each other and are likely to mate only with those on the same island. Choice (A) implies that the finches are able to cross between islands, and Choice (B) refers to a natural disaster. Neither is mentioned in the information provided. Choice (C) describes the finches developing similar features, but they developed different types of beaks. Only Choice (D) references the isolated populations associated with allopatric speciation.

11.   **B**   If meiosis occurred as it should, there should be 4 haploid gametes, each with one long and one short chromosome. None of the gametes look correct. There was a nondisjunction at Meiosis I, and the two long homologous chromosomes failed to separate. This caused two gametes to get no copies of the long chromosome and two gametes to get two copies. If the nondisjunction was in Meiosis II, then two of the gametes would look normal and two of the gametes would have incorrect amounts of chromosomes. Barr bodies are small and unviable, but they still have the correct number of chromosomes in them.

12.   **A**   The location appears to be upstream of the transcription start site. This is the promoter region and is likely important for polymerase binding and initiation of transcription. Without it, there would be no transcription and subsequently there could not be any translation.

13.   **B**   Dark fur would be caused by high expression of fursilla. The triggers for that high expression would be an underexpression of DRKFR and high expression of LSFR.

14.   **D**   If something competes with the polymerase, then it is likely an inhibitor and high amounts of it would decrease transcription. DRKFR is the only protein that significantly decreases Fursilla transcript levels when overexpressed.

15.   **B**   Two proteins fitting together would not occur from similarities in their nucleotide sequences, which is referenced in Choices (A) and (C). As proteins, they should not be formed from lipids, so Choice (D) is incorrect. Proteins are formed from amino acids, and complementary conformations would allow them to bind together.

16.   **A**   DRKFR is a transcriptional inhibitor, and it significantly decreases Fursilla transcript levels when overexpressed. If BLD binds to the transcriptional inhibitor, DRKFR, then it would reduce inhibition and increase the transcript levels of Fursilla.

17.   **D**   Solvents dissolve solutes that are similar to them, so polar solutes would no longer dissolve in water if water turned nonpolar. Choice (B) is incorrect, because nonpolar molecules are more likely be found as solids. Less liquid water would be found on Earth due to the lack of strong bonds holding it together, which also ties into the melting and boiling points decreasing instead of increasing. The polarity of water allows it to hydrogen bond and form a less dense solid, so Choice (D) describes what would happen if the polarity changed.

18. **C** The wind speed, tree height, and altitude do not tie directly into the function of the jaw and the size of the opening. Only the size of seeds that the mice eat would directly relate to the mouth size.

19. **A** It says in the passage that the mice are the same species. This means they should be able to mate with each other. As we do not know anything about the genetics and inheritance of jaw angle, it is not possible to predict how the jaw angle would be affected.

20. **A** Since the jaw angle on island B is now large, the smaller jaw angle must have disappeared. The small jaw angle is likely selected against, and only mice with large jaws survive on Island B. The survival on the other islands is not directly relevant.

21. **B** Punctuated equilibrium is defined as an isolated episode of rapid development of a population. A slow change, described in Choices (A) and (C), or a fluctuation, described in Choice (D), do not suggest a quick change.

22. **D** The percent change in mass will be closest to 0% when the solutions are isotonic. In other words, the potato cells should stay the same mass when they are soaked in a sodium chloride solution with a similar concentration as the cells. The percent change crosses over from a gain to a loss between 0.3M and 0.5M, so the isotonic concentration must occur between these two molarities. Only Choice (D) describes a number in between these concentrations.

23. **B** HRIET1 has a signal sequence AND a targeting signal, so it must be part of an organelle that is in the secretory pathway that helps things exit the cell. The Golgi apparatus is the only organelle among these choices that functions in this pathway.

24. **D** The receptor must be in the membrane on the surface of the cell. It should have a signal sequence, but it should not have a targeting signal or a localization sequence. The only protein that meets these requirements is K8TE.

25. **B** The electron transport chain in humans occurs in mitochondria. Proteins destined for the mitochondria should have a localization sequence. The only two proteins that have a localization sequence are NUH8 and LNACT.

26. **B** HAZL2 has none of the sequences or signals represented in the data table. Proteins without these features would remain in the cytosol where it is assembled, so a scientist would have to collect samples that contain cytosol.

27. **D** The raccoon population seems steady, with a slight dip in 2008. If raccoons rely on berries as a primary food source, then it is likely that the berry population would have also remained stable.

28. **B** The cougar population is shown to be decreasing over time. One possible explanation for the decline may be due to a disappearing food source, so the food source should also show a decrease over time. The only animal that also shows a decreasing population is deer.

29. **D** There is no information about what species is a predator of skunks, and the population seems to be stable over time, so this is impossible to determine. Even though the skunks and raccoons are both stable, there is no indication that they are related to each other.

30. **A** If the chipmunk is dependent on the acorns, then the acorns affect the carrying capacity. Since the quantity of acorns would affect a large population differently than a small population, this makes it a density-dependent factor.

31. **D** None of the populations demonstrate exponential growth because they each have a carrying capacity that sets a limit on their growth. Only populations growing as fast as they can reproduce at a maximal amount will grow at an exponential rate.

32. **B** The bases at the 5' side of the transcript are in the untranslated region because this is generally the site for ribosome binding. Without this region, the ribosome couldn't bind, and there would be no translation.

33. **A** Before the transcripts are completed and leave the nucleus, they undergo certain modifications including the removal of introns and the addition of a 5' GTP cap and a 3' poly-A tail. The transcript in the nucleus would not have undergone these modifications yet, so it would contain introns.

34. **B** The coding strand is the strand that is identical to the mRNA transcript, except that the mRNA has the nucleotide uracil in place of the nucleotide thymine. It is the partner strand to the coding strand and used as a template during transcription. Therefore, the answer should match the transcript shown in the image with "U's" in place of any "T's."

35. **D** A transcription factor binds to DNA before RNA is created during transcription. Since this is already a map of an RNA transcript, transcription has already happened and no transcription factors will bind at any location.

36. **D** At each level of the food chain, approximately 10% of energy is used to build biomass for the next level. The primary consumers would have approximately 298.7 g/m2 of biomass, and the secondary consumers would have approximately 29.87 g/m2 of biomass.

37. **A** The question states that the change affects the place where the substrate binds, which is defined as the active site. Position A is identified as the active site.

38. **D** Position B has nonpolar residues and is not the active site, since Position A is the active site. Therefore, Choice (A) is not correct since the substrate does not bind at Position B. It doesn't look like it is where the homodimer is, so Choice (C) is incorrect. There is no information to indicate Choice (B). Nonpolar residues are known to be helpful for inserting into the lipid bilayer, which makes (D) a strong answer.

39. **D** A homodimer is formed when more than one polypeptide subunit come together. This is defined as a quaternary structure, (D).

40. **C** The protein is known to be a homodimer, so it must have two polypeptides in it. Each polypeptide has one N-terminus, so this dimer protein would have two. They would not occur at the active site, so position A should not be in the answer.

41. **B** If fats are insoluble in water, figuring out what type of mixture it forms when mixed with water would help target this property of lipids. The only option that discusses mixing with water is Choice (B).

42. **A** While Choice (A) is true, the question is asking about stored energy. Choice (C) describes glycerol, which is a component of cell membranes, and glucagon's function as a hormone will not affect energy storage.

43. **B** Organic compounds are defined as molecules that contain carbon and are found in living organisms. Choice (B) is the only option that references carbon.

44. **A** Figure 1 is adenine, which is a nucleotide. Nucleotides are the monomers of nucleic acids, so the image represents Choices (B) and (C). According to the information in the passage, nucleic acids relay genetic information, so Choice (A) is the only answer that incorrectly describes the figure.

45. **A** A true-breeding white nettle would be homozygous dominant (*WW*). A true breeding pink nettle would be homozygous recessive (*ww*). All offspring would inherit one dominant allele and one recessive allele, so they would all be heterozygous (*Ww*). Heterozygous nettles would display the dominant white phenotype.

46. **C** The F1 generation refers to the first generation of offspring. The previous question determined that all of the offspring from the first cross would be heterozygous. If the heterozygote offspring was taken and crossed with a true-breeding pink nettle, which would be homozygous recessive (*ww*), half of the offspring would be *Ww* and half would be *ww*. The ww offspring would have pink flowers.

47. **B** The heterozygote (*WwSs*) would be crossed with a nettle recessive for both traits (*wwss*). If the genes were not linked, the flowers should be equally white/ruffled, white/slotted, pink/ruffled, and pink/slotted. If the genes are linked, then some combinations of color/texture should be more common. Since a pattern is found in the offspring, and only some of those combinations are present, the traits are likely linked.

48. **B** The equations of Hardy-Weinberg ($p + q = 1$ and $p^2 + 2pq + q^2 = 1$) should be used. There are 5,300 plants, and 400 of them are pink plants (homozygous recessive). Therefore, the frequency of homozygous recessive alleles = 400/5,300 = q^2 = 0.075 and thus q = 0.27.

49. **B** If you correctly solved the previous problem, you're halfway there, as you've already found that q = 0.27. However, here's that information again. The equations of Hardy-Weinberg ($p + q = 1$ and $p^2 + 2pq + q^2 = 1$) should be used. There are 5,300 plants, and 400 of them are pink plants (homozygous recessive). Therefore, the frequency of homozygous recessive alleles = 400/5,300 = q^2 = 0.075 and thus q = 0.27. If $p + q = 1$, the equation becomes p + 0.27 = 1, or p = 0.73. The frequency of heterozygotes is $2pq$, which becomes 2(0.73)(0.27) = 0.3942 = 39%.

50. **D** G-protein coupled receptors (GPCRs) are necessary to bring a signal from outside the cell to inside the cell. If the ligand is a small nonpolar molecule, then it can go through the membrane without trouble and does not need an extracellular receptor.

51. **B** According to the figure, the first step in the activation pathway was the binding of epinephrine, the external ligand.

52. **C** The G-protein coupled receptor (GPCR) allows a message to be transmitted inside the cell without the ligand actually entering the cell. This is most like a drive-through window where the order is communicated to someone inside a restaurant, but the patron does not actually enter the building.

53. **B** The conformational change allows a message to be passed from the outside of the cell to the inside. The conformational change is a sign that the ligand is bound. The ligand does not bind to the membrane, so Choice (A) is incorrect. The ligand triggers binding of the GTP-protein, so Choice (D) is incorrect. A conformational change would not make an attachment more solid, which rules out Choice (C).

54. **C** According to the table, the average carrot mass was stable around 60 grams, but the carrot length decreased by almost 4 cm over time.

55. **C** If a weevil lived at 17 cm, the carrots that are longer than 17 cm would not survive. The shorter carrots would be untouched by the weevils and have more evolutionary fitness. The shorter carrots would be naturally selected for over time and the short trait would become more prevalent in the population.

56. **B** By removing the weevil population, any long carrots would have the chance to thrive. Additionally, by crossing the longest carrots, the genotypes that promote carrot length will be selected for. Since all of the carrots are longer than 10 cm, a predator at 10 cm would still be able to prey on the carrots.

57. **B** If the carrot became shorter, the weevil population would still survive as long as the radish supply was sufficient. Without the radishes, the weevils that lived at those depths would die, and only the weevils that can live at depths where some food is found will survive, (B). Carrots grow to a depth of 17 cm, so the population of weevils might be selected to live at 17 cm. If only a small percentage of weevils died from a virus, this shouldn't change the preferred depth of the entire population.

58. **B** The table shows that the weevil predation is a significant selective pressure on carrot length. Length determines whether or not a carrot is able to prosper in its environment and display a high level of fitness.

59. **D** Sexual selection is defined as natural selection specifically for a mate of the opposite sex. The dance would increase mate attraction, but the predator would also decrease the health of the same mates. The sexual selection and the predation are selective pressures acting in opposing ways at the same time, and both are important.

60. **B** Bacteria reproduce through asexual reproduction, so they don't have "parents" as mentioned in Choice (A). Other species are not mentioned as in Choice (C), and the bacteria still fill the same niche before and after the resistance trait is developed as mentioned in Choice (D).

Section II: Free Response

Short student-style responses have been provided for each of the questions. These samples indicate an answer that would get full credit, so if you're checking your own response, make sure that the actual answers to each part of the question are similar to your own. The structure surrounding them is less important, although we've modeled it as a way to help organize your own thoughts and to make sure that you actually respond to the entire question.

Note that the rubrics used for scoring periodically change based on the College Board's analysis of the previous year's test takers. This is especially true as of the most recent Fall 2019 changes to the AP Biology exam! We've done our best to approximate their structure, based on our institutional knowledge of how past exams have been scored and on the information released by the test makers. However, the 2020 exam's free-response questions will be the first of their kind.

Our advice is to over-prepare. Find a comfortable structure that works for you, and really make sure that you're providing all of the details required for each question. Also, continue to check the College Board's website, as they may release additional information as the test approaches. For some additional help, especially if you're worried that you're not being objective in scoring your own work, ask a teacher or classmate to help you out. Good luck!

Question 1

(a) Bacterial communities containing different species of bacteria are no different than other communities with multiple species. **Explain** what factors contribute to a carrying capacity. (2 points)

Carrying capacity is determined by the available resources including food/nutrients, safe, healthy space to live in, water, sunlight, etc. Anything that is required but is in short supply contributes to the carrying capacity. When populations are competing for resources, there is a finite population that the environment can sustain, and that is the carrying capacity.

(b) **Identify** the two important controls that are used in this experiment. In addition, the experiment includes a 7-day waiting period between when the bacterial transplant takes place. **Justify** why this is necessary. (3 points)

One important control is the use of a clean swab. Another control is to deposit the interdigital swab onto another interdigital space. These are important to confirm that the process of applying the swab to the new area is not a tainted or contaminated process. The 7-day waiting period is important to allow the new bacteria to compete a little with the existing bacteria. Then we can better judge which bacteria is going to stick around in the new site.

(c) **Interpret** the effect of the transplants between interdigital web space and the buttock, external auditory canal, and plantar heel areas. (3 points)

When deposited on the buttock, the bacteria community already on the buttock did not get changed much. The existing bacteria still made up the community. This means the interdigital bacteria didn't survive well there or could not compete with the existing bacteria.

When deposited on the external auditory canal and the plantar heel areas it looked like the new bacteria made an impact and were competing with the existing bacteria. The plantar heel area, in particular, seemed to be recolonized by the bacteria or the interdigital web space.

(d) If each deposition site is pretreated with known antibiotics, **explain** how this would likely affect the results. (2 points)

Antibiotics destroy bacteria. If the new sites had been treated with antibiotics this would have wiped out the existing bacteria. If the antibiotics were still present, then they would kill the newly deposited bacteria as well. Either way, the skin sites would be clear of the existing bacteria and ready to receive the newly deposited bacteria, like a clean slate. This would make the results show a population more like the interdigital than the existing bacteria at those sites.

Question 2

(a) **Describe** how CO_2 is a measure of respiration. **Explain** why temperature might affect the rate of respiration. (2 points)

CO_2 is a byproduct of cellular respiration, which is why we need to get rid of it when we exhale. Living things, like crickets, make more CO_2 when they are performing more cellular respiration (with the processes glycolysis and the Krebs cycle and the electron transport chain). CO_2 is a byproduct of the Krebs cycle.

Temperature might affect respiration because the crickets change their behavior levels with the temperature. This makes them use more oxygen and energy and need more respiration when they are more active. Also, certain enzymes work better at different temperatures, so this might increase the rate of respiration or decrease the rate of respiration depending on how the temperature impacts enzyme efficiency.

(b) On the axes provided, **construct** an appropriately labeled graph to demonstrate the CO_2 levels with changing temperature. Please include all temperatures, timepoints, and error bars. (4 points)

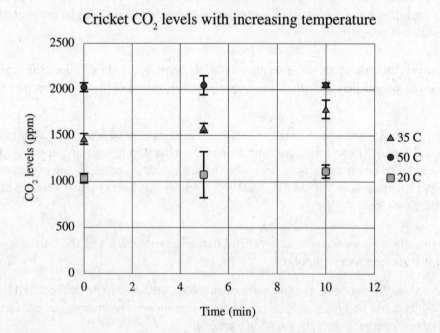

Cricket CO_2 levels with increasing temperature

(c) **Identify** which temperature led to having the highest respiration rate in the crickets. (1 point)

The highest respiration rate was found at 35°C as that showed the greatest increase over time.

(d) **Predict** what would happen to the CO_2 levels in the crickets with an additional 5°C increase above the highest temperature. **Justify** your prediction with evidence from the data. (2 points)

With a 5°C increase in temperature, the respiration rate would increase to approximately 2100-2200ppm. This is justified because the rate increased by approximately 500ppm when the temperature increased by 15°C. This means that a 5°C increase in temp would increase the respiration rate by about 166ppm.

Question 3

(a) **Explain** how changing a single nucleotide can change the sequence of the protein.

Changing a nucleotide will change the codon, which is the group of three nucleotides that gets read by the ribosome when the protein is made. If the codon changes, it might change the amino acid that gets added to the protein. It could also make the chain shorter if it changes to a stop codon.

(b) Given what is known about the neuraminidase enzyme, **describe** what is measured to provide the results in Figure 2 and Figure 3.

The neuraminidase enzyme cleaves a tether that holds the viral particle to the membrane. To measure the "neuraminidase activity" for figure 2 and figure 3, the cleavage action of the enzyme must be measured.

(c) Will an NA_H virus with an N234Y substitution be a rapid spreading or slow spreading virus?

It will be a slow spreading virus.

(d) **Justify** your prediction.

The N234Y causes the enzyme activity to decrease, and this would mean that the viral particles do not get released. This means that the virus should not spread quickly. NA_H has high enzyme activity and spreads quickly, but the N234Y version should not spread quickly.

Question 4

(a) **Identify** which cellular process thylakoids are associated with.

Thylakoids are associated with photosynthesis.

(b) **Explain** why ions in the hydroponic tank cannot pass into the roots by simple diffusion.

The roots, like all cells, have a phospholipid bilayer membrane. This membrane is not permeable to polar or charged things. The inner hydrophobic space prevents things like ions from diffusing through. Ions must enter by facilitated diffusion or sometimes active transport using a pump.

(c) **Predict** the immediate effect this would have on the cell's ability to produce reactants necessary for carbon fixation.

The reactants for the carbon fixation in the Calvin cycle would not be produced.

(d) **Justify** your prediction.

The permeability of the thylakoid membrane would make the hydrogen ions able to flow freely from one side to the other. Without the ability to create a concentration gradient with hydrogen ions, the electron transport chain would fail. This would prevent production of ATP and NADPH, both of which are required for carbon fixation in the Calvin cycle.

Question 5

(a) **Describe** what is represented by a node in a phylogenetic tree.

A node represents a common ancestor. This means that the species that branch from that point had a common ancestor and were once the same species. That species got split and each group evolved differently.

(b) **Identify** two of the viruses that are LEAST similar to GGE.

Two viruses with the least similarity to GGE are YF and SRE. Actually, YF and TYU and SRE are all distantly related to GGE.

(c) In 50 years, new data have identified that TBEwest(Hypr) is no longer in circulation, but it seems to have evolved in two directions: TBEwest1 and TBEwest2. Using the empty tree below, correctly **indicate** TBEwest(neu), TBEwest(Kuml), TBEwest1m and TBEwest2.

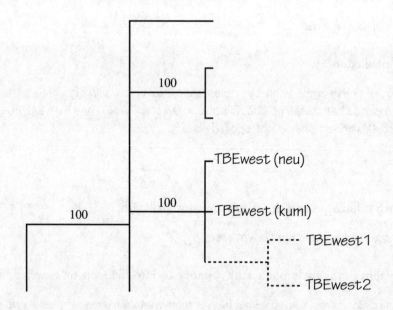

(d) This phylogenetic tree was created using nucleotide sequences. If protein sequences were used instead, **explain** why this might increase the similarity between the viruses?

If protein sequences were used, the sequences might show more similarity because there is more than one codon for some amino acids. This means that the nucleotide sequence might vary, but the amino acid sequence would be the same. These are called silent mutations.

Question 6

(a) **Identify** which color moth was the most plentiful between 1969 and 1995?

Although the two moth colors take turns being the most plentiful, most of the years showed the beige colored moth being the most plentiful.

(b) **Explain** why the lines of the graph never exceed 100?

The graph is showing the percentages of moths of each color. 100% would represent the entire population of moths. It cannot be higher than 100%. The two colors of moths should always add up to 100% since they are the only possible colors of moth.

(c) **Predict** during which years there was a drought?

There was likely a drought from approximately 1976-1988. The beginning of the drought seemed to happen quickly in 1976 when the green moths disappeared and the beige moths rose in numbers. It is hard to know exactly when the drought ended, but seems like maybe the green moths are making a comeback starting around 1988, and by the end of the graph the numbers of each were approaching 50% again, suggesting the drought conditions might be disappearing.

(d) **Explain** how natural selection led to the rise of the beige moth in the late 1970s?

Natural selection caused the beige moths to become more plentiful because they survive better when the grass turns brown from the drought. This is probably because the birds cannot find them to eat them as well as the green moths, which are not camouflaged against the grass because the grass is dry and brown and not green. As the green moths get eaten, they do not reproduce, and the population becomes filled with a higher percentage of beige moths.

Part III
About the AP Biology Exam

- The Structure of the AP Biology Exam
- How the AP Biology Exam Is Scored
- Overview of Content Topics
- Breakdown of Free Response Questions
- How AP Exams Are Used
- Other Resources
- Designing Your Study Plan

THE STRUCTURE OF THE AP BIOLOGY EXAM

As of 2018, students may use a four-function, scientific, or graphing calculator.

The AP Biology Exam is three hours long and is divided into two sections: Section I (multiple-choice questions) and Section II (free-response questions).

Section I consists of 60 questions. You will have 90 minutes to complete this section.

Section II involves free-response questions. You'll be presented with two long-form free-response questions and four short-form free-response questions touching upon key issues in biology. You'll be given 90 minutes to answer all six questions.

If you're thinking that this sounds like a heap of work to try to finish in three hours, you're absolutely right. How can you possibly tackle so much science in so little time? Fortunately, there's absolutely no need to. As you'll soon see, we're going to ask you to leave a small chunk of the test blank. Which part? The parts you don't like. This selective approach to the test, which we call "pacing," is probably the most important part of our overall strategy. But before we talk strategy, let's look at the topics that are covered by the AP Biology Exam.

HOW THE AP BIOLOGY EXAM IS SCORED

AP scores are calculated from your scores on the multiple-choice and free-response sections. The final score is reported on a scale from 1 to 5. The following table explains what that final score means:

Score (Meaning)	Number of test-takers receiving this score*	Percentage of test-takers receiving this score*	Equivalent grade in a first-year college course	Credit granted for this score?
5 (extremely qualified)	18,800	7.2%	A	Most schools grant credit.
4 (well qualified)	57,795	22.2%	A–, B+	Most schools grant credit.
3 (qualified)	92,073	35.3%	B–, C	Some schools grant credit; some do not.
2 (possibly qualified)	69,312	26.6%	C, D	Very few schools grant credit.
1 (no recommendation)	22,836	8.8%	D	No schools grant credit.

*The data above is from the College Board website and based on the May 2019 test administration.

Remember that colleges' rules may vary when it comes to granting credit for AP courses. You should contact the individual admissions departments to find out what score you need on the exam to ensure you'll be given credit.

For the AP Biology Exam, your scores on the multiple-choice section and free-response section are each worth 50 percent of your final score. On the multiple-choice section, your total score is based on the number of questions answered correctly, and you do not lose any points for incorrect answers. Unanswered questions do not receive points. The free-response section is graded on a separate point system. Your scores are tallied to determine your total free-response score. The free-response score is then combined with your multiple-choice score and weighted to figure out where your score falls within the standard AP scoring scale (1 to 5).

OVERVIEW OF CONTENT TOPICS

The AP Biology Exam covers these four Big Ideas.

These are the *official* Big Ideas:

- **Big Idea 1: Evolution**—The process of evolution drives the diversity and unity of life.
- **Big Idea 2: Energetics**—Biological systems use energy and molecular building blocks to grow, to reproduce, and to maintain dynamic homeostasis.
- **Big Idea 3: Information Storage and Transmission**—Living systems store, retrieve, transmit, and respond to information essential to life processes.
- **Big Idea 4: Systems Interactions**—Biological systems interact, and these systems and their interactions exhibit complex properties.

As you read this book, think about how these themes fit with various areas of biology:

- Building blocks/hierarchy
- Responses to stimuli
- Organization
- Structure/function
- Communication
- Relationships
- Disruptions and consequences
- Critical thinking

To fully understand the four big ideas, a solid grasp of eight content units is required. The following is a list of units/topics and the weight of each unit on the exam.

- Chemistry of Life (8–11%)
  - Important properties of water
  - pH
  - Carbohydrates
  - Proteins
  - Lipids
  - Nucleic acids
  - Origins of life

- Cell Structure and Function (10–13%)
  - Prokaryotic and eukaryotic cells
  - Organelles
  - Membranes and transport
  - Cell junctions
  - Cell communication

- Cellular Energetics (12–16%)
  - Change in free energy
  - Enzymes
  - Coupled reactions and ATP
  - Photosynthesis
  - Cellular respiration (glycolysis, Krebs, oxidative phosphorylation)
  - Fermentation

- Cell Communication and Cell Cycle (10–15%)
  - Cell cycle
  - Mitosis
  - Meiosis

- Heredity (8–11%)
  - Mendelian genetics
  - Inheritance patterns

- Gene Expression and Regulation (12–16%)
  - DNA and genome structure
  - Transcription
  - Translation
  - Gene regulation
  - Mutation
  - Biotechnology

- Natural Selection (13–20%)
    o Natural selection
    o Evidence of evolution
    o Phylogenetic trees
    o Impact of genetic variation
    o Speciation
    o Hardy-Weinberg equilibrium

- Ecology (10–15%)
    o Behavior and communication
    o Food webs and energy pyramids
    o Succession
    o Communities and ecosystems
    o Global issues

This might seem like an awful lot of information, but for each topic, there are just a few key facts you'll need to know. Your biology textbooks may go into far greater detail about some of these topics than we do. That's because they're trying to teach you "correct science," whereas we're aiming to improve your scores. Our science is perfectly sound; it's just cut down to size. We've focused on crucial details and given you only what's important. Moreover, as you'll soon see, our treatment of these topics is far easier to handle.

The AP Biology Exam not only tests your content knowledge, but it also tests how you apply that knowledge during scientific inquiry. Simply put, the test's authors are testing whether you can design and/or think critically about experiments and the hypotheses, evidence, math, data, conclusions, and theories therein. There are six broad science practices that are tested. The weight that each practice holds within the multiple-choice section is given.

- **Science Practice 1: Concept Explanation:** Explain biological concepts, processes, and models presented in written format. (25–33%)
- **Science Practice 2: Visual Representations:** Analyze visual representations of biological concepts and processes. (16–24%)
- **Science Practice 3: Questions and Methods:** Determine scientific questions and methods. (8–14%)
- **Scientific Practice 4: Representing and Describing Data:** Represent and describe data. (8–14%)
- **Scientific Practice 5: Statistical Tests and Data Analysis:** Perform statistical tests and mathematical calculations to analyze and interpret data. (8–14%)
- **Science Practice 6: Argumentation:** Develop and justify scientific arguments using evidence. (20–26%)

BREAKDOWN OF FREE RESPONSE QUESTIONS

Each of the six free-response questions will follow a slightly different outline.

Questions 1 and 2 are long questions worth 8–10 points each.

Question 1: Interpreting and Evaluating Experimental Results

Students will be provided with a scenario and an accompanying graph and/or table.

- Part A (1–2 points): Describe and explain biological concepts, processes, or models.
- Part B (3–4 points): Identify experimental design procedures.
- Part C (1–3 points): Analyze data.
- Part D (2–4 points): Make and justify predictions.

Question 2: Interpreting and Evaluating Experimental Results with Graphing

This question is similar to question 1, except the student will be required to construct a data representation.

- Part A (1–2 points): Describe and explain biological concepts, processes, or models.
- Part B (4 points): Construct a graph, plot, or chart and use confidence intervals or error bars.
- Part C (1–3 points): Analyze data.
- Part D (1–3 points): Make and justify predictions.

Questions 3–6 are short questions that are each worth 4 points.

Question 3: Scientific Investigation

Students will be provided with a lab investigation scenario.

- Part A (1 point): Describe biological concepts or processes.
- Part B (1 point): Identify experimental procedures.
- Part C (1 point): Predict results.
- Part D (1 point): Justify predictions.

Question 4: Conceptual Analysis

Students are provided a scenario describing a biological phenomenon with a disruption.

- Part A (1 point): Describe biological concepts or processes.
- Part B (1 point): Explain biological concepts or processes.
- Part C (1 point): Predict the causes or effects of a change in a biological system.
- Part D (1 point): Justify predictions.

Question 5: Analyze Model or Visual Representation

Students are provided a scenario accompanied by a visual model or representation.

- Part A (1 point): Describe characteristics of a biological concept, process, or model represented visually.
- Part B (1 point): Explain relationships between different characteristics of a biological concept or process represented visually.
- Part C (1 point): Represent results within a biological model.
- Part D (1 point) Explain how a biological concept or process represented visually relates to a larger biological principle, concept, process, or theory.

Question 6: Analyze Data

Students are provided data in a graph, table, or other visual representation.

- Part A (1 point): Describe data.
- Part B (1 point): Describe data.
- Part C (1 point): Use data to evaluate a hypothesis or prediction.
- Part D (1 point): Explain how experimental results relate to biological principles, concepts, processes, or theories.

HOW AP EXAMS ARE USED

Different colleges use AP Exams in different ways, so it is important that you go to a particular college's website to determine how it uses AP Exams. The three items below represent the main ways in which AP Exam scores can be used:

- **College Credit.** Some colleges will give you college credit if you score well on an AP Exam. These credits count toward your graduation requirements, meaning that you can take fewer courses while in college. Given the cost of college, this could be quite a benefit, indeed.
- **Satisfy Requirements.** Some colleges will allow you to "place out" of certain requirements if you do well on an AP Exam, even if they do not give you actual college credits. For example, you might not need to take an introductory-level course, or perhaps you might not need to take a class in a certain discipline at all.
- **Admissions Plus.** Even if your AP Exam will not result in college credit or allow you to place out of certain courses, most colleges will respect your decision to push yourself by taking an AP Course or an AP Exam outside of a course. A high score on an AP Exam shows mastery of more difficult content than is taught in many high school courses, and colleges may take that into account during the admissions process.

OTHER RESOURCES

There are many resources available to help you improve your score on the AP Biology Exam, not the least of which are your **teachers**. If you are taking an AP class, you may be able to get extra attention from your teacher, such as obtaining feedback on your essays. If you are not in an AP course, reach out to a teacher who teaches AP Biology and ask if the teacher will review your essays or otherwise help you with content.

The AP Students home page address for this course is apstudents.collegeboard.org/courses/ap-biology. Here you'll find the following:

- a course description, which includes details on what content is covered and sample questions
- sample questions from the AP Biology Exam
- free-response question prompts and multiple-choice questions from previous years

Finally, The Princeton Review offers tutoring and small group instruction. Our expert instructors can help you refine your strategic approach and add to your content knowledge. For more information, call 1-800-2REVIEW.

If there are any late-breaking changes from the College Board after the printing of this book, we'll let you know about them in your free online student tools.

Go online to access two more AP Bio practice exams. Head over to PrincetonReview.com and register this book for a host of test prep resources, including extra AP Bio practice tests and SAT and ACT practice!

DESIGNING YOUR STUDY PLAN

In Part I, you identified some areas of potential improvement. Let's now delve further into your performance on Practice Test 1 with the goal of developing a study plan appropriate to your needs and time commitment.

Read the answers and explanations associated with the multiple-choice questions (starting at page 35). After you have done so, respond to the following questions:

- Review the bulleted list of topics on pages 50 and 51. Next to each topic, indicate your rank of the topic as follows: 1 means "I need a lot of work on this," 2 means "I need to beef up my knowledge," and 3 means "I know this topic well."

- How many days/weeks/months away is your exam?

- What time of day is your best, most focused study time?

- How much time per day/week/month will you devote to preparing for your exam?

- When will you do this preparation? (Be as specific as possible: Mondays and Wednesdays from 3:00 to 4:00 P.M., for example.)

- Based on the answers above, will you focus on strategy (Part IV), content (Part V), or both?

- What are your overall goals in using this book?

Use those answers to create a study plan. Start with the topics that need the most work and map out when you will study and what you will study.

Remember, your schedule may evolve along the way. If a certain time/location is not working for you, then try mixing it up. If you are struggling with a topic, perhaps try tackling it with a teacher, a tutor, or a classmate.

Part IV
Test-Taking Strategies for the AP Biology Exam

- Preview
- 1 How to Approach Multiple-Choice Questions
- 2 How to Approach Free-Response Questions
- 3 Using Time Effectively to Maximize Points
- Reflect

PREVIEW

Review your responses to the three questions on page 2 and then respond to the following questions:

- How many multiple-choice questions did you miss even though you knew the answer?

- On how many multiple-choice questions did you guess blindly?

- How many multiple-choice questions did you miss after eliminating some answers and guessing based on the remaining answers?

- Did you find any of the free-response questions easier/harder than the others—and, if so, why?

HOW TO USE THE CHAPTERS IN THIS PART

For the following Strategy chapters, think about what you are doing now before you read the chapters. As you read and engage in the directed practice, be sure to appreciate the ways you can change your approach. At the end of Part IV, you will have the opportunity to reflect on how you will change your approach.

Chapter 1
How to Approach
Multiple-Choice
Questions

SECTION I

Section I consists of 60 multiple-choice questions that deal with an experiment, set of data, or with general knowledge of biology. For a breakdown of the average percentage of each unit/topic, please refer back to pages 50–51. These questions will also each test a specific Science Practice topic, the list of which can be found on page 52. On the following pages, we'll provide an example of how that topic might be tested in multiple-choice form. These are just model questions, so don't worry about the answers—we just want you to see how these might look in context, and with figures, diagrams, and/or charts.

Concept Explanation

1. Electrophoresis is used to separate DNA fragments by their size and charge. DNA fragments are pipetted into an agarose gel well and covered with a buffer. An electrical current is applied to the gel and buffered solution. Which of the following best explains why DNA fragments migrate towards the positive pole through the agarose gel?

   (A) The nitrogenous bases found in DNA have a positive charge.
   (B) The phosphate groups found in DNA have a negative charge.
   (C) The histone proteins which are intertwined with the DNA have a negative charge.
   (D) The polar bonds on the deoxyribose molecules found in DNA have a slightly negative charge.

Concept Explanation questions require you to explain biological concepts, processes, and written models. These are your standard content knowledge questions, where you may be asked to apply prior knowledge to a specific context or situation. These questions typically provide a specific example, then ask you to identify or explain something more general about the example.

The most obvious feature of these questions is that they start with a few sentences of information, usually containing a specific example or phenomenon. Many of the questions start with "which of the following best describes/explains…" and ask you to identify something's function or explain the general science behind a situation.

Visual Representation

2. Marfan syndrome is a genetic disorder that affects tissues in the body. People with Marfan syndrome are typically tall and thin and have scoliosis and high levels of flexibility in their joints. The pattern of incidence of Marfan syndrome in a specific family is represented in Figure 1.

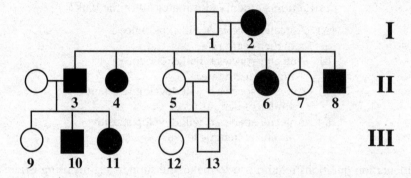

Figure 1. Family pedigree of Marfan syndrome

Which of the following best describes the inheritance pattern for the disorder represented in Figure 1?

(A) X-linked recessive
(B) Autosomal recessive
(C) Autosomal dominance
(D) Incomplete dominance

Visual Representations questions require you to analyze pictorial depictions of biological concepts and processes. These questions are similar to Concept Explanation questions in that you may be asked to apply prior knowledge to a specific context or situation, except they will include a model, diagram, flow chart, or other representation instead of, or in addition to, a written description. These questions may also ask you to explain relationships between different topics or between specific concepts and larger biological principles and theories.

The most obvious feature of these questions is that they come with a graphic. Many of the questions start with "which of the following best describes/explains…" and reference a figure. You may be asked to identify the trend represented on a figure or for an explanation behind something in the figure.

Argumentation

3. A mosquito species that is a carrier for dengue fever lives in a rainforest in which two species of anteaters cohabit. Species A is immune to dengue fever but species B is not. The mosquito species is the primary food source for a specific species of bat that also lives in the rainforest. Which of the following statements best predicts the observable consequence if roost destruction suddenly eliminates all of the bats?

   (A) Anteater species A will experience a higher death rate.
   (B) Anteater species B will experience a higher death rate.
   (C) Anteater species B will develop resistant strains of dengue fever.
   (D) Anteater species A will develop sensitive strains of dengue fever.

Argumentation questions require you to make and support claims using evidence, biological principles and theories, and/or evidence. You may be asked to provide reasoning to justify a claim, explain the relationship between experimental results and larger concepts, and predict the effects of a disruption in a biological system or model. Evidence could be provided as a general biological concept or process, a visual representation, and/or data. These questions typically provide a specific example, then ask you to make a claim or predict something about the example.

The most obvious feature of Argumentation questions is that they reference a source of data or a piece of information. While Concept Explanation questions also reference information, they are more focused on memorized biological facts instead of justification and evidence. Many of the questions start with "which of the following best predicts the consequence…" or "which of the following best supports the claim that…"

Questions 4–6 refer to the following material.

Table 1. Effect of temperature on glucose production.

Temperature	0°C	20°C	40°C	60°C	80°C
Rate of glucose production (mg/dL)/min	0	68.6	105	6.6	0
Color of solution	yellow	light green	dark green	very light green	yellow

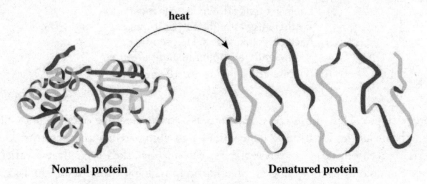

Figure 1. Effect of heat on the structure of lactase

Lactase is an enzyme found in many mammals that breaks down lactose, a sugar found in milk, into glucose through a hydrolysis reaction. Scientists hypothesized that lactase's efficiency is affected by temperature. To test this hypothesis, the scientists dissolved lactase tablets in 200 mL of water and mixed 1 mL of the resulting enzyme solution with 2 mL of milk at different temperatures. After 20 minutes, the scientists tested for the presence of glucose using test strips. The glucose test strips are yellow in the absence of glucose and turn increasingly darker shades of green when glucose is present. The results are shown in Figure 1.

Questions & Methods

4. Which of the following would have provided the best control group for the investigation?

(A) A sample of milk containing no enzyme solution
(B) A sample of milk at a pH of 8
(C) A sample containing orange juice instead of milk
(D) A sample of milk at 100°C

Questions & Methods questions ask you to determine the methods best suited for an experiment. You may be asked to develop a testable question to prove or disprove an idea and/or develop hypotheses for an experiment. You may also be asked to identify experimental design components for experiments, such as independent and dependent variables and controls. You may be asked to propose the next steps for an experiment based on the methods used or the data collected.

The most obvious feature of Questions & Methods questions is the use of experimental design vocabulary, such as question, independent variable, dependent variable, and control. Many of the questions start with "which of the following questions will best guide the scientists towards…" or "which of the following best describes the independent variable/dependent variable/control of the experiment…"

Representing & Describing Data

5. Which of the following claims is best supported by the data?

    (A) Lactase breaks down lactose most efficiently at low temperatures.
    (B) Lactase breaks down lactose most efficiently at high temperatures.
    (C) Lactase breaks down lactose most efficiently at moderate temperatures.
    (D) Lactase efficiency is not affected by temperature.

Representing & Describing Data questions ask you to describe data from a table or graph, including identifying data points, describing trends and patterns, and describing relationships between variables. You may be asked to analyze a variety of representations, such as line graphs, bar graphs, logarithmic data, dual axes, histograms, box-and-whisker plots, and pie charts.

The most obvious feature of Representing & Describing Data questions is that they include an accompanying graph or data table. Many of the questions start with "which of the following claims is best supported by…" or "which of the following best describes the relationship between…"

Statistical Tests & Data Analysis

6. The average rate of glucose production from 20°C to 60°C is closest to which of the following?

    (A)   0 (mg/dL)/min
    (B)   20 (mg/dL)/min
    (C)   60 (mg/dL)/min
    (D) 100 (mg/dL)/min

Statistical Tests & Data Analysis questions ask you to perform calculations, such as finding averages, rates, ratios, and percentages, to draw conclusions from data. You may be asked to determine whether information is statistically significant or different based on confidence intervals or error bars, and you may be asked to perform chi-square statistics. While Argumentation questions also ask you to support claims with evidence, Statistical Tests & Data Analysis questions will focus more on the data than the scientific principles and theories behind the data.

The most obvious feature of Statistical Tests & Data Analysis questions is the associated numerical data and numbers in the answer choices. Many of the questions start with "The average/ratio/percent … is closest to which of the following…"

PACE YOURSELF

When you take a test in school, how many questions do you answer? Naturally, you try to answer all of them. You do this for two reasons: (1) your teacher told you to, and (2) if you left a question blank, your teacher would mark it wrong. However, that's not the case when it comes to the AP Biology Exam. In fact, finishing the test is the worst thing you can do. Before we explain why, let's talk about timing.

One of the main reasons that taking the AP Biology Exam can be so stressful is the time constraint we discussed earlier—90 seconds per multiple-choice question, 25 minutes per long essay, and 10 minutes per short essay. If you had all day, you would probably ace the test. We can't give you all day, but we can do the next best thing: we can give you more time for each question. How? By having you slow down and answer fewer questions.

Slowing down and doing well on the questions you do answer is the best way to improve your score on the AP Biology Exam. Rushing through questions in order to finish, on the other hand, will always hurt your score. When you rush, you're far more likely to make careless errors, misread, and fall into traps. Keep in mind that blank answers are not counted against you.

In case you were curious, the answers to the questions modeled on these pages are: 1 (B), 2 (C), 3 (B), 4 (A), 5 (C), 6 (C).

THE THREE-PASS SYSTEM

The AP Biology Exam covers a broad range of topics. There's no way, even with our extensive review, that you will know everything about every topic in biology. So what should you do?

Do the Easiest Questions First

The best way to rack up points is to focus on the questions you find the easiest. If you know the answer, nail it and move on. Other questions, however, will be a little more complicated. As you read each question, decide if you think it's easy, medium, or hard. During a first pass, do all of your easy questions. If you come across a problem that seems time-consuming or completely incomprehensible, skip it. Remember:

3 is the Magic Number
Using the Three-Pass System will help you earn more points by doing the questions you know first and saving the ones you don't know for last.

> Questions that you find easy are worth just as many points as the ones that stump you, so your time is better spent focusing on the ones in the former group.

Save the medium questions for the second pass. These questions are either time-consuming or require you to analyze all the answer choices (i.e., the correct answer doesn't pop off the page). If you come across a question that makes no sense from the outset, save it for the last pass. You're far less likely to fall into a trap or settle on a silly answer.

Watch Out for Those Bubbles!

Since you're skipping problems, you need to keep careful track of the bubbles on your answer sheet. One way to accomplish this is by answering all the questions on a page and then transferring your choices to the answer sheet. If you prefer to enter them one by one, make sure you double-check the number beside the ovals before filling them in. We'd hate to see you lose points because you forgot to skip a bubble!

So then, what about the questions you don't skip?

Outsmart the Tricky Questions
EXCEPT and NOT questions are considered more difficult than other question types. Here's where POE comes in handy!

PROCESS OF ELIMINATION (POE)

On most tests, you need to know your material backward and forward to get the right answer. In other words, if you don't know the answer beforehand, you probably won't answer the question correctly. This is particularly true of fill-in-the-blank and essay questions. We're taught to think that the only way to get a question right is by knowing the answer. However, that's not the case on Section I of the AP Biology Exam. You can get a perfect score on this portion of the test without knowing a single right answer, provided you know all the wrong answers!

What are we talking about? This is perhaps the single most important technique in terms of the multiple-choice section of the exam. Let's take a look at the example below.

7. Trees and shrubs have hard, woody stems and grow buds that persist above ground throughout the colder months. Which of the following best explains how gas is exchanged in a woody stem?

(A) Oxygen is taken in by lungs and carbon dioxide is released into the air.
(B) Oxygen, carbon dioxide, and water vapor pass through small pores called lenticels.
(C) Hormones and neurotransmitters are transferred between adjacent ganglia.
(D) Lentil beans provide small openings for gases to pass into the woody stem.

Now if this were a fill-in-the-blank-style question, you might be in a heap of trouble. But let's take a look at what we've got. You see "woody stems" and "buds" in the question, which leads you to conclude that the question is about plants. Right away, you know the answer is not (A) or (C) because plants don't have lungs or ganglia. Now you've narrowed it down to (B) and (D). Notice that (B) and (D) look very similar. Obviously, one of them is a trap. At this point, if you don't know what "lentil beans" are, you have to guess. However, even if you don't know precisely what they are, it's safe to say that you probably know that lentil beans have nothing to do with gas exchange in plants. Therefore, the correct answer is (B).

> **Process of Elimination (POE)** is the best way to approach
> the multiple-choice questions.

Even if you don't know the answer right off the bat, you'll surely know that two or three of the answer choices are not correct. What then?

AGGRESSIVE GUESSING

The testing board, the service that develops and administers the exam, tells you that random guessing will not affect your score. This is true. There is no guessing penalty on the AP Biology Exam. For each correct answer you'll receive one point and you will not lose any points for each incorrect answer.

Although you won't lose any points for wrong answers, you should guess aggressively by getting rid of the incorrect answer choices. The moment you've eliminated a couple of answer choices, your odds of getting the question right, even if you guess, are far greater. If you can eliminate as many as two answer choices, your odds improve enough that it's in your best interest to guess.

WORD ASSOCIATIONS

Another way to rack up the points on the AP Biology Exam is by using word associations in tandem with your POE skills. Make sure that you memorize the words in the Key Terms lists throughout this book. Know them backward and forward. As you learn them, make sure you group them by association, since the testing board is bound to ask about them on the AP Biology Exam. What do we mean by "word associations"? Let's take the example of mitosis and meiosis.

You'll soon see from our review that there are several terms associated with mitosis and meiosis. Synapsis, crossing-over, and tetrads, for example, are words associated with meiosis but not mitosis. We'll explain what these words mean later in this book. For now, just take a look.

8. In mitosis, plant cytokinesis differs from animal cytokinesis. Scientists studying the process have hypothesized that the Golgi apparatus sends cellulose in vesicles to form a cell plate between plant daughter cells. Which of the following questions will help guide the scientists toward a direct test of their hypothesis?

(A) In what types of cells do homologous chromosomes exchange DNA through crossing-over?

(B) Are tetrads formed in the center of the cell as a result of plant cytokinesis?

(C) When homologous chromosomes pair up through the process of synapsis, does a cell plate form?

(D) Does cytokinesis still occur in plant cells when Golgi vesicles are inhibited from travelling to the center of the cell?

This might seem like a difficult problem. But let's think about the associations we just discussed. The question asks us about mitosis. However, (A), (B), and (C) all mention events that we've associated with meiosis. Therefore, they are out. Without even racking your brain, you've managed to find the correct answer: (D). Not bad!

Once again, don't worry about the science for now. We'll review it later. What is important to recognize is that by combining the associations we'll offer throughout this book and your aggressive POE techniques, you'll be able to rack up points on problems that might have seemed difficult at first.

MNEMONICS—OR THE BIOLOGY NAME GAME

One of the big keys to simplifying biology is the organization of terms into a handful of easily remembered packages. The best way to accomplish this is by using mnemonics. Biology is all about names: the names of chemical structures, processes, theories, and so on. How are you going to keep them all straight? A mnemonic, as you may already know, is a convenient device for remembering something.

For example, one important issue in biology is taxonomy—that is, the classification of life-forms, or organisms. Organisms are classified in a descending system of similarity, leading from domains (the broadest level) to species (the most specific level). The complete order runs as follows: domain, kingdom, phylum, class, order, family, genus, and species. Don't freak out yet. Use a mnemonic to help you:

King Philip of Germany decided to walk to America. What do you think happened?

Dumb	→	**D**omain
King	→	**K**ingdom
Philip	→	**P**hylum
Came	→	**C**lass
Over	→	**O**rder
From	→	**F**amily
Germany	→	**G**enus
Soaked	→	**S**pecies

Learn the mnemonic and you'll never forget the science!

Mnemonics can be as goofy as you like, as long as they help you remember. Throughout this book, we'll give you mnemonics for many of the complicated terms we'll be seeing. Use ours, if you like them, or feel free to invent your own. Be creative! Remember: the important thing is that you remember the information, not how you remember it.

IDENTIFYING EXCEPT QUESTIONS

You might encounter some EXCEPT/NOT/LEAST questions. With this type of question, you must remember that you're looking for the wrong (or the least correct) answer. The best way to go about these is by using POE.

More often than not, the correct answer is a true statement, but is wrong in the context of the question. However, the other three tend to be pretty straightforward. Cross off the three that apply, and you're left with the one that does not. Here's a sample question.

> All of the following are true statements about gametes EXCEPT
>
> (A) they are haploid cells
> (B) they are produced only in the reproductive structures
> (C) they bring about genetic variation among offspring
> (D) they develop from polar bodies

If you don't remember anything about gametes and gametogenesis, or the production of gametes, this might be a particularly difficult problem. We'll see these again later on, but for now, remember that gametes are the sex cells of sexually reproducing organisms. As such, we know that they are haploid and are produced in the sexual organs. We also know that they come together to create offspring.

Practice Makes Perfect
This question type may not appear on the newly designed AP Biology exam. However, because the POE skills they teach remain valuable for solving all other problems, our sample tests will, for the time being, continue to feature these questions for your benefit.

Outsmart the Tricky Questions
EXCEPT and NOT questions are considered more difficult than other question types. Here's where POE comes in handy!

Remember Your Resources!

The AP Biology Equations and Formulas sheet is included at the end of this book and in your online Student Tools. Print it and remember to use it when you take the Practice Tests.

From this very basic review, we know immediately that (A) and (B) are not our answers. Both of these are accurate statements, so we eliminate them. That leaves us with (C) and (D). If you have no idea what (D) means, focus on (C). In sexual reproduction, each parent contributes one gamete, or half the genetic complement of the offspring. This definitely helps vary the genetic makeup of the offspring. Choice (C) is a true statement, so it can be eliminated. The correct answer is (D).

Don't sweat it if you don't recall the biology. We'll be reviewing it in detail soon enough. For now, remember that the best way to answer these types of questions is to spot all the right statements and cross them off. You'll wind up with the wrong statement, which happens to be the correct answer.

Chapter 2
How to Approach
Free-Response
Questions

Remember the Questions

For a full list of the six types of questions and how they're scored, refer to page 52.

THE ART OF THE FREE-RESPONSE QUESTION

Section II of the exam is made up of six free-response questions. Although each of these questions generally has four parts, labeled (a), (b), (c), and (d), each of which is to be answered separately, the first two questions tend to be longer, and usually contain multiple bolded task verbs. You're largely expected to answer in complete sentences, although in a few instances, you may be asked instead to show your mathematical work, or to create a diagram, graph, or table that models something from the question.

This particular free-response section is new to the May 2020 AP Biology exam. It provides far more context for each question—in the form of data or a passage—and asks you to synthesize that information in a wider variety of formats. Because this version of the free-response section has never been administered before, we can't tell you *exactly* what sort of rubric they're going to use. However, we can take our decades of expertise with the AP exams and provide you with some firm strategies and a model for what you might expect the College Board to do, based on what they've done before.

Ultimately, the best way to rack up points in this section is to give the test readers exactly what they're looking for. However, unlike the multiple-choice, where there is only one valid response—a letter that either matches or doesn't—these free-response readers will accept any answer that accurately fits the checklist that they have been given. (The "accurate" part is important. If you just list a bunch of incorrect things along with the correct response, the reader may assume you're just guessing.) In short, be short! Don't waste time including material that hasn't been asked for. Support your statements as needed, and write in full sentences (as opposed to outlines or bullet points). Use the following strategies to focus on providing specific, complete answers.

KNOW WHAT YOU NEED

The first thing you should do is take less than a minute to skim all of the questions and put them into your own personal order of difficulty from easiest to toughest. Once you've decided the order in which you will answer the questions (easiest first, hardest last), you can begin to formulate your responses. Your first step should be a more detailed assessment of each question.

The most important advice we can give you is to read each question at least twice. As you read the question, focus on key words, especially the bolded direction words. Almost every essay question begins with a direction word. Make sure to read closely: some questions contain more than one direction word and you need to address each of these in your answer. As you read through those outlines of what each question contains, you will recognize various direction words. Some examples of direction words are *calculate*, *explain*, *identify*, and *predict*. These words tell you exactly what the question is looking for, and they're exactly what you should do. For instance, a question that asks you to support a claim expects you to actually provide reasoning and/or evidence that backs it up. By contrast, a question that asks you to construct, draw, or represent data may just want you to include a graph or build a table, no further explanation necessary.

Many students lose points because they either misread the question or fail to do what's asked of them.

FIND WHAT YOU NEED

Now that you know what you need, you have to figure out where your answers are coming from. There are only two options, so this isn't as tricky as it might sound. Some parts can be answered based entirely on what you've learned in the course thus far; others, however, will require you to cite evidence or use information from the passage and/or graph that precedes the question.

Spend about two minutes per question re-familiarizing yourself with any key terms, definitions, or processes that you'll need to answer the question. You can scribble these next to the questions, and you can underline or circle the relevant parts of the figures that you'll need to focus on.

USE WHAT YOU NEED

The multiple parts of each question may connect to each other, but they're answered individually. Label the information that you've just set aside so that you know you've got enough to get all the available points for each part. If you've brainstormed more examples than the question is asking you for, choose only those that you feel strongest talking about.

Spend about a minute here organizing your thoughts, making sure that you're addressing as many of the bolded verbs as you can. You want to answer every part of a question so as to maximize your score, but if there's something you just don't know, first focus on the other parts. Then, time permitting, go back and look at any unanswered parts: a well-written, educated guess may get you partial credit, even if you're not sure about it.

ANSWER EACH PART OF THE QUESTION SEPARATELY

Unless one part specifically references an earlier one, treat each part as its own question. This shouldn't be difficult to remember. Each part should already have its own space in which to write or draw a response, and if not, you can feel free to break between parts by inserting headers—Part A, Part B, etc.

The key here is to make sure you answer as many parts as you can, and to not get bogged down in any one part, because each part has a maximum value. For the first two long questions, you can score at most 4 points for a part (and if this is the case, the others will be worth 1 to 2 points). For the four short questions, each is worth a single point. Ask yourself: is the extra point you might be able to squeak out of a long response worth the individual points you'll miss out on elsewhere if you run out of time on the short responses?

You don't need a perfectly written response to get the maximum number of points, either, so don't waste your time with fancy introductions like "It was the best of experiments, it was the worst of experiments." Just leap right into your response. Don't even sweat grammar or spelling errors; these can hurt you only if they're so bad that they seriously impair the reader's ability to understand you.

Know How to Label Diagrams and Figures

Some responses of the **represent** variety may require you to answer with a figure as opposed to text. (The same is true for any math-based **calculate** parts.) The key here is to clearly and correctly show your work and understanding, so let's briefly discuss the important elements in setting up a graph. The favorite type of graph on the AP Biology Exam is the **coordinate graph**. The coordinate graph has a horizontal axis (x-axis) and a vertical axis (y-axis).

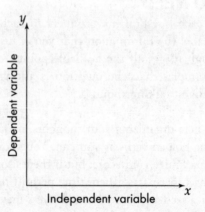

The x-axis usually contains the **independent variable**—the thing that's being manipulated or changed. The y-axis contains the **dependent variable**—the thing that's affected when the independent variable is changed.

Now let's look at what happens when you put some points on the graph. Every point on the graph represents both an independent variable and a dependent variable.

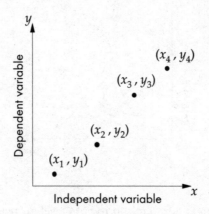

Once you draw both axes and label the axes as *x* and *y*, you can plot the points on the graph. We're just going to look at (b), which asks us to design an experiment.

> Enzymes are biological catalysts.
>
> (b) **Construct** an experiment that investigates the influence of temperature, substrate concentration, or pH on the activity of an enzyme.

Let's set up a graph that shows the results of an experiment examining the relationship between pH and enzyme activity. Notice that we've chosen only one factor here, pH. We could have chosen any of the three. Why did we choose pH and not temperature or substrate concentration? Well, perhaps it's the one we know the most about.

What is the independent variable? It is pH. In other words, pH is being manipulated in the experiment. We'll therefore label the *x*-axis with pH values from 0 through 14.

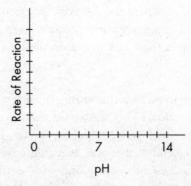

What is the dependent variable? It's the enzyme activity—the thing that's affected by pH. Let's label the *y*-axis "Rate of Reaction." Now we're ready to plot the values on the graph. Based on our knowledge of enzymes, we know that for most enzymes the functional range of pH is narrow, with optimal performance occurring at or around a pH of 7.

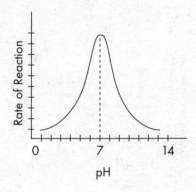

Now you should interpret your graph. If the pH level decreases from a neutral pH of 7, the reaction rate of the enzyme will decrease. If the pH level increases, the rate of reaction will also decrease. Don't forget to include a simple explanation of your graph.

Additional quantitative and graphing skills will be discussed at length in a later chapter of this text. There you will learn how to include statistics and hypothesis testing in your free-response questions and how best to present data using graphs.

UNDERSTAND CONTROLS IN EXPERIMENTS

Almost every experiment will have at least one group that represents the normal/unchanged/neutral version of the independent variable. This is called the *control*. A control is simply a standard of comparison. What does a control do? It enables the biologist to verify that the outcome of the study is due to changes in the independent variable and nothing else.

Let's say the principal of your school thinks that students who eat breakfast do better on the AP Biology exam than those who don't eat breakfast. He gives a group of 10 students from your class free breakfast every day for a year. When the school year is over, he administers the AP Biology exam and they all score brilliantly! Did they do well because they ate breakfast every day? We don't know for sure. Maybe those kids were the smartest kids in the class and would have scored well anyway.

In this case, the best way to be sure that eating breakfast made a difference is to have a control group. In other words, he would need to also follow for a year a group of students of equal intelligence to the first group, BUT who are known to never eat breakfast. At the end of that year, he could have them take the AP Biology exam. If they do just as well as the group with similar intelligence who ate breakfast, then we can probably conclude that eating breakfast wasn't necessarily leading to higher AP scores. The group of students who didn't eat breakfast is called the control group because those students were not "exposed" to the variable of interest—in this case, breakfast. Control groups are generally any group that you include just for the sake of comparison.

Put It All Together
Use the techniques discussed in this chapter and practice writing your free responses. Additional sample free-response questions are available on the AP Students page for the AP Biology Exam.

Chapter 3
Using Time
Effectively to
Maximize Points

BECOMING A BETTER TEST TAKER

Very few students stop to think about how to improve their test-taking skills. Most assume that if they study hard, they will test well and that if they do not study, they will do poorly. Most students continue to believe this even after experience teaches them otherwise. Have you ever studied really hard for an exam and then blown it on test day? Have you ever aced an exam for which you thought you weren't well-prepared? Most students have had one, if not both, of these experiences. The lesson should be clear: factors other than your level of preparation influence your final test score. This chapter will provide you with some insights that will help you perform better on the AP Biology Exam and on other exams as well.

PACING AND TIMING

A big part of scoring well on an exam is working at a consistent pace. The worst mistake made by inexperienced or less savvy test-takers is that they come to a question that stumps them and rather than just skip it, they panic and stall. Time seems to stand still when you're working on a question you cannot answer, and it is not unusual for students to waste five minutes on a single question (especially a question involving a graph or the word EXCEPT) because they are too stubborn to cut their losses. It is important to be aware of how much time you have spent on a given question or section. There are several ways to improve your pacing and timing for the test:

- **Know your average pace.** While you prepare for your test, try to gauge how long you take on 5, 10, or 20 questions. Knowing how long you spend on average per question will help you identify how many questions you can answer effectively and how best to pace yourself for the test.

- **Have a watch or clock nearby.** You are permitted to have a watch or clock nearby to help you keep track of time. However, it's important to remember that constantly checking the clock is in itself a waste of time and can be distracting. Devise a plan. Try checking the clock every fifteen or thirty questions to see if you are keeping the correct pace or whether you need to speed up or slow down. This will ensure that you're cognizant of the time but will not permit you to fall into the trap of dwelling on it.

- **Know when to move on.** Since all questions are scored equally, investing appreciable amounts of time on a single question is inefficient and can potentially deprive you of the chance to answer easier ones later on. You should eliminate answer choices if you are able to, but don't feel bad about picking a random answer and moving on if you cannot find the correct answer. Remember, tests are like marathons:

you do best when you work through them at a steady pace. You can always come back to a question you don't know. When you do, very often you will find that your previous mental block is gone and you will wonder why the question perplexed you the first time around (as you gleefully move on to the next question). Even if you still don't know the answer, you will not have wasted valuable time you could have spent on easier questions.

- **Be selective.** You don't have to do any of the questions in a given section in order. If you are stumped by an essay or multiple-choice question, skip it or choose a different one. In the next section, you will see that you may not have to answer every question correctly to achieve your desired score. Select the questions or essays that you can answer and work on them first. This will make you more efficient and give you the greatest chance of getting the most questions correct.

- **Use Process of Elimination (POE) on multiple-choice questions.** Many times, you can eliminate one or more answer choices. Every answer choice that you eliminate increases the odds that you will answer the question correctly.

Remember, when all the questions on a test are of equal value, no one question is that important, and your overall goal for pacing is to get the most questions correct. Finally, you should set a realistic goal for your final score.

GETTING THE SCORE YOU WANT

Depending on the score you need, it may be in your best interest not to try to work through every question. Check with the schools to which you are applying to find out what scores they will accept for credit.

AP Exams in all subjects no longer include a "guessing penalty" of a quarter of a point for every incorrect answer. Instead, students are assessed only on the total number of correct answers. A lot of AP materials, even those you receive in your AP class, may not include this information. It's really important to remember that if you are running out of time, you should fill in all the bubbles before the time for the multiple-choice section is up. Even if you don't plan to spend a lot of time on every question or even if you have no idea what the correct answer is, you need to fill something in.

Answer Every Question
Here's something worth repeating: there is no penalty for guessing. Don't leave a question blank; there's at least a 25% chance you might have gotten it right!

TEST ANXIETY

Everybody experiences anxiety before and during an exam. To a certain extent, test anxiety can be helpful. Some people find that they perform more quickly and efficiently under stress. If you've ever pulled an all-nighter to write a paper and ended up doing good work, you know the feeling.

However, too much stress is definitely a bad thing. Hyperventilating during the test, for example, almost always leads to a lower score. If you find that you stress out during exams, here are a few preemptive actions you can take:

- **Take a reality check.** Evaluate your situation before the test begins. If you have studied hard, remind yourself that you are well-prepared. Remember that many others taking the test are not as well-prepared, and (in your classes, at least) you are being graded against them, so you have an advantage. If you didn't study, accept the fact that you will probably not ace the test. Make sure you get to every question you know something about. Don't stress out or fixate on how much you don't know. Your job is to score as high as you can by maximizing the benefits of what you do know. In either scenario, it's best to think of a test as if it were a game. How can you get the most points in the time allotted to you? Always answer questions you can answer easily and quickly before tackling those that will take more time.

- **Try to relax.** Slow, deep breathing works for almost everyone. Close your eyes, take a few slow, deep breaths, and concentrate on nothing but your inhalation and exhalation for a few seconds. This is a basic form of meditation that should help you to clear your mind of stress and, as a result, concentrate better on the test. If you have ever taken yoga classes, you probably know some other good relaxation techniques. Use them when you can (obviously, anything that requires leaving your seat and, say, assuming a handstand position won't be allowed by any but the most free-spirited proctors).

- **Eliminate as many surprises as you can.** Make sure you know where the test will be given, when it starts, what type of questions are going to be asked, and how long the test will take. You don't want to be worrying about any of these things on test day or, even worse, after the test has already begun.

The best way to avoid stress is to study both the test material and the test itself. Congratulations! By reading this book, you are taking a major step toward a stress-free AP Biology Exam.

REFLECT

Respond to the following questions:

- How long will you spend on multiple-choice questions?

- How will you change your approach to multiple-choice questions?

- What is your multiple-choice guessing strategy?

- How much time will you spend on the free-response questions?

- What will you do before you begin writing your free-response answers?

- Will you seek further help outside of this book (such as a teacher, tutor, or AP Students) on how to approach the questions that you will see on the AP Biology Exam?

Part V
Content Review for the AP Biology Exam

Chapter 4
Chemistry of Life

ELEMENTS

Although life-forms exist in many diverse forms, they all have one thing in common: they are all made up of matter. Matter is made up of elements. **Elements,** by definition, are substances that cannot be broken down into simpler substances by chemical means.

The Essential Elements of Life

Although there are 92 natural elements, 96% of the mass of all living things is made up of just four of them: **oxygen** (O), **carbon** (C), **hydrogen** (H), and **nitrogen** (N). Other elements such as calcium (Ca), phosphorus (P), potassium (K), sulfur (S), sodium (Na), Chlorine (Cl), and magnesium (Mg) are also present, but in smaller quantities. These elements make up most of the remaining four percent of a living thing's weight. Some elements are known as **trace elements** because they are required by an organism only in very small quantities. Trace elements include iron (Fe), iodine (I), and copper (Cu).

SUBATOMIC PARTICLES

The smallest unit of an element that retains its characteristic properties is an **atom**. Atoms are the building blocks of the physical world.

Within atoms, there are even smaller subatomic particles called **protons**, **neutrons**, and **electrons**. Let's take a look at a typical atom.

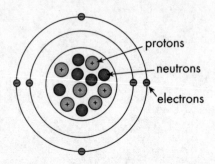

Protons and neutrons are particles that are packed together in the core of an atom called the **nucleus**. You'll notice that protons are positively charged (+) particles, whereas neutrons are uncharged particles.

Electrons, on the other hand, are negatively charged (–) particles that spin around the nucleus. Electrons are pretty small compared to protons and neutrons. In fact, for our purposes, electrons are considered massless. Most atoms have the same number of protons and electrons, making them electrically neutral. Some atoms have the same number of protons but differ in the number of neutrons in the nucleus. These atoms are called **isotopes**.

COMPOUNDS

When two or more individual elements are combined in a fixed ratio, they form a chemical **compound**. You'll sometimes find that a compound has different properties from those of its elements. For instance, hydrogen and oxygen exist in nature as gases. Yet when they combine to make water, they often pass into a liquid state. When hydrogen atoms get together with oxygen atoms to form water, we've got a **chemical reaction**.

$$2H_2\ (g) + O_2\ (g) \rightarrow 2H_2O\ (l)$$

The atoms of a compound are held together by **chemical bonds**, which may be **ionic bonds**, **covalent bonds**, or **hydrogen bonds**.

An ionic bond is formed between two atoms when one or more electrons are transferred from one atom to the other. In order for this to occur, first, one atom *loses* electrons and becomes positively charged, and the other atom *gains* electrons and becomes negatively charged. The charged forms of the atoms are called **ions**. An ionic bond results from the attraction between the two oppositely charged ions. For example, when Na reacts with Cl, the charged ions Na^+ and Cl^- are formed.

A covalent bond is formed when electrons are *shared* between atoms. If the electrons are shared equally between the atoms, the bond is called **nonpolar covalent**. If the electrons are shared unequally, the bond is called **polar covalent**. When one pair of electrons is shared between two atoms, the result is a single covalent bond. When two pairs of electrons are shared, the result is a double covalent bond. When three pairs of electrons are shared, the result is a triple covalent bond.

WATER: THE VERSATILE MOLECULE

One of the most important substances in nature is water. Water is considered a unique molecule because it plays an important role in chemical reactions.

Let's take a look at one of the properties of water. Water has two hydrogen atoms joined to an oxygen atom.

In water molecules, the electrons are not shared equally in the bonds between hydrogen and oxygen. This means that the hydrogen atoms have a partial positive charge and the oxygen atom has a partial negative charge. Molecules that have partially positive and partially negative charges are said to be **polar**. Water is therefore a polar molecule. The positively charged elements of the water molecules strongly attract the negatively charged ends of other polar compounds (including water). Likewise, the negatively charged ends strongly attract the positively charged ends of polar compounds. These forces are most readily apparent in the tendency of water molecules to stick together, as in the formation of water beads or raindrops.

Water Weight
Did you know that more than 60 percent of your body weight consists of water?

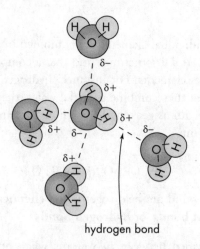

hydrogen bond

These types of intermolecular attractions are called hydrogen bonds. Hydrogen bonds are weak chemical bonds that form when a hydrogen atom that is covalently bonded to one electronegative atom is also attracted to another electronegative atom. Water molecules are held together by hydrogen bonds.

Although hydrogen bonds are individually weak, they are strong when present in large numbers. Because water reacts well with other polar substances, it makes a great solvent; it can dissolve many kinds of substances. The hydrogen bonds that hold water molecules together contribute to a number of special properties, including **cohesion**, **adhesion**, **surface tension**, **high heat capacity**, and **expansion on freezing**.

Cohesion and Adhesion

As mentioned above, water molecules have a strong tendency to stick together. That is, water exhibits cohesive forces. These forces are extremely important to life. For instance, during transpiration, water molecules evaporate from a leaf, "pulling" on neighboring water molecules. These, in turn, draw up the molecules immediately behind them, and so on, all the way down the plant vessels. The resulting chain of water molecules enables water to move up the stem.

Water molecules also like to stick to other substances—that is, they're **adhesive**. Have you ever tried to separate two glass slides stuck together by a film of water? They're difficult to separate because of the water sticking to the glass surfaces.

These two forces taken together—cohesion and adhesion—account for the ability of water to rise up the roots, trunks, and branches of trees. Since this phenomenon occurs in thin vessels, it's called **capillary action**.

Surface Tension

The cohesion of water molecules contributes to another property of water, its surface tension. Like a taut trampoline, the surface of water has a tension to it. The water molecules are stuck together and light things like leaves and water striders can sit atop the surface without sinking.

High Heat Capacity

Another remarkable property of water is its high heat capacity. What's heat capacity? Your textbook will give you a definition something like this: "heat capacity is the quantity of heat required to change the temperature of a substance by 1 degree." What does that mean? In plain English, heat capacity refers to the ability of a substance to resist temperature changes. For example, when you heat an iron kettle, it gets hot pretty quickly. Why? Because it has a *low* specific heat. It doesn't take much heat to increase the temperature of the kettle. Water, on the other hand, has a high heat capacity. You have to add a lot of heat to get an increase in temperature. The boiling point of water is pretty darn high. Liquid water is essential for cellular structures, and it is important that it doesn't boil away if someone goes out on a warm day. Water's ability to resist temperature changes is one of the things that helps keep the temperature in our oceans fairly stable. It's also why organisms that are mainly made up of water, like us, are able to keep a constant body temperature.

Expansion on Freezing

One of the most important properties of water is yet another result of hydrogen bonding. When four water molecules are bound in a solid lattice of ice, the hydrogen bonds actually cause solid water to expand on freezing. In most materials, when the molecules lose kinetic energy and cool from liquid to solid, the molecules get more dense, moving closer together. However, in liquid water, the molecules are slightly more dense than in solid water. Because water expands on freezing, becoming slightly less dense than liquid water, ice can crack lead pipes in the winter, or a soda can will pop if it is left in your cold car. The important consequence of this property of water is that ice floats on the top of lakes or streams, allowing animal life to live underneath the ice. If ice was denser than water, it would sink to the bottom, freezing the body of water solid. All aquatic life would be unable to survive.

So let's review the unique properties of water:

- Water is polar and can dissolve other polar substances.
- Water has cohesive and adhesive properties.
- Water has a high heat capacity.
- Water expands when it freezes (or, ice floats).

ACIDS AND BASES

We just said that water is important because most reactions occur in watery solutions. Well, there's one more thing to remember: reactions are also influenced by whether the solution in which they occur is **acidic**, **basic**, or **neutral**.

What makes a solution acidic or basic? A solution is acidic if it contains a lot of *hydrogen ions* (H^+). That is, if you dissolve an acid in water, it will release a lot of hydrogen ions. When you think about acids, you usually think of substances with a sour taste, like lemons. For example, if you squeeze a little lemon juice into a glass of water, the solution will become acidic. That's because lemons contain citric acid, which releases a lot of H^+ into the solution.

Important Formulas
Don't forget to check out what formulas will be given on the equations and formulas sheet.

Bases, on the other hand, do not release hydrogen ions when added to water. They release a lot of *hydroxide ions* (OH⁻). These solutions are said to be **alkaline** (the fancy name for a basic solution). Bases usually have a slippery consistency. Common soap, for example, is composed largely of bases.

The acidity or alkalinity of a solution can be measured with a **pH scale**. The pH scale is numbered from 1 to 14. The midpoint, 7, is considered neutral pH. The concentration of hydrogen ions in a solution will indicate whether it is acidic, basic, or neutral. If a solution contains a lot of hydrogen ions, then it will be acidic and have a low pH. Here's the trend:

> An increase in H⁺ ions causes a decrease in the pH.
>
> $$pH = -\log [H^+]$$

One more thing to remember: the pH scale is not a linear scale—it's logarithmic. That is, a change of *one* pH number actually represents a *tenfold* change in hydrogen ion concentration. For example, a pH of 3 is actually ten times more acidic than a pH of 4. This is also true in the reverse direction: a pH of 4 represents a tenfold decrease in acidity compared to a pH of 3.

Therefore, as the concentration of H⁺ ions increases by a factor of 10, the pH becomes one number smaller. For example, stomach acid has a pH of 2, and if we use the equation, we discover that the concentration of H⁺ ions in stomach acid is 10^{-2} *M*. This is pretty high when you consider the other extreme is lye, which has a pH of 14, a concentration of H⁺ ions around 10^{-14} *M*! Use your calculator to double-check that these numbers are correct.

You'll notice from the scale below that stronger acids have lower pHs. If a solution has a low concentration of hydrogen ions, it will have a high pH.

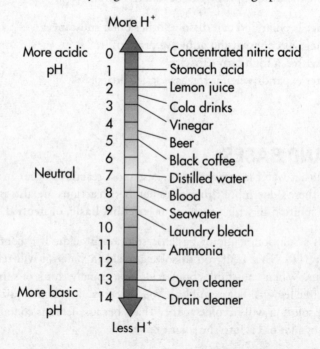

The equation for pH is listed on the AP Biology Equations and Formulas sheet. However, you will not be expected to perform calculations using this equation. Instead, you should understand how the equation works and when pH calculations are useful.

ORGANIC MOLECULES

Now that we've discussed chemical compounds in general, let's talk about a special group of compounds. Most of the chemical compounds in living organisms contain a skeleton of carbon atoms surrounded by hydrogen atoms and often other elements. These molecules are known as **organic compounds**. By contrast, molecules that do not contain carbon atoms are called **inorganic compounds**. For example, salt (NaCl) is an inorganic compound.

Carbon is important for life because it is a versatile atom, meaning that it has the ability to bind not only with other carbons but also with a number of other elements including nitrogen, oxygen, and hydrogen. The resulting molecules are key in carrying out the activities necessary for life.

To recap:

> - Organic compounds contain carbon atoms AND hydrogen atoms (and sometimes other things too).
> - Inorganic compounds don't contain both carbon atoms and hydrogen atoms.

Now let's focus on the following four classes of organic compounds central to life on Earth:

- carbohydrates
- proteins
- lipids
- nucleic acids

Most macromolecules are chains of building blocks, called **polymers**. The individual building blocks of a polymer are called **monomers**.

Carbohydrates

Organic compounds that contain carbon, hydrogen, and oxygen are called **carbohydrates**. They usually contain these three elements in a ratio of appoximately 1:2:1, respectively. We can represent the proportion of elements within carbohydrate molecules by the formula $C_nH_{2n}O_n$, which can be simplified as $(CH_2O)_n$.

Most carbohydrates are categorized as either **monosaccharides**, **disaccharides**, or **polysaccharides**. The term *saccharides* is a fancy word for "sugar." The prefixes *mono-*, *di-*, and *poly-* refer to the number of sugars in the molecule. *Mono-* means "one," *di-* means "two," and *poly-* means "many." A monosaccharide is therefore a carbohydrate made up of a single saccharide.

Monosaccharides: The Simplest Sugars

Monosaccharides, the simplest sugars, serve as an energy source for cells. The two most common sugars are (1) **glucose** and (2) **fructose**. Galactose, ribose, and deoxyribose are also monosaccharides.

Both glucose and fructose are six-carbon sugars with the chemical formula $C_6H_{12}O_6$. Glucose, the most abundant monosaccharide, is the most popular sugar around. Glucose is an important part of the food we eat and it is the favorite food made by plants during photosynthesis. Living organisms break down glucose in order to provide cells with energy. Almost all living organisms can perform this ancient biochemical process (which is called cell respiration; you'll learn more about this process in Chapter 6). In fact, the evolution of glycolysis, photosynthesis, and oxidative metabolism is thought to have occurred about 3 billion years ago! Fructose, the other monosaccharide you need to know for the test, is a common sugar in fruits.

Glucose and fructose can be depicted as either "straight" or "rings." Both of them are pretty easy to spot; just look for the carbon molecules with a lot of OHs and Hs attached to them.

Here are the two different forms.

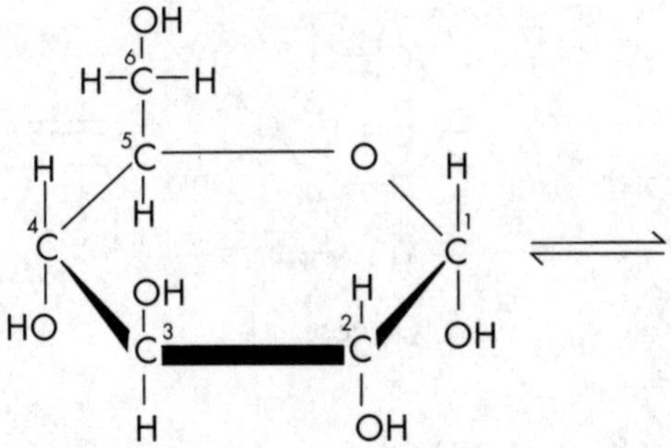

Ring form of glucose

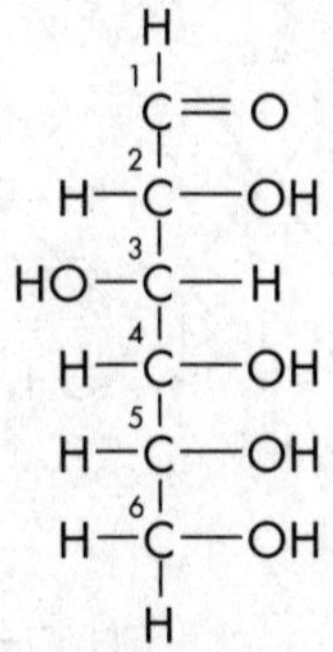

Straight-chain
form of glucose

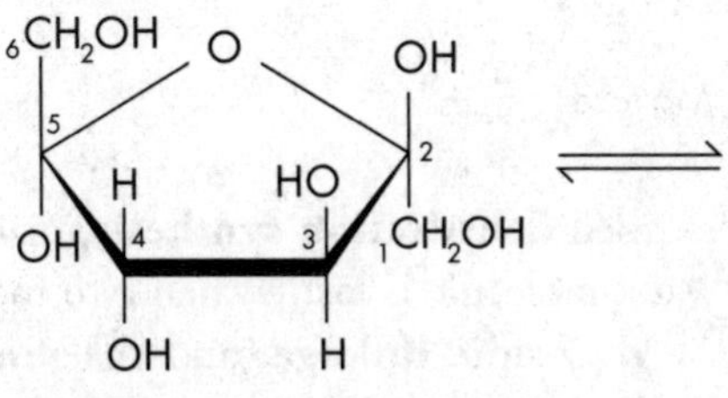

Ring form of fructose

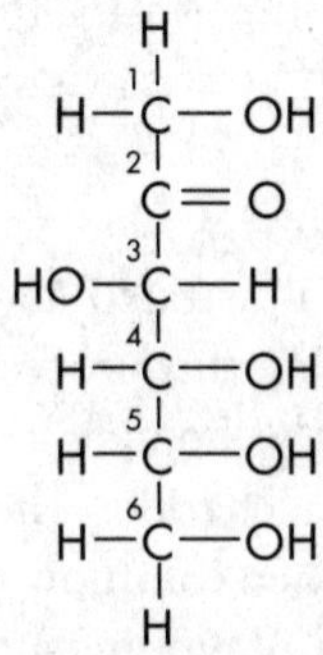

Straight-chain
form of fructose

Disaccharides

What happens when two monosaccharides are brought together? The hydrogen (–H) from one sugar molecule combines with the hydroxyl group (–OH) of another sugar molecule. What do H and OH add up to? Water (H_2O)!

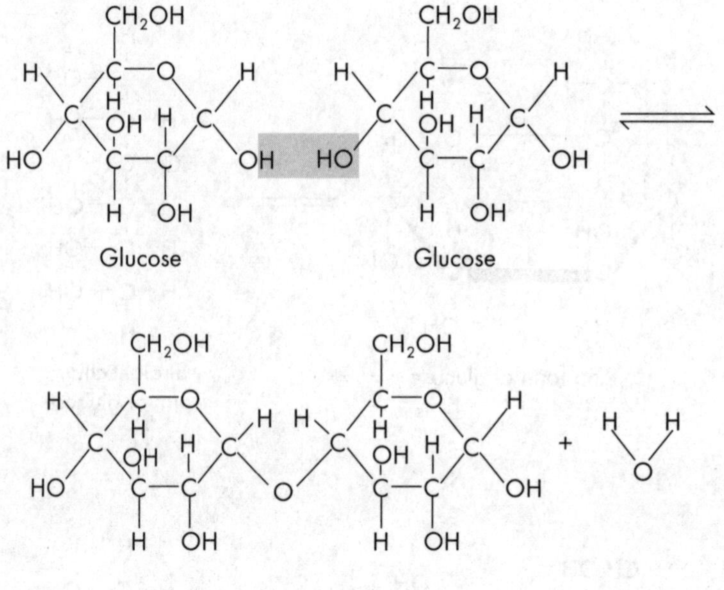

Glucose   Glucose

Maltose

The process illustrated above is called **dehydration synthesis**, or **condensation**, because during this process, a water molecule is lost. When two monosaccharides are joined, the bond is called a **glycosidic linkage,** and the resulting sugar is called a disaccharide. The disaccharide formed from two glucose molecules is maltose. Other common disaccharides include sucrose (table sugar) and lactose (found in dairy products).

Now, what if you want to break up the disaccharide and form two monosaccharides again? Just add water. That's called **hydrolysis**, which means "water" (*hydro-*) and "breaking" (*-lysis*). The water breaks the bond between the two monomers. Like condensation, hydrolysis is a common reaction in biology.

Maltose

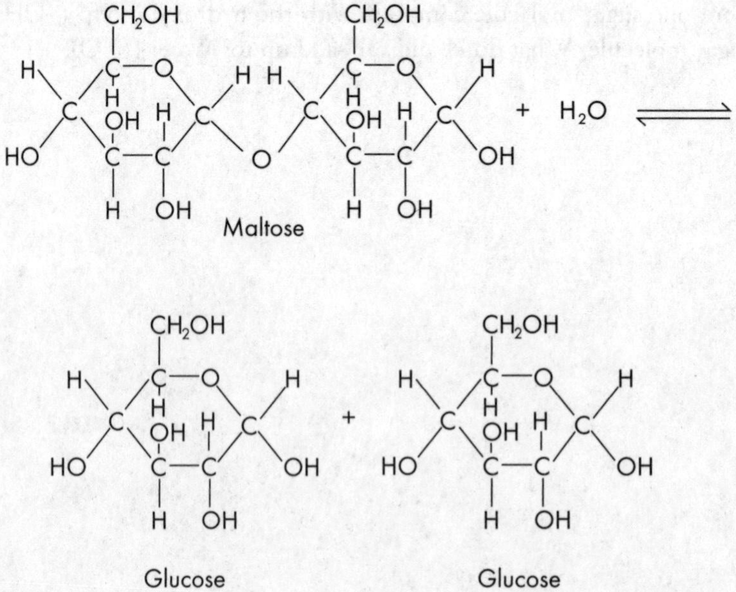

Glucose   Glucose

Polysaccharides

Polysaccharides are made up of many repeated units of monosaccharides. Polysaccharides can consist of branched or unbranched chains of monosaccharides. The most common polysaccharides you'll need to know for the test are **starch**, **cellulose**, and **glycogen**. Glycogen and starch are sugar storage molecules. Glycogen stores sugar in animals and starch stores sugar in plants. Cellulose, on the other hand, is made up of β-glucose and is a major part of the cell walls in plants. Its function is to lend structural support. Chitin, a polymer of β-glucose molecules, serves as a structural molecule in the walls of fungus and in the exoskeletons of arthropods.

Proteins

Proteins perform most of the work in your cells, and are important for structure, function, and regulation of your tissues and organs. There are thousands of different types of proteins. **Amino acids** are organic molecules that serve as the building blocks of proteins. They contain carbon, hydrogen, oxygen, and nitrogen atoms. There are 20 different amino acids commonly found in proteins. Fortunately, you don't have to memorize all 20 amino acids. However, you do have to remember that every amino acid has four important parts around a central carbon: an **amino group** (–NH$_2$), a **carboxyl group** (–COOH), a hydrogen, and an **R-group**.

Here's a typical amino acid.

Amino acids differ only in the R-group, which is also called the **side chain**. The R-group associated with an amino acid could be as simple as a hydrogen atom (as in the amino acid *glycine*) or as complex as a charged carbon skeleton (as in the amino acid *arginine*).

Glycine

Arginine

Digest This Tidbit!
Did you ever look at a pile of firewood and think, Yummy!? Probably not. Wood is almost entirely made of cellulose, which contains β-linked glucose that humans can't digest. When we eat plants, we can get nutrients from their stored starch, but not from the cellulose of their cell walls. This passes through us as fiber roughage. Fiber is important in the diet for other reasons, but you should probably leave the firewood alone.

When it comes to spotting an amino acid, simply keep an eye out for the amino group (NH_2); then look for the carboxyl molecule (COOH). The side chains for each amino acid vary greatly. You don't have to memorize the structures of the side chains for individual amino acids, but you should know that they can vary in composition, polarity, charge, and shape depending on the side chain that they have.

> How do the side chains of the amino acids differ?
>
> - Composition of elements (C, H, O, N, and S)
> - Polarity (polar, nonpolar)
> - Charge (neutral, positive, negative)
> - Shape (long chain, short chain, ring-shape)

Side chain polarity affects whether an animo acid is more hydrophobic or more hydrophilic.

You can see all 20 amino acids on the next page. We've put them in some common categories: nonpolar, polar uncharged, and polar charged. In addition to the three-letter abbreviations we've listed, each amino acid is also given a one-letter name. For example, histidine can be referred to as His or just H. Don't worry about memorizing these abbreviations for the test.

Check out the bottom of the chart on the next page: at physiological pH (which is about 7.4), two of the amino acids (glutamic acid and aspartic acid) donate a proton, making them negatively charged. In contrast, lysine and arginine accept protons at physiological pH, so they are usually positively charged.

Finally, notice that only two amino acids contain the atom sulfur: methionine and cysteine.

Naturally Occurring Amino Acids

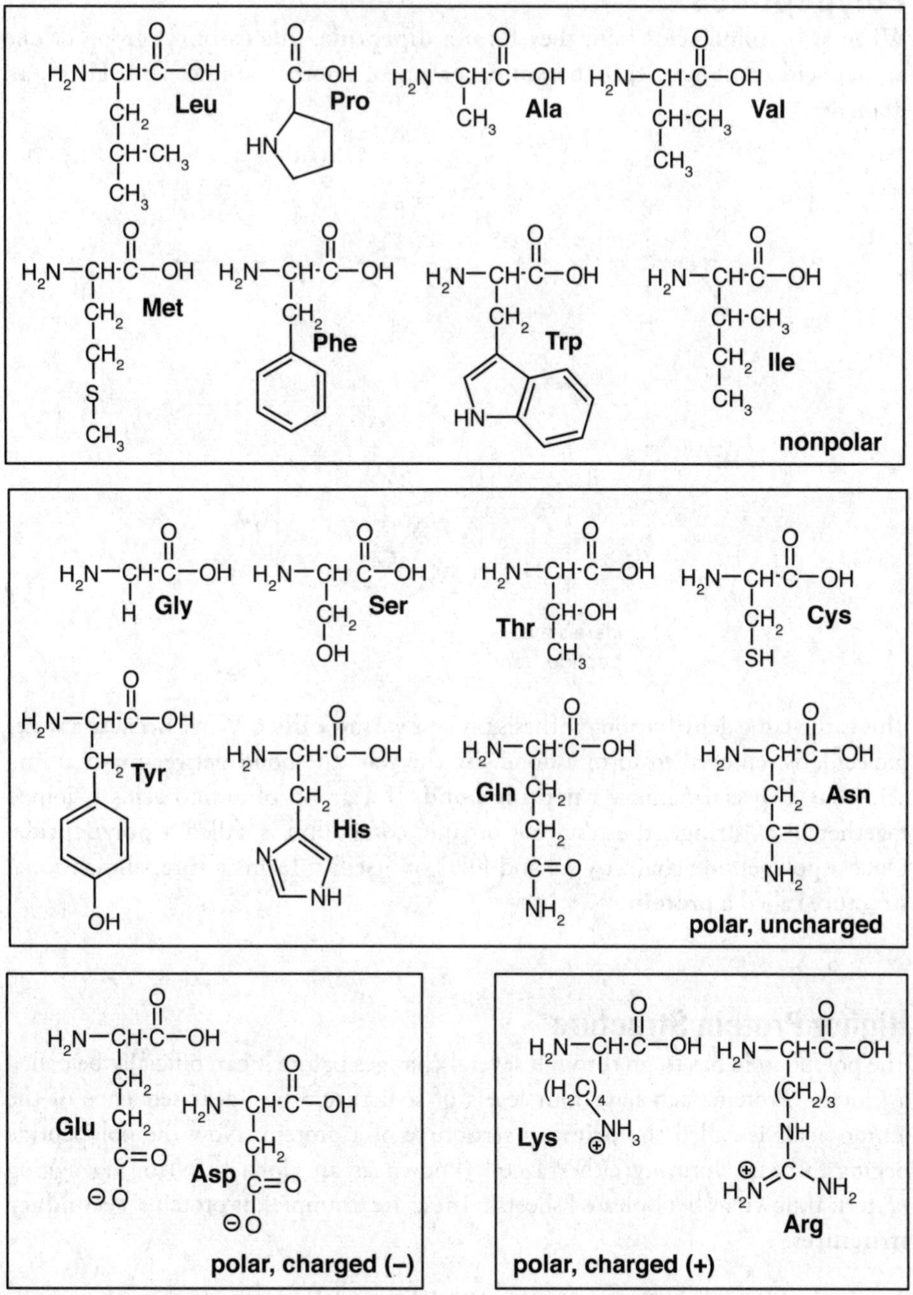

Leu
Pro
Ala
Val
Met
Phe
Trp
Ile
nonpolar
Gly
Ser
Thr
Cys
Tyr
His
Gln
Asn
polar, uncharged
Glu
Asp
polar, charged (–)
Lys
Arg
polar, charged (+)

We've included this reference page so you can familiarize yourself with these common acids. But don't worry! You don't have to memorize all these structures.

Polypeptides

When two amino acids join, they form a **dipeptide**. The carboxyl group of one amino acid combines with the amino group of another amino acid. Here's an example:

Here's the
peptide bond

Building a Chain
The new amino acid is always added on the carboxyl end of the existing chain.

This is the same dehydration synthesis process we saw earlier. Why? Because a water molecule is removed to form a bond. By the way, the bond between two amino acids has a special name—a **peptide bond**. If a group of amino acids is joined together in a "string," the resulting organic compound is called a **polypeptide**. Once a polypeptide chain twists and folds on itself, it forms a three-dimensional structure called a **protein**.

Higher Protein Structure

The polypeptide has to go through several changes before it can officially be called a protein. Proteins can have four levels of structure. The linear sequence of the amino acids is called the **primary structure** of a protein. Now the polypeptide begins to twist, forming either a coil (known as an alpha helix) or zigzagging pattern (known as beta-pleated sheets). These are examples of proteins' **secondary structures**.

A polypeptide folds and twists because the different R-groups of the amino acids are interacting with each other. Remember, each R-group is unique and has a particular size, shape, charge, and so on. that allows it to react to things around it. Depending on which amino acids are in a protein and the order that they are in, the protein can twist and fold in very different ways. This is why proteins can form so many different shapes.

The secondary structure is formed by amino acids that interact with other amino acids close by in the primary structure. However, after the secondary structure reshapes the polypeptide, amino acids that were far away in the primary structure arrangement can now also interact with each other. This is called the **tertiary structure**. Because most proteins are found in an aqueous environment, hydrophilic amino acids and regions of the peptide chain are often located on the exterior of the protein. Hydrophobic amino acids and regions are usually found on the interior of proteins. Sometimes, two cysteine amino acids can react with each other to form a covalent disulfide bond that stabilizes the tertiary structure. Tertiary structure often minimizes the free energy of the molecule and locks it into a stable 3D shape.

Lastly, several different polypeptide chains sometimes interact with each other to form a **quaternary structure**. The different polypeptide chains that come together are often called subunits of the final whole protein.

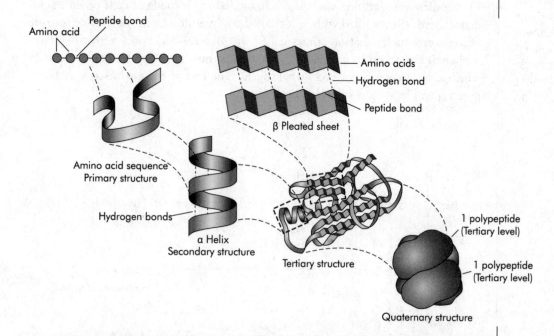

Only proteins that have folded correctly into their specific three-dimensional structure can perform their function properly. Mistakes in the amino acid chain can create oddly shaped proteins that are nonfunctional. One more thing: in some cases, the folding of proteins involves other proteins known as **chaperone proteins** (**chaperonins**). They help the protein fold properly and make the process more efficient.

Lipids

Like carbohydrates, **lipids** consist of carbon, hydrogen, and oxygen atoms, but not in the 1:2:1 ratio typical of carbohydrates. The most common examples of lipids are **triglycerides, phospholipids**, and **steroids.** Lipids are important because they function as structural components of cell membranes, sources of insulation, signaling molecules, and a means of energy storage.

Triglycerides

Our bodies store fat in tissue called adipose, which is made of cells called adipocytes; these cells are filled with lipids called triglycerides. Each triglyceride is made of a glycerol molecule (sometimes called the glycerol backbone) with three fatty acid chains attached to it. A fatty acid chain is mostly a long chain of carbons in which each carbon is covered in hydrogen. One end of the chain has a carboxyl group (just like we saw in an amino acid).

Let's take a look.

To make a triglyceride, each of the carboxyl groups (–COOH) of the three fatty acids must react with one of the three hydroxyl groups (–OH) of the glycerol molecule. This happens by the removal of a water molecule. So, the creation of a fat requires the removal of three molecules of water. Once again, what have we got? You probably already guessed it—dehydration synthesis! The linkage now formed between the **glycerol** molecule and the fatty acids is called an **ester linkage**. A fatty acid can be **saturated** with hydrogens along its long carbon chain or it can have a few gaps where double bonds exist instead of a hydrogen. If there is a double bond in the chain, then it is an **unsaturated** fatty acid. A **polyunsaturated** fatty acid has many double bonds within the fatty acid.

Phospholipids

Another special class of lipids is known as phospholipids. Phospholipids contain two fatty acid "tails" and one negatively charged phosphate "head." Take a look at a typical phospholipid.

Phospholipids are extremely important, mainly because of some unique properties they possess, particularly with regard to water.

The two fatty acid tails are **hydrophobic** ("water-hating"). In other words, just like oil and vinegar, fatty acids and water don't mix. The reason for this is that fatty acid tails are nonpolar, and nonpolar substances don't mix well with polar ones, such as water.

On the other hand, the phosphate "head" of the lipid is **hydrophilic** ("water-loving"), meaning that it does mix well with water. Why? It carries a negative charge, and this charge draws it to the positively charged end of a water molecule. Since a phospholipid has both a hydrophilic region and a hydrophobic region, it is an **amphipathic molecule**. One side of a phospholipid loves to hang out with water, and the other side hates to.

Thus, the two fatty acid chains orient themselves away from water, while the phosphate portion orients itself toward the water. Keep these properties in mind. We'll see later how this orientation of phospholipids in water relates to the structure and function of cell membranes.

Cholesterol

Cholesterol is another important type of lipid. Cholesterol is a four-ringed molecule that is found here and there in membranes. It generally increases membrane fluidity, except at very high temperatures when it helps to hold things together instead. Cholesterol is also important for making certain types of hormones and for making vitamin D. Here are some examples:

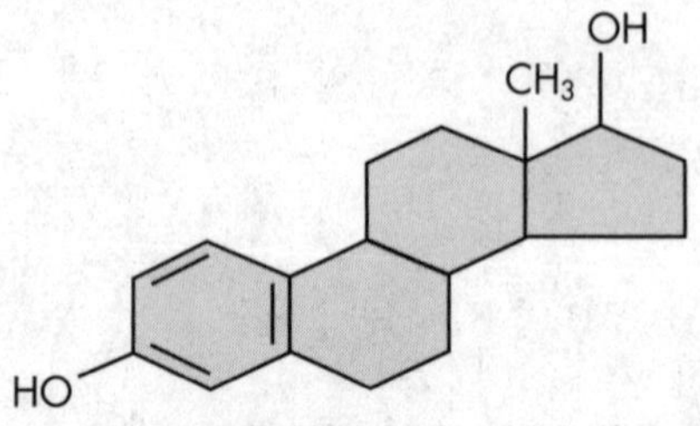

Estradiol

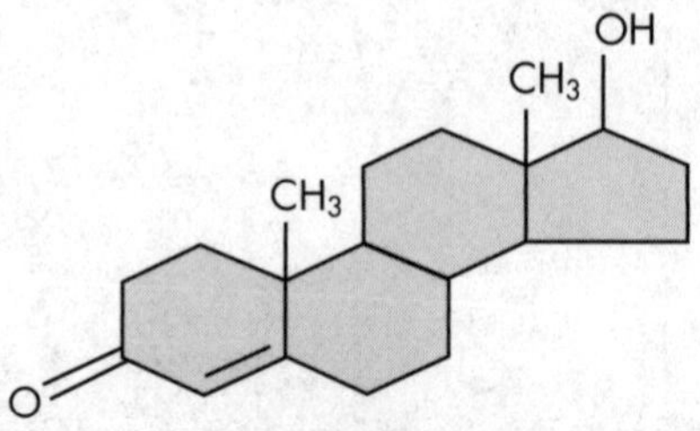

Testosterone

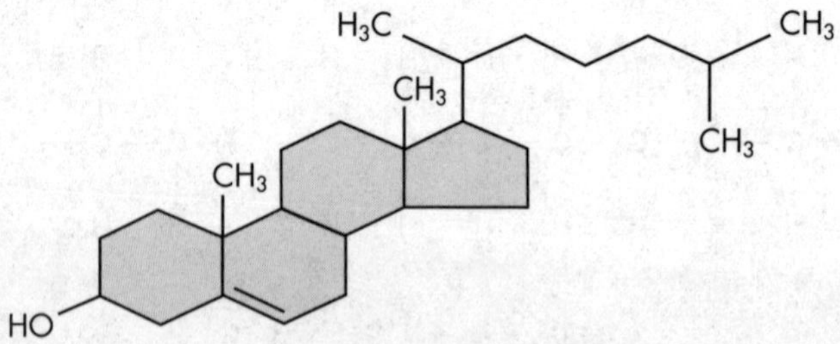

Cholesterol

Nucleic Acids

The fourth class of organic compounds is the **nucleic acids.** Like proteins, nucleic acids contain carbon, hydrogen, oxygen, and nitrogen, but nucleic acids also contain phosphorus. Nucleic acids are molecules that are made up of simple units called **nucleotides.** For the AP Biology Exam, you'll need to know about two kinds of nucleic acids: **deoxyribonucleic acid (DNA)** and **ribonucleic acid (RNA).**

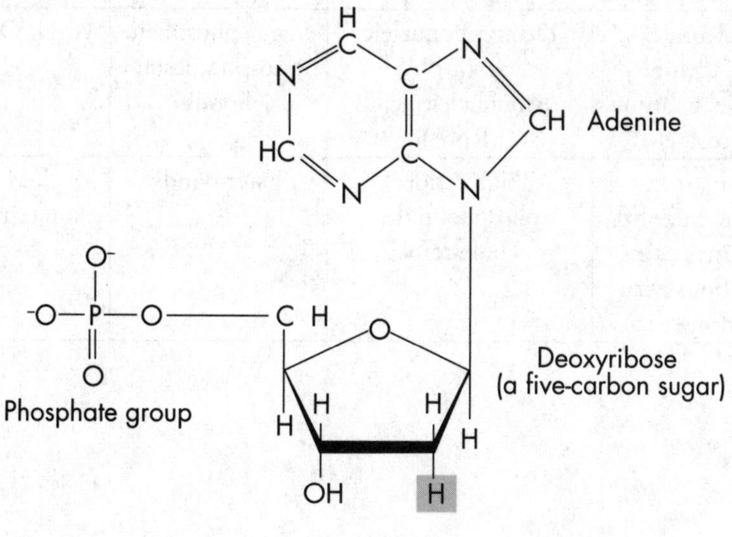

DNA

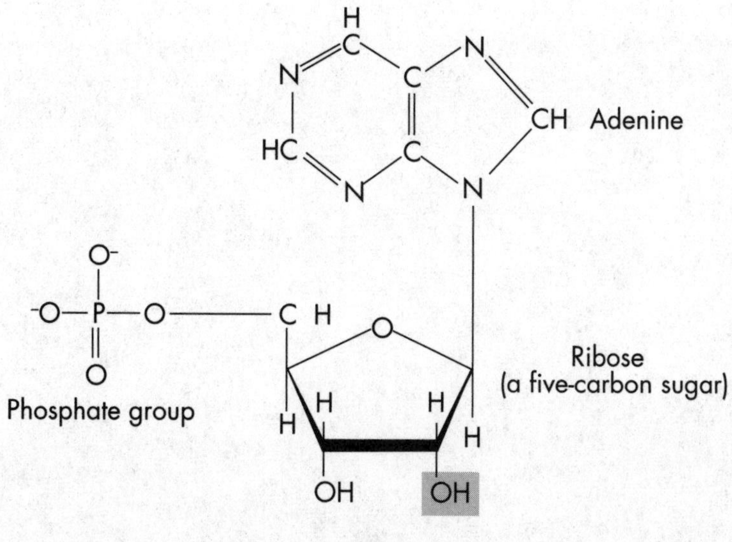

RNA

Need-to-Know Nucleic Acids

For the AP Biology Exam, you'll be expected to know about two types of nucleic acids, DNA and RNA, which are shown here.

DNA is important because it contains the hereditary "blueprints" of all life. RNA is important because it's essential for protein synthesis. We'll discuss DNA and RNA in greater detail when we discuss heredity (see Chapter 8). The following table summarizes the important macromolecules discussed here.

Important Biological Macromolecules

Macromolecule	Monomer	Polymer	Linkage Bond	Elements
Carbohydrates	Monosaccharide (e.g. Glucose)	Polysaccharide (e.g. Starch, glycogen, cellulose)	Glycosidic linkage	C, H, O,
Proteins	Amino Acid (e.g. Glycine)	Polypeptide (e.g. Actin)	Peptide bond	C, H, O, N, S
Nucleic Acids	Nucleotides (e.g. Adenine, thymine, guanine, cytosine)	Deoxyribonucleic acid (DNA) Ribonucleic acid (RNA)	Sugar-phosphate phosphodiester bonds	C, H, O, N, P
Lipids	Not a true polymer, but often contains chains of carbons with hydrogens	Triglycerides, phospholipids, cholesterol	Ester bonds	C, H, O, sometimes P

KEY TERMS

elements
oxygen
carbon
hydrogen
nitrogen
trace elements
atom
protons
neutrons
electrons
nucleus
isotopes
radiometric dating
compound
chemical reaction
chemical bond
ionic bond
covalent bond
hydrogen bond
ions
nonpolar covalent
polar covalent
polar molecule
cohesion
adhesion
surface tension
heat capacity
expansion on freezing
adhesive
capillary action
acidic
basic
neutral
alkaline
pH scale
organic compounds
inorganic compounds
polymer
monomer
carbohydrates
monosaccharides
disaccharides

polysaccharides
glucose
fructose
dehydration synthesis
  (condensation)
glycosidic linkage
hydrolysis
starch
cellulose
glycogen
amino acids
amino group
carboxyl group
R-group
side chain
dipeptide
peptide bond
polypeptide
protein
primary structure
secondary structure
tertiary structure
quaternary structure
chaperone proteins (chaperonins)
lipids
triglycerides
phospholipids
steroids
glycerol
ester linkage
saturated
unsaturated
polyunsaturated
hydrophobic
hydrophilic
amphipathic molecule
cholesterol
nucleic acids
nucleotides
deoxyribonucleic acid (DNA)
ribonucleic acid (RNA)

Summary

o Compounds are composed of two or more types of elements in a fixed ratio.

o Molecules are held together by ionic bonds and covalent bonds.

o Hydrogen bonding is essential for the following emergent properties of water:
  - cohesion
  - adhesion
  - surface tension
  - high heat capacity
  - expansion on freezing
  - polar structure
  - ability to dissolve other polar (hydrophilic) substances but inability to dissolve nonpolar (hydrophobic) substances

o pH is important for biological reactions. Solutions may be acidic, basic or alkaline, or neutral.

o Organic molecules contain carbon atoms. Many molecules in biology are composed of several monomers bound together into polymers. The creation of these polymers occurs through dehydration synthesis reactions. The breakdown of these polymers occurs through hydrolysis reactions.

o Important organic macromolecules in biology include:
  - Carbohydrates: monosaccharides link by glyosidic bonds to form di- and polysaccharides. Examples are glucose, fructose, glycogen, cellulose, and starch.
  - Proteins: 20 different amino acids, each with a different R-group, link together via peptide bonds to form polypeptides. The chain further folds up into special shapes to make the final protein structure. The structure can be divided into primary (amino acid sequence), secondary (alpha helices and beta-pleated sheets), tertiary (disulfide bonds), and quaternary (multiple polypeptides).
  - Lipids: chains called fatty acids (either saturated or unsaturated) combine on a glycerol backbone to form phospholipids and triglycerides. Cholesterol is another type of lipid. Lipids are hydrophobic.
  - Nucleic acids: chains of either DNA or RNA nucleotides. They carry the genetic recipes in the body.

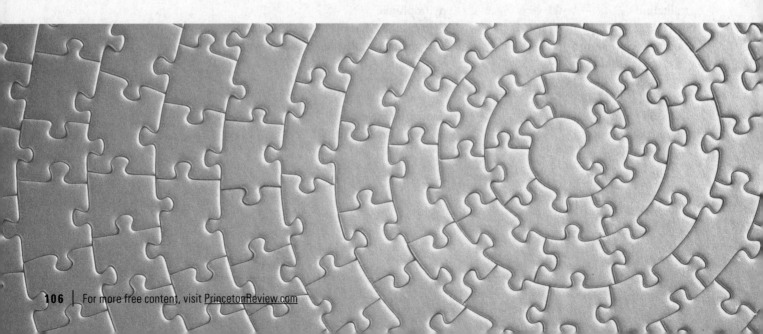

Chapter 4 Drill

Answers and explanations can be found in Chapter 15.

1. Water is a critical component of life due to its unique structural and chemical properties. Which of the following does NOT describe a way that the exceptional characteristics of water are used in nature to sustain life?

   (A) The high heat capacity of water prevents lakes and streams from rapidly changing temperature and freezing completely solid in the winter.
   (B) The high surface tension and cohesiveness of water facilitates capillary action in plants.
   (C) The low polarity of water prevents dissolution of cells and compounds.
   (D) The high intermolecular forces of water, such as hydrogen bonding, result in a boiling point which exceeds the tolerance of most life on the planet.

2. The intracellular pH of human cells is approximately 7.4. Yet, the pH within the lumen (inside) of the human stomach averages 1.5. Which of the following accurately describes the difference between the acidity of the cellular and gastric pH?

   (A) Gastric juices contain approximately 6 times more H^+ ions than the intracellular cytoplasm of cells and are more acidic.
   (B) Gastric juices contain approximately 1,000,000-fold more H^+ ions than the intracellular cytoplasm of cells and are more acidic.
   (C) The intracellular cytoplasm of cells contain approximately 6 times more H^+ ions than gastric juices and is more acidic.
   (D) The intracellular cytoplasm of cells contains approximately 1,000,000-fold more H^+ ions than gastric juices and is more acidic.

3. Amino acids are the basic molecular units which compose proteins. All life on the planet forms proteins by forming chains of amino acids. Which labeled component of the amino acid structure of phenylalanine shown below will vary from amino acid to amino acid?

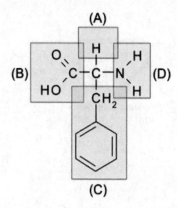

Questions 4–6 refer to the following paragraph and diagram.

In 1953, Stanley Miller and Harold Urey performed an experiment at the University of Chicago to test the hypothesis that the conditions of the early Earth would have favored the formation of larger, more complex organic molecules from basic precursors. The experiment, as shown below, consisted of sealing basic organic chemicals (representing the atmosphere of the primitive Earth) in a flask, which was exposed to electric sparks (to simulate lightning) and water vapor.

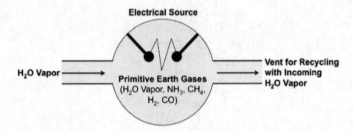

After one day of exposure, the mixture in the flask had turned pink in color, and later analysis showed that at least 10% of the carbon had been transformed into simple and complex organic compounds including at least 11 different amino acids and some basic sugars. No nucleic acids were detected in the mixture.

4. Which of the following contradicts the hypothesis of the experiment that life may have arisen from the formation of complex molecules in the conditions of the primitive Earth?

   (A) Complex carbon-based compounds were generated after only one day of exposure to simulated primitive Earth conditions.
   (B) Nucleic acid compounds such as DNA and RNA were not detected in the mixture during the experiment.
   (C) Over half of known amino acids involved in life were detected in the mixture during the experiment.
   (D) Basic sugar molecules were generated and detected in the mixture during the experiment.

5. Some amino acids, such as cysteine (shown below) and methionine, could not be formed in this experiment. Which of the following best explains why these molecules could not be detected?

   (A) The chemical reactions necessary to create amino acids such as cysteine and methionine require more energy than the simulated lightning provided in the experiment.
   (B) The chemical reactions necessary to create amino acids such as cysteine and methionine require enzymes for catalysis to occur, which were not included in the experiment.
   (C) Sulfur-based compounds were not included in the experiment.
   (D) Nitrogen-based compounds were not included in the experiment.

6. A scientist believes that the Miller-Urey experiment failed to yield the remaining amino acids and the nucleic acids because of the absence of critical chemical substrates that would have existed on the primordial Earth due to volcanism. Which of the following basic compounds, which are associated with volcanism, would NOT need to be added in a follow-up Miller-Urey experiment?

   (A) H_2S (gas)
   (B) SiO_2 (silica)
   (C) SO_2
   (D) H_3PO_4 (phosphoric acid)

7. Catabolism refers to breaking down complex macromolecules into their basic components. Many biological processes use hydrolysis for catabolism. Hydrolysis of proteins could directly result in

    (A) free water
    (B) adenine
    (C) cholesterol
    (D) dipeptides

8. Which of the following contain both hydrophilic and hydrophobic properties and are often found in cell plasma membranes?

    (A) Nucleotides
    (B) Phospholipids
    (C) Water
    (D) Amino acids

9. Maltotriose is a trisaccharide composed of three glucose molecules linked through α-1,4 glycosidic linkages formed via dehydration synthesis. What would the formula be for maltotriose?

    (A) $C_{18}H_{36}O_{18}$
    (B) $C_{18}H_{10}O_{15}$
    (C) $C_{18}H_{32}O_{16}$
    (D) $C_3H_6O_3$

10. A radioactive isotope of hydrogen, ^3H, is called tritium. Tritium differs from the more common form of hydrogen because

    (A) it contains two neutrons and one proton in its nucleus
    (B) it contains one neutron and two protons in its nucleus
    (C) it differs by its atomic number
    (D) it is radioactive and therefore gives off one electron

REFLECT

Respond to the following questions:

- Which topics from this chapter do you feel you have mastered?

- Which content topics from this chapter do you feel you need to study more before you can answer multiple-choice questions correctly?

- Which content topics from this chapter do you feel you need to study more before you can effectively compose a free response?

- Was there any content that you need to ask your teacher or another person about?

Chapter 5
Cell Structure
and Function

LIVING THINGS

All living things are composed of **cells**. According to the cell theory, the cell is life's basic unit of structure and function. This simply means that the cell is the smallest unit of living material that can carry out all the activities necessary for life.

It may seem strange that we are still made of tiny little cells. Why not just be a living thing that is one giant cell?

One reason for this is because compartments allow more specialization. As we will see, compartmentalization is an important part of the organization of unicellular and multicellular organisms.

The second reason we can't just be giant cells is the **surface area-to-volume ratio**. There is always lots of exchange going on between the inside of things and the outside. This ratio must be kept large so that there is lots of space to do these exchanges. Some parts of living things are special because they have lots of folds to increase the surface area.

Therefore, even the largest things are still made of cells.

Surface area-to-volume ratio
This is a concept that comes up often. Use the volume formulas on the Equations and Formulas sheet to practice calculating the surface area-to-volume ratios of different-sized cubes and spheres.

MICROSCOPES AND THE STUDY OF CELLS

Cells are studied with different types of microscopes. **Light microscopes** (for example, the compound microscopes commonly found in labs) are used to study stained or living cells. They can magnify the size of an organism up to 1,000 times. **Electron microscopes** are used to study detailed structures of a cell that cannot be easily seen or observed by light microscopy. They are capable of resolving structures as small as a few nanometers in length, such as individual virus particles or the pores on the surface of the nucleus.

TYPES OF CELLS AND ORGANELLES

For centuries, scientists have known about cells. However, it wasn't until the development of the electron microscope that scientists were able to figure out what cells do. We now know that there are two distinct types of cells: **prokaryotic cells** and **eukaryotic cells**.

A prokaryotic cell, which is a lot smaller than a eukaryotic cell, is relatively simple. Bacteria and archaea are examples of prokaryotes. The inside of the cell is filled with a substance called **cytoplasm**. The genetic material in a prokaryote is one continuous, circular DNA molecule that is found free in the cell in an area called the **nucleoid** (this is not the same as a nucleus!). Most prokaryotes have a **cell wall** composed of peptidoglycans that surrounds a lipid layer called the **plasma membrane**. Prokaryotes also have ribosomes (though smaller than those found in eukaryotic cells). Some bacteria may also have one or more **flagella**, which are long projections used for motility (movement) and they might have a thick **capsule** outside their cell wall to give them extra protection.

Eukaryotic cells are more complex than prokaryotic cells. Fungi, protists, plants, and animals are eukaryotes. Eukaryotic cells are organized into many smaller structures called **organelles**. Some of these organelles are the same structures seen in prokaryotic cells, but many are uniquely eukaryotic. A good way to remember

Endosymbiosis
It is believed that chloroplasts and mitochondria used to be independent prokaryotic cells. They have since become a permanent part of eukaryotic cells.

the difference is that prokaryotes do not have any membrane-bound organelles. Their only membrane is the plasma membrane.

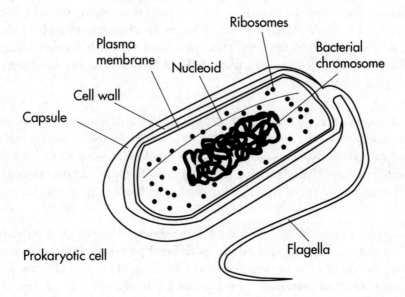

Organelles

A eukaryotic cell is like a microscopic factory. It's filled with organelles, each of which has its own special tasks. Let's take a tour of a eukaryotic cell and focus on the structure and function of each organelle. Here's a picture of a typical animal cell and its principal organelles.

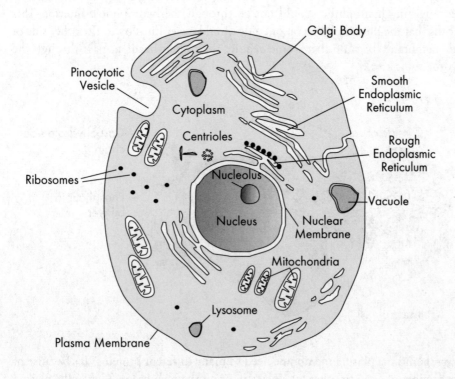

The surface area of a cell must be large enough to allow sufficient exchange of materials. In general, smaller cells have a more favorable surface area-to-volume ratio.

Plasma Membrane

The cell has an outer envelope known as the plasma membrane. Although the plasma membrane appears to be a simple thin layer surrounding the cell, it's actually a complex, double-layered structure made up of mostly phospholipids and proteins. The *hydrophobic* fatty acid tails face inward and the *hydrophilic* phosphate heads face outward. It is called a **phospholipid bilayer** since two lipid layers are forming a hydrophobic sandwich.

The plasma membrane is important because it regulates the movement of substances into and out of the cell. The membrane itself is semipermeable, meaning that only certain substances, namely small hydrophobic molecules (such as O_2 and CO_2), pass through it unaided. Anything large and/or hydrophilic can pass through the membrane only via special tunnels (discussed more later in this chapter).

Many proteins are associated with the cell membrane. Some of these proteins are loosely associated with the lipid bilayer (**peripheral proteins**). They are located on the inner or outer surface of the membrane. Others are firmly bound to the plasma membrane (**integral proteins**). These proteins are amphipathic, which means that their hydrophilic regions extend out of the cell or into the cytoplasm, while their hydrophobic regions interact with the tails of the membrane phospholipids. Some integral proteins extend all the way through the membrane (**transmembrane proteins**).

This arrangement of phospholipids and proteins is known as the **fluid-mosaic model**. This means that each layer of phospholipids is flexible, and it is a mosaic because it is peppered with different proteins and carbohydrate chains. Remember, anything hydrophilic should not go through the hydrophobic interior. This means that the phospholipids on one side should never flip-flop to the other side of the membrane (because that would require their polar heads to pass through the hydrophobic area).

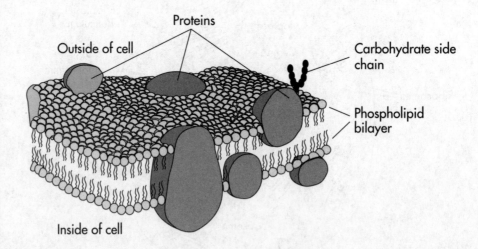

Why should the plasma membrane need so many different proteins? It's because of the number of activities that take place in or on the membrane. Generally, plasma membrane proteins fall into several broad functional groups. Some membrane proteins (**adhesion proteins**) form junctions between adjacent cells. Others (**receptor proteins**), such as hormones, serve as docking sites for arrivals at the cell. Some proteins (**transport proteins**) form pumps that use ATP to actively transport solutes across the membrane. Others (**channel proteins**) form channels that selectively allow the passage of certain ions or molecules. Finally, some proteins (**cell surface markers**), such as glycoproteins, are exposed on the extracellular surface and play a role in cell recognition and adhesion.

Attached to the surface of some proteins are **carbohydrate side chains**. They are found only on the outer surface of the plasma membrane. As mentioned above, cholesterol molecules are also found in the phospholipid bilayer because they help stabilize membrane fluidity in animal cells.

The Nucleus

The **nucleus** is usually the largest organelle in the cell. The nucleus not only directs what goes on in the cell, but is also responsible for the cell's ability to reproduce. It's the home of the hereditary information—DNA—which is organized into large structures called **chromosomes**. The most visible structure within the nucleus is the **nucleolus**, which is where rRNA is made and ribosomes are assembled.

Ribosomes

The **ribosomes** are sites of protein synthesis. Their job is to manufacture all the proteins required by the cell or secreted by the cell. Ribosomes are round structures composed of two subunits, the large subunit and the small subunit. The structure is composed of rRNA and proteins. Ribosomes can be either free floating in the cell or attached to another structure called the **endoplasmic reticulum (ER)**.

Endoplasmic Reticulum (ER)

The endoplasmic reticulum (ER) is a continuous channel that extends into many regions of the cytoplasm. The region of the ER that is attached to the nucleus and "studded" with ribosomes is called the **rough ER** (RER). Proteins generated in the rough ER are trafficked to or across the plasma membrane, or they are used to build Golgi bodies, lysosomes, or the ER. The region of the ER that lacks ribosomes is called the **smooth ER** (SER). The smooth ER makes lipids, hormones, and steroids and breaks down toxic chemicals.

Golgi Bodies

The **Golgi bodies**, which look like stacks of flattened sacs, also participate in the processing of proteins. Once the ribosomes on the rough ER have completed synthesizing proteins, the Golgi bodies modify, process, and sort the products. They're the packaging and distribution centers for materials destined to be sent out of the cell. They package the final products in little sacs called **vesicles**, which carry products to the plasma membrane. Golgi bodies are also involved in the production of lysosomes.

Ribosomes are found in all forms of life, although prokaryotic and eukaryotic ribosomes differ slightly in size. Eukaryotic ribosomes (80s) are slightly larger than prokaryotic ribosomes (70s). The "s" is a unit of measurement used to compare sizes of small things.

The flattened membrane disks of the endoplasmic reticulum and Golgi apparatus are called cisternae. A Golgi stack may contain anywhere from 3 to 20 cisternae, and these are divided into cis (close to the RER), medial (in the middle), and trans (farthest from the RER). Each cisternae type contains different types of enzymes, and molecules can move between cisternae via vesicles.

There are lots of examples in biology in which folding is used to increase surface area. This is similar to the surface area-to-volume ratio concept we mentioned earlier. Some common examples are inner mitochondrial membrane, ER membrane, convoluted membranes in chloroplasts, brain tissue, small intestine (villi and microvilli), alveoli in the lungs, or root hairs in plants.

Mitochondria

Another important organelle is the mitochondrion. **Mitochondria** are often referred to as the "powerhouses" of the cell. They're power stations responsible for converting energy from organic molecules into useful energy for the cell. The most common energy molecule in the cell is **adenosine triphosphate (ATP)**.

The mitochondrion is usually an easy organelle to recognize because it has a unique oblong shape and a characteristic double membrane consisting of an inner portion and an outer portion. The inner mitochondrial membrane forms folds known as **cristae** and separates the innermost area called the matrix from the intermembrane space. The outer membrane separates the intermembrane space from the cytoplasm. As we'll see later, most of the production of ATP is done on the cristae. Having folds in the membrane increases the surface area for making ATP.

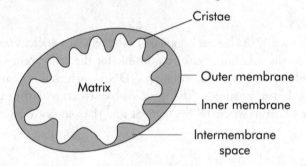

Lysosomes

Throughout the cell are small, membrane-enclosed structures called **lysosomes**. These tiny sacs carry digestive enzymes, which they use to break down old, worn-out organelles, debris, or large ingested particles. Lysosomes make up the cell's clean-up crew, helping to keep the cytoplasm clear of unwanted flotsam. Lysosomes contain hydrolytic enzymes that function only at acidic pH, which is enclosed inside the lumen of the lysosome. Lysosomes are made when vesicles containing specific enzymes from the trans Golgi fuse with vesicles made during endocytosis (which you'll learn about in a bit). Lysosomes are also essential during programmed cell death called apoptosis.

Centrioles

The **centrioles** are small, paired, cylindrical structures that are often found within **microtubule organizing centers (MTOCs)**. Centrioles are most active during cellular division. When a cell is ready to divide, the centrioles produce microtubules, which pull the replicated chromosomes apart and move them to opposite ends of the cell. Although centrioles are common in animal cells, they are not found in most types of plant cells.

Vacuoles

In Latin, the term *vacuole* means "empty cavity." But **vacuoles** are far from empty. They are fluid-filled sacs that store water, food, wastes, salts, or pigments.

Peroxisomes

Peroxisomes are organelles that detoxify various substances, producing hydrogen peroxide (H_2O_2) as a byproduct. They also contain enzymes that break down hydrogen peroxide into oxygen and water. In animals, they are common in liver and kidney cells.

Cytoskeleton

Have you ever wondered what actually holds the cell together and enables it to keep its shape? The shape of a cell is determined by a network of protein fibers called the **cytoskeleton**. The most important fibers you'll need to know are **microtubules** and **microfilaments**.

Microtubules, which are made up of the protein **tubulin**, participate in cellular division and movement. These small fibers are an integral part of three structures: centrioles, **cilia,** and flagella. We've already mentioned that centrioles help chromosomes separate during cell division.

Microfilaments, like microtubules, are important for movement. These thin, rodlike structures are composed of the protein actin. Actin monomers are joined together and broken apart as needed to allow microfilaments to grow and shrink. Microfilaments assist during cytokinesis, muscle contraction, and formation of pseudopodia extensions during cell movement.

Cilia and Flagella

Cilia and flagella are threadlike structures best known for their locomotive properties in single-celled organisms. The beating motion of cilia and flagella structures propels these organisms through their watery environments.

The two classic examples of organisms with these structures are the *Euglena*, which gets about using its whiplike flagellum, and the *Paramecium*, which is covered in cilia. The rhythmic beating of the *Paramecium*'s cilia enables it to motor about in waterways, ponds, and microscope slides in your biology lab. You may have seen these in lab, but here's what they look like:

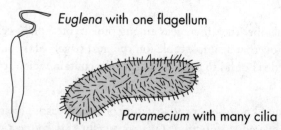

Euglena with one flagellum

Paramecium with many cilia

Though we usually associate such structures with microscopic or unicellular organisms, they aren't the only ones with cilia and flagella. As you probably know, these structures are also found in certain human cells. For example, the cells lining your respiratory tract possess cilia that sweep constantly back and forth (beating up to 20 times per second), helping to keep dust and unwanted debris from descending into your lungs. And every sperm cell has a type of flagellum, which enables it to swim through the female reproductive organs to fertilize the waiting ovum.

Plant Cells Versus Animal Cells

Plant cells contain most of the same organelles and structures seen in animal cells, with several key exceptions. Plant cells, unlike animal cells, have a protective outer covering called the **cell wall** (made of cellulose). A cell wall is a rigid layer just outside the plasma membrane that provides support for the cell. It is found in plants, protists, fungi, and bacteria. Cell walls are important for protection against osmotic changes as well. In fungi, the cell wall is usually made of **chitin**, a modified polysaccharide. Chitin is also a principle component of an arthropod's exoskeleton. In addition, plant cells possess **chloroplasts**. Chloroplasts contain chlorophyll, the light-capturing pigment that gives plants their characteristic green color. Because chloroplasts are involved in photosynthesis, we will discuss them in more detail in Chapter 6.

Did You Know?
Chloroplasts are found in plants and photosynthetic algae.

Another difference between plant and animal cells is that most of the cytoplasm within a plant cell is usually taken up by a large vacuole—the **central vacuole**—that crowds the other organelles. In mature plants, this vacuole contains the **cell sap**. A full vacuole in a plant is a sign that it is not dehydrated. Dehydrated plants cannot fill their vacuoles and they wilt. Plant cells also differ from animal cells in that plant cells do not contain centrioles.

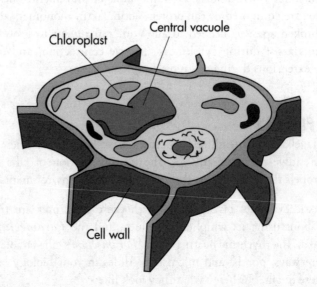

To help you remember the differences among prokaryotes, plant cells, and animal cells, we've put together a simple table (on the next page). Make sure you learn it! The testing board is bound to ask you which cells contain which structures.

Why do we need to know about the structure of cells? Because biological structure is often closely related to function. (Watch out for this connection: it's a favorite theme for the AP Biology Exam.) And, more importantly, because the testing board likes to test you on it!

Structural Characteristics of Different Cell Types			
Structure	**Prokaryote**	**Plant Cell**	**Animal Cell**
Cell Wall	Yes	Yes	No
Plasma Membrane	Yes	Yes	Yes
Membrane-Bound Organelles	No	Yes	Yes
Nucleus	No	Yes	Yes
Centrioles	No	No	Yes
Ribosomes	Yes	Yes	Yes

Transport: Traffic Across Membranes

We've talked about the structure of cell membranes; now let's discuss how molecules and fluids pass through the plasma membrane. What are some of the patterns of membrane transport? The ability of molecules to move across the cell membrane depends on two things: (1) the semipermeability of the plasma membrane and (2) the size and charge of particles that want to get through.

Since the plasma membrane is composed primarily of phospholipids, small, lipid-soluble substances cross the membrane without any resistance. Why? Because "like dissolves like." The lipid bilayer is a hydrophobic sandwich (hydrophilic outside and hydrophobic interior), and only hydrophobic things can pass that central zone. If a substance is hydrophilic, the bilayer won't let it pass without assistance, called **facilitated transport**.

Facilitated transport depends upon a number of proteins that act as tunnels through the membrane. Channels are very specialized types of tunnels that let only certain things through.

The most famous of these are **aquaporins**, which are water-specific channels. Although water is polar, there are typically sufficient aquaporins for water to traverse the membrane whenever it wishes without forming traffic jams. However, without aquaporins, no water would be able to cross the membrane. Glucose and ions such as Na^+ and K^+ are also transported across the plasma membrane via membrane proteins.

Passive Transport: Simple and Facilitated Diffusion

Now we know how things cross the membrane, but let's talk about why they cross the membrane. The simple answer is "To get to the other side." If there is a high concentration of something in one area, it will move to spread out and diffuse into an area with a lower concentration, even if that means entering or exiting a cell. In other words, the substance moves *down* a concentration gradient. This is called diffusion. When the molecule that is diffusing is hydrophobic, the diffusion is called **simple diffusion** because the small nonpolar molecule can just drift right through the membrane without trouble. When the diffusion requires the help of a channel-type protein, it is called **facilitated diffusion**. Anytime that a substance is moving by diffusion, it is called **passive transport** because there is no outside

Homeostasis
Moving things across membranes is important for growth and homeostasis. Osmoregulation is the movement of water and solute across membranes to maintain homeostasis.

Passive transport does not require the input of metabolic energy. It is important because it helps the cell import resources and export wastes.

energy required to power the movement. It's like riding a bicycle downhill. The molecules are just "going with the flow."

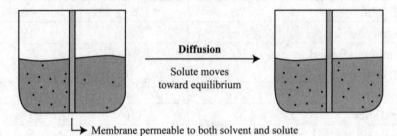

Membrane permeable to both solvent and solute

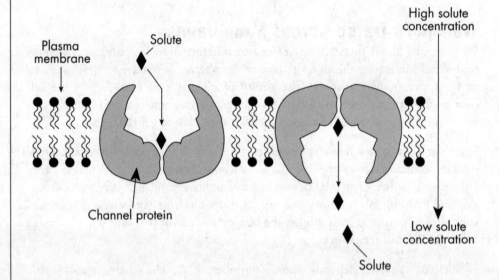

Osmosis

When the molecule that is diffusing is water, the process is called **osmosis**. Water always wants to move from an area where it is most concentrated to an area where it is least concentrated. Water is not usually pure water; since it is so great at dissolving things, it usually exists as a solution. A solution is made when a liquid solvent dissolves solute particles. So, if we rethink water moving down its concentration gradient, this is the same as saying that it wants to go from where there is less solute (ie. watery solution) to where there is more solute (ie. concentrated solution). It's as if the water is moving to dilute the concentrated solute particles.

Often, situations discussing osmosis set it up so that there are two areas and the water can flow freely, but the solute cannot. For example, if a chamber containing water and a chamber containing a sucrose solution are connected by a semipermeable membrane that allows water but not sucrose to cross, diffusion of sucrose between the chambers cannot occur. In this case, osmosis draws water into the sucrose chamber to reduce the sucrose concentration. This will reduce the total volume of the water chamber. Water will flow into the sucrose chamber until the concentration is the same across the membrane.

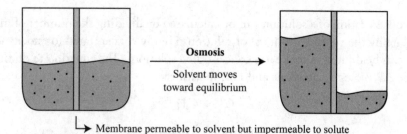

Osmosis

Solvent moves
toward equilibrium

Membrane permeable to solvent but impermeable to solute

In both diffusion and osmosis, the final result is that solute concentrations are the same on both sides of the membrane. The only difference is that in diffusion the membrane is usually permeable to solute, and in osmosis it is not.

In plants, the cell wall is important to protect it against osmotic changes. The cell wall stays relatively the same size all the time. On the other hand, the cell membrane can shrink away from the wall (a process called plasmolysis) if it loses water and can expand and squeeze tightly against the cell wall if it takes in water.

The term **tonicity** is used to describe osmotic gradients. If an environment is **isotonic** to the cell, the solute concentration is the same inside and outside. A **hypertonic** solution has more total dissolved solutes than the cell, while a **hypotonic** solution has less. The tendency of water to move down its concentration gradient (into or out of cells) can be a powerful force; it can cause cells to explode and can overcome gravity. You may also hear the terms *isosmotic*, *hyperosmotic*, and *hypoosmotic*. These words are generally used when comparing two solutions, while tonicity terms are used when comparing a solution to a cell. Other than this, the definitions are similar; isotonic and isosmotic both describe situations in which the concentration is the same on either side of a membrane, for example. The only difference is what type of membrane.

Water potential (Ψ) is the measure of potential energy in water and describes the eagerness of water to flow from an area of high water potential to an area of low water potential. It is affected by two factors: pressure potential (Ψ_p) and solute potential (Ψ_s). The equation for **solute (osmotic) potential** is listed on the AP Biology Equations and Formulas sheet and describes the effect of solute concentration on water flow. Adding a **solute** lowers the water potential of a solution, causing water to be less likely to leave this solution and more likely to flow into it. This makes sense because the added solute makes the solution more concentrated, and the water is now unlikely to diffuse away. The more solute molecules present, the more negative the solute potential is.

A red blood cell dropped into a hypotonic solution (such as distilled water) will expand because water will move into the cell, an area of lower water potential. Eventually, the red blood cell will pop. If a similar experiment is done with a plant cell, water will still move into the cell, but the cell wall will exert pressure, increasing the water potential and limiting the gain of water. Water potential also explains how water moves from soil into plant roots and how plants transport water from roots to leaves to support photosynthesis.

Don't Forget the Formula Sheet!
Be sure to look at the Equations and Formulas sheet to be prepared for questions about water potential and solute potential.

Remember It Like This
An area of high water potential is an area of high water concentration (e.g., low solute).

The concentration of a solution can be calculated by dividing the number of moles of solute by the volume (in liters) of solution. Highly concentrated (or stock) solutions can be diluted to make less concentrated solutions. The equation to do this is on the AP Biology Equations and Formulas sheet.

$$C_i V_i = C_f V_f$$

This equation tells you how much (V_i) of a concentrated solution (at C_i) is required to make a more dilute solution (C_f) of a certain volume (V_f). The amount of solvent added will be the final volume of the solution (V_f) minus whatever amount of stock solution you need to add. For example, if you have a bottle of 2 M solution of Tris buffer and would like to make 50 mL of 0.1 M Tris:

$$C_i V_i = C_f V_f$$
$$(2\ M)(V_i) = (0.1\ M)(0.050\ \text{L})$$
$$V_i = 0.0025\ \text{L, or 2.5 mL}$$

Thus, you would add 2.5 mL of the concentrated (stock) Tris solution to a new tube or bottle and then add water (the solvent) up to a final volume of 50 mL. In other words, you should add 47.5 mL of water.

Active Transport

Suppose a substance wants to move in the opposite direction—from a region of lower concentration to a region of higher concentration. A transport protein can help usher the substance across the plasma membrane, but it's going to need energy to accomplish this. This time it's like riding a bicycle uphill. Compared with riding downhill, riding uphill takes a lot more work. Movement against the natural flow is called **active transport**.

But where does the protein get this energy? Some proteins in the plasma membrane are powered by ATP. The best example of active transport is a special protein called the **sodium-potassium pump**. It ushers out three sodium ions (Na⁺) and brings in two potassium ions (K⁺) across the cell membrane. This pump depends on ATP to get ions across that would otherwise remain in regions of higher concentration. Primary active transport occurs when ATP is directly utilized to transport something. Secondary active transport occurs when something is actively transported using the energy captured from the movement of another substance flowing down its concentration gradient.

We've now seen that small substances can cross the cell membrane by:

- simple diffusion
- facilitated transport
- active transport

Resting Membrane Potential

The sodium-potassium pump is important for setting up the typical charge within the cell, and the inside is always a little bit negative. This resting membrane potential (difference in charge between inside and outside the cell) is important for passing signals in the body.

Endocytosis

When the particles that want to enter a cell are just too large, the cell uses a portion of the cell membrane to engulf the substance. The cell membrane forms a pocket, pinches in, and eventually forms either a vacuole or a vesicle. This process is called **endocytosis**.

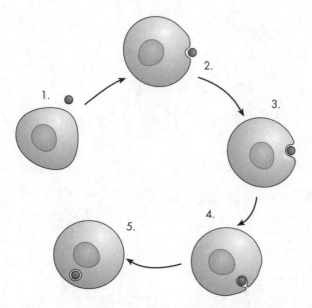

Three types of endocytosis exist: **pinocytosis**, **phagocytosis**, and **receptor-mediated endocytosis**. In pinocytosis, the cell ingests liquids ("cell-drinking"). In phagocytosis, the cell takes in solids ("cell-eating"). A special type of endocytosis, receptor-mediated endocytosis, involves cell surface receptors that work in tandem with endocytic pits that are lined with a protein called clathrin. When a particle, or ligand, binds to one of these receptors, the ligand is brought into the cell by the invagination, or "folding in" of the cell membrane. A vesicle then forms around the incoming ligand and carries it into the cell's interior.

Bulk Flow

Other substances move by **bulk flow**. Bulk flow is the one-way movement of fluids brought about by *pressure*. For instance, the movement of blood through a blood vessel and the movement of fluids in xylem and phloem of plants are examples of bulk flow.

Dialysis

Dialysis is the diffusion of *solutes* across a selectively permeable membrane. Special membranes that have holes of a certain size within them can be used to sort substances by using the process of diffusion. Kidney dialysis is a specialized process by which the blood is filtered by using machines and concentration gradients. Things present at high levels will naturally diffuse out of the blood; dialysis just gives them the opportunity.

Exocytosis

Sometimes large particles are transported *out* of the cell. In **exocytosis**, a cell ejects waste products or specific secretion products, such as hormones, by the fusion of a vesicle with the plasma membrane, which then expels the contents into the extracellular space. Think of exocytosis as reverse endocytosis.

KEY TERMS

cells
surface area-to-volume ratio
light microscopes
electron microscopes
prokaryotic cells
eukaryotic cells
cytoplasm
nucleoid
cell wall
plasma membrane
flagella
capsule
organelles
phosholipid bilayer
peripheral proteins
integral proteins
transmembrane proteins
fluid-mosaic model
adhesion proteins
receptor proteins
transport proteins
channel proteins
cell surface markers
carbohydrate side chains
nucleus
chromosomes
nucleolus
ribosomes
endoplasmic reticulum (ER)
rough ER
smooth ER
Golgi bodies
vesicles
mitochondria
adenosine triphosphate (ATP)
cristae
lysosomes
centrioles

microtubule organizing centers
 (MTOCs)
vacuoles
peroxisomes
cytoskeleton
microtubules
microfilaments
tubulin
cilia
Euglena
Paramecium
cell wall
chitin
chloroplasts
central vacuole
cell sap
facilitated transport
aquaporins
simple diffusion
facilitated diffusion
passive transport
osmosis
tonicity
isotonic
hypertonic
hypotonic
water potential
solute (osmotic) potential
solutes
active transport
sodium-potassium pump
endocytosis
pinocytosis
phagocytosis
receptor-mediated endocytosis
bulk flow
dialysis
exocytosis

Summary

o Living things are composed of cells, the smallest unit of living organisms. Cells may be categorized into prokaryotes, which do not have a nucleus or membrane-bound organelles, and eukaryotes, which have a nucleus and membrane-bound organelles. Features of prokaryotic cells (like bacteria, for example) can include the following:

- plasma membrane
- flagella and cilia
- cell wall
- nucleoid region
- ribosomes
- circular chromosome
- cytoplasm

o Components of the eukaryotic cells include the following:

- nucleus
- nucleolus
- rough endoplasmic reticulum
- smooth endoplasmic reticulum
- Golgi body
- ribosomes
- mitochondria
- vacuole
- lysosome
- peroxisomes
- centrioles (animals only)
- cytoplasm
- plasma membrane
- cytoskeleton (microtubules, microfilaments)
- central vacuole (plants only)
- cell wall (plants, fungi, and some protists)

o The plasma membrane is composed of a phospholipid bilayer. The hydrophobic nature of the inside of the membrane is responsible for its selective permeability. Transport through the membrane is dependent on size and polarity of the molecules and concentration gradients. There are a few different types of cellular transport:

- simple diffusion
- facilitated transport
- active transport
- bulk flow
- dialysis
- endocytosis (pinocytosis, phagocytosis, receptor-mediated endocytosis)
- exocytosis

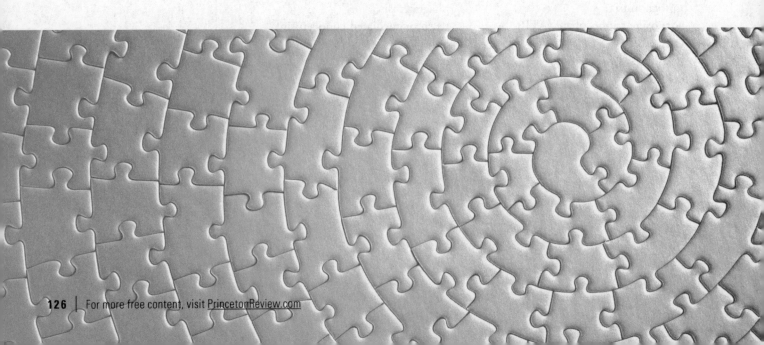

Chapter 5 Drill

Answers and explanations can be found in Chapter 15.

1. Movement of substances into the cell is largely dependent on the size, polarity, and concentration gradient of the substance. Which of the following represents an example of active transport of a substance into a cell?

   (A) Diffusion of oxygen into erythrocytes (red blood cells) in the alveolar capillaries of the lungs
   (B) Influx of sodium ions through a voltage-gated ion channel in a neuron cell during an action potential
   (C) The sodium-potassium pump, which restores resting membrane potentials in neurons through the use of ATP
   (D) Osmosis of water into an epithelial cell lining the lumen of the small intestine

2. The development of electron microscopy has provided key insights into many aspects of cellular structure and function, which had previously been too small to be seen. All of the following would require the use of electron microscopy for visualization EXCEPT

   (A) the structure of a bacteriophage
   (B) the matrix structure of a mitochondrion
   (C) the shape and arrangement of bacterial cells
   (D) the pores on the nuclear membrane

3. *Vibrio cholerae* (shown below) are highly pathogenic bacteria that are associated with severe gastrointestinal illness and are the causative agent of cholera. In extreme cases, antibiotics are prescribed that target bacterial structures that are absent in animal cells. Which of the following structures is most likely targeted by antibiotic treatment?

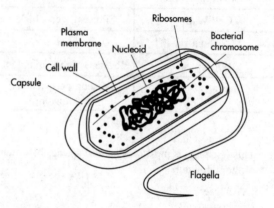

   (A) Cytoplasm
   (B) Plasma membrane
   (C) Ribosomes
   (D) Cell wall

Question 4 is based on the information and table below.

A new unicellular organism has recently been identified living in thermal pools in Yellowstone National Park. The thermal pools have average temperatures of 45°C, a pH of approximately 3.2, and high concentrations of sulfur-containing compounds. To identify the organism, a microbiologist performs a series of tests to evaluate its structural organization. The table below summarizes the microscopy data of the newly identified organism.

Cellular Structure	Analysis
Plasma Membrane	Present
Cell Wall	Present, very thick
Mitochondria	Absent
Ribosomes	Present, highly abundant
Flagella	Present, peritrichous organization

4. This organism is most likely a new species of which of the following?

   (A) Algae
   (B) Protozoa
   (C) Bacteria
   (D) Fungi

5. In a eukaryotic cell, which of the following organelles directly work together?

   (A) Nuclear envelope, nucleolus, vacuoles, centrioles
   (B) Ribosomes, rough endoplasmic reticulum, Golgi bodies, plasma membrane
   (C) Mitochondria, ribosomes, lysosomes, chloroplasts
   (D) Centrioles, nucleolus, smooth endoplasmic reticulum, lysosomes

6. A lipid-soluble hormone, estrogen, is secreted from the ovaries. This molecule travels through the body via the bloodstream. A researcher was interested in reducing estrogen's effect in order to determine the response of decreased estrogen on the organism. Which of the following is a valid strategy for reducing effects of estrogen on the whole research organism?

   (A) Treat with a competitive inhibitor drug that blocks all receptors at the plasma membrane
   (B) Treat with lipid-soluble testosterone
   (C) Treat with a lipid-soluble noncompetitive inhibitor that specifically reduces estrogen binding to the intracellular receptor
   (D) Remove ovaries of the organism

7. Which type of cell would be the most useful for studying the rough endoplasmic reticulum?

   (A) Neurons firing action potentials
   (B) Insulin-making cells in the pancreas
   (C) Bacterial cells in a colony
   (D) White blood cells

8. *Paramecium* is a single-celled protist that lives in freshwater habitats. In these conditions, *Paramecium* has evolved strategies to handle the potential consequences of inhabiting this hypotonic environment. One of these strategies could be

   (A) contractile vacuoles, which expel water forcefully
   (B) increased aquaporins in its cellular membrane
   (C) many cilia covering its surface
   (D) salt receptors on its surface to seek out less concentrated areas

9. A cotransporter is something that moves two substances across a membrane, one passively and the other actively. The Na^+K^+ ATPase transports sodium and potassium ions across the plasma membrane against their concentration gradients. This pump is not considered a cotransporter because

   (A) ATP is produced through this transporter
   (B) both ions are moved via active transport
   (C) both ions are moved via passive transport
   (D) ATP hydrolysis does not occur during transport

REFLECT

Respond to the following questions:

- Which topics from this chapter do you feel you have mastered?

- Which content topics from this chapter do you feel you need to study more before you can answer multiple-choice questions correctly?

- Which content topics from this chapter do you feel you need to study more before you can effectively compose a free response?

- Was there any content that you need to ask your teacher or another person about?

Chapter 6
Cellular Energetics

BIOENERGETICS

In Chapter 4, we discussed some of the more important organic molecules. But what makes these molecules so important? Glucose, starch, and fat are all energy-rich. However, the energy is packed in chemical bonds holding molecules together. To carry out the processes necessary for life, cells must find a way to release the energy in these bonds when they need it and store it away when they don't. The study of how cells accomplish this is called **bioenergetics**. Generally, bioenergetics is the study of how energy from the sun is transformed into energy in living things.

During chemical reactions, such bonds are either broken or formed. This process involves energy, no matter which direction we go. Every chemical reaction involves a change in energy.

Thermodynamics

We've all heard the expression *"nothing in life is free."* The same holds true in nature.

> Energy cannot be created or destroyed. In other words, the sum of energy in the universe is constant.

This rule is called the **First Law of Thermodynamics**. As a result, the cell cannot take energy out of thin air. Rather, it must harvest it from somewhere.

The **Second Law of Thermodynamics** states that energy transfer leads to less organization. That means the universe tends toward disorder (**entropy**).

Types of Reactions

Exergonic reactions are those in which the products have *less* energy than the reactants. Simply put, energy is given off during the reaction.

Let's look at an example. The course of a reaction can be represented by an **energy diagram**. Here's an energy diagram for an exergonic reaction.

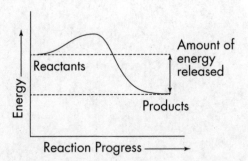

You'll notice that energy is represented along the *y*-axis. Based on the diagram, we see that our reaction released energy. An example of an exergonic reaction is the oxidizing of molecules in mitochondria of cells, which then releases the energy stored in the chemical bonds.

Reactions that require an input of energy are called **endergonic reactions**. You'll notice that the products have *more* energy than the reactants. An example is plants' use of carbon dioxide and water to form sugars.

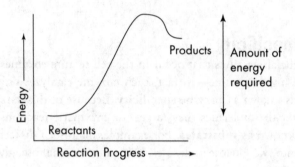

Activation Energy

Even though exergonic reactions release energy, the reaction might not occur naturally without a little bit of energy to get things going. This is because the reactants must first turn into a high energy molecule, called the **transition state**, before turning into the products. The transition state is sort of a reactants-products hybrid state that is difficult to achieve. In order to reach this transition state, a certain amount of energy is required. This is called the **activation energy**.

Be careful: a transition state is not the same thing as a reaction intermediate. Transition states occur in each step of every reaction, between the reactant(s) and the product(s). Intermediates are formed between each step of a multistep reaction.

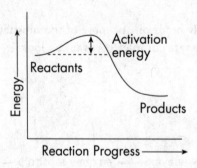

You'll notice that we needed a little energy to get us going. That's because chemical bonds must be broken before new bonds can form. This energy barrier—the hump in the graph—is called the activation energy. Once a set of reactants has reached its activation energy, the rest of the reaction is all downhill. Reaching the transition state is the tough part.

ENZYMES

A catalyst is something that speeds something up. **Enzymes** are biological catalysts that speed up reactions. They accomplish this by lowering the activation energy and helping the transition state to form. Thus, the reaction can occur more quickly because the tricky transition state is not as much of a hurdle to overcome. Enzymes do NOT change the energy of the starting point or the ending point of the reaction. They only lower the activation energy.

Enzyme Specificity

Most of the crucial reactions that occur in the cell require enzymes. Yet enzymes themselves are highly specific—in fact, each enzyme catalyzes only one kind of reaction. This is known as **enzyme specificity**. Because of this, enzymes are usually named after the molecules they target. In enzymatic reactions, the targeted molecules are known as **substrates**. For example, maltose, a disaccharide, can be broken down into two glucose molecules. Our substrate, maltose, gives its name to the enzyme that catalyzes this reaction: maltase.

Many enzymes are named simply by replacing the suffix of the substrate with –*ase*. Using this nomenclature, malt*ose* becomes malt*ase*.

Enzyme-Substrate Complex

Enzymes have a unique way of helping reactions along. As we just saw, the reactants in an enzyme-assisted reaction are known as substrates. During a reaction, the enzyme's job is to bring the transition state about by helping the substrate(s) get into position. It accomplishes this through a special region on the enzyme known as an **active site**.

The enzyme temporarily binds one or more of the substrates to its active site and forms an **enzyme-substrate complex**. Let's take a look:

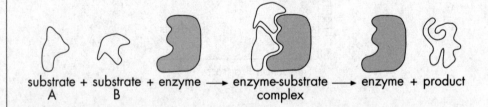

substrate + substrate + enzyme ⟶ enzyme-substrate ⟶ enzyme + product
A   B   complex

Once the reaction has occurred and the product is formed, the enzyme is released from the complex and restored to its original state. Now the enzyme is free to react again with another bunch of substrates.

By binding and releasing substrates over and over again, the enzyme speeds the reaction along, enabling the cell to release much-needed energy from various molecules. Here is a quick review on the function of enzymes.

Enzymes Do:
- increase the rate of a reaction by lowering the reaction's activation energy
- form temporary enzyme-substrate complexes
- remain unaffected by the reaction

Enzymes Don't:
- change the reaction
- make reactions occur that would otherwise not occur at all

Induced-fit

However, scientists have discovered that enzymes and substrates don't fit together quite so seamlessly. It appears that the enzyme has to change its shape slightly to accommodate the shape of the substrates. This is called **induced-fit**. Sometimes certain factors are involved in activating the enzyme and making it capable of binding the substrate. Because the fit between the enzyme and the substrate must be perfect, enzymes operate only under a strict set of biological conditions.

Enzymes Don't Always Work Alone

Enzymes sometimes need a little help in catalyzing a reaction. Those factors are known as **cofactors**. Cofactors can be either organic molecules called **coenzymes** or inorganic molecules or ions. Inorganic cofactors are usually metal ions (Fe^{2+}, Mg^{2+}). Vitamins are examples of organic coenzymes. Your daily dose of vitamins is important for just this reason: vitamins are active and necessary participants in crucial chemical reactions.

Factors Affecting Reaction Rates

Enzymatic reactions can be influenced by a number of factors, such as temperature and pH. The concentrations of enzyme and substrate will also determine the speed of the reaction.

Temperature

The rate of a reaction increases with increasing temperature, up to a point, because an increase in the temperature of a reaction increases the chance of collisions among the molecules. But too much heat can damage an enzyme. If a reaction is conducted at an excessively high temperature (above 42°C), the enzyme loses its three-dimensional shape and becomes inactivated. Enzymes damaged by heat and deprived of their ability to catalyze reactions are said to be **denatured**.

Here's one thing to remember: all enzymes operate at an ideal temperature. For most human enzymes, this temperature is body temperature, 37°C.

Break Down Words
To denature is to "un-nature" something, to change its nature, or to remove the characteristics that make it unique. Without proper and special folding, a protein loses its nature.

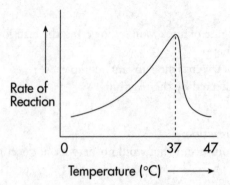

Not all organisms have a constant body temperature. For example, ectotherms depend on the environment to control their varying body temperatures. Q_{10} is a measure of temperature sensitivity of a physiological process or enzymatic reaction rate. The equation for Q_{10} is included on the AP Biology Equations and Formulas sheet.

$$Q_{10} = \left(\frac{k_2}{k_1}\right)^{\frac{10}{T_2 - T_1}}$$

The temperature unit must be in either Celsius or Kelvin, and the same unit must be used for T_1 and T_2. The two reaction rates (k_1 and k_2) must also have the same unit. Q_{10} has no unit; it is the factor by which the rate of a reaction increases due to a temperature increase. The more temperature-dependent a reaction is, the higher Q_{10} will be. Reactions with $Q_{10} = 1$ are temperature independent.

pH

Enzymes also function best at a particular pH. At an incorrect pH, the hydrogen bonds can be disrupted and the structure of the enzyme can be altered. For most enzymes, the optimal pH is at or near a pH of 7.

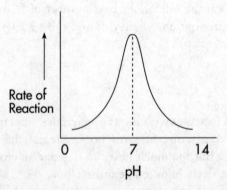

Other enzymes operate at a low pH. For instance, pepsin, the digestive enzyme found in the stomach, is most effective at an extremely acidic pH of 2. The enzymes in the lysosome also function best in an acidic pH.

Enzyme Regulation

We know that enzymes control the rates of chemical reactions. But what regulates the activity of enzymes? It turns out that a cell can control enzymatic activity by regulating the conditions that influence the shape of the enzyme. Enzymes can be turned on/off by things that bind to them. Sometimes these things can bind at the active site, and sometimes they bind at other sites, called **allosteric sites**.

If the substance has a shape that fits the active site of an enzyme (i.e., similar to the substrate or the transition state), it can compete with the substrate and block the substrate from getting into the active site. This is called **competitive inhibition**. If there was enough substrate available, the substrate would out-compete the inhibitor and the reaction would occur. You can always identify a competitive inhibitor based on what happens when you flood the system with lots of substrate.

If the inhibitor binds to an allosteric site, it is an **allosteric inhibitor** and it is **noncompetitive inhibition**. A noncompetitive inhibitor generally distorts the enzyme shape so that it cannot function. The substrate can still bind at the active site, but the enzyme will not be able to catalyze the reaction.

Not all molecules that bind to the allosteric site of an enzyme are inhibitors. Some enzymes are activated by allosteric regulators.

REACTION COUPLING AND ATP

As we just saw, almost everything an organism does requires energy. How, then, can the cell acquire the energy it needs without becoming a major mess? Fortunately, it's through adenosine triphosphate (ATP).

ATP, as the name indicates, consists of a molecule of adenosine bonded to three phosphates. The great thing about ATP is that an enormous amount of energy is packed into those phosphate bonds.

The body can perform difficult endergonic reactions by coupling them to ATP hydrolysis, which is very exergonic.

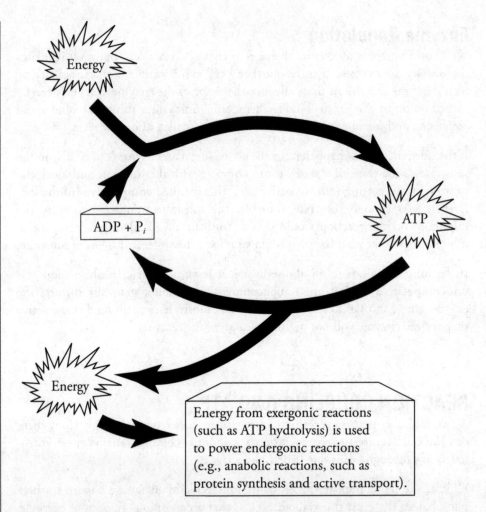

Energy from exergonic reactions (such as ATP hydrolysis) is used to power endergonic reactions (e.g., anabolic reactions, such as protein synthesis and active transport).

When a cell needs energy, it takes one of these potential-packed molecules of ATP and splits off the third phosphate, forming adenosine diphosphate (ADP) and one loose phosphate (P_i), while releasing energy in the process.

$$ATP \rightarrow ADP + P_i + energy$$

The energy released from this reaction can then be put to whatever use the cell pleases. Of course, this doesn't mean that the cell is above the laws of thermodynamics. But within those constraints, ATP is the best source of energy the cell has available. It is relatively neat (only one bond needs to be broken to release that energy) and relatively easy to form. Organisms can use exergonic processes that increase energy, like breaking down ATP, to power endergonic reactions, like building organic macromolecules.

Sources of ATP

But where does all this ATP come from? It can be formed in several ways, but the bulk of it comes from a process called **cellular respiration**. Cellular respiration is a process of breaking down sugar and making ATP.

In autotrophs, the sugar is made during **photosynthesis**. In heterotrophs, glucose comes from the food we eat. We will start by going over photosynthesis, since in plants it is a precursor to cellular respiration.

PHOTOSYNTHESIS

Plants, algae, and cyanobacteria are producers. The earliest photosynthesis was conducted by prokaryotic cyanobacteria, and later eukaryotes developed photosynthetic abilities. All they do is bask in the sun, churning out the glucose necessary for life. In this section, we'll focus on how plants conduct photosynthesis.

As we discussed earlier, photosynthesis is the process by which light energy is converted to chemical energy. Here's an overview of photosynthesis.

$$6CO_2 + 6H_2O \longrightarrow C_6H_{12}O_6 + 6O_2$$

You'll notice from this equation that carbon dioxide and water are the raw materials used to manufacture simple sugars. But remember, there's much more to photosynthesis than what's shown in the simple reaction above.

There are two stages in photosynthesis: the **light reactions** (also called the light-dependent reactions) and the **dark reactions** (also called the light-independent reactions). The whole process begins when **photons** (energy units) of sunlight strike a leaf, activating chlorophyll and exciting electrons. The activated chlorophyll molecule then passes these excited electrons down to a series of electron carriers, ultimately producing ATP and NADPH. Both of these products, along with carbon dioxide, are then used in the dark reactions (light-independent) to make carbohydrates. Along the way, water is also split and oxygen gets released.

Algae are a diverse group of aquatic, eukaryotic organisms that can perform photosynthesis. Most algae are protists. Some common examples are seaweed (such as kelp or phytoplankton), pond scum, or algal blooms in lakes.

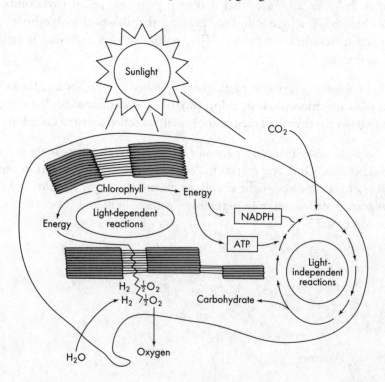

We'll soon see that this beautifully orchestrated process occurs thanks to a whole host of special enzymes and pigments. But before we turn to the stages in photosynthesis, let's talk about where photosynthesis occurs.

Chloroplast Structure

The leaves of plants contain lots of chloroplasts, which are the primary sites of photosynthesis.

Now let's look at an individual chloroplast. If you split the membranes of a chloroplast, you'll find a fluid-filled region called the **stroma**. Inside the stroma are structures that look like stacks of coins. These structures are the **grana**.

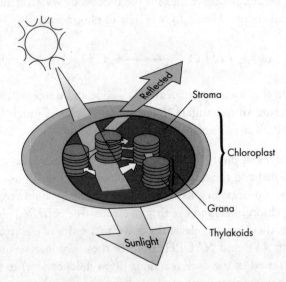

The many disk-like structures that make up grana are called **thylakoids.** They contain chlorophyll, a light-absorbing pigment that drives photosynthesis, as well as enzymes involved in the process. The very inside of a thylakoid is called the thylakoid lumen.

Many light-absorbing pigments participate in photosynthesis. Some of the more important ones are **chlorophyll *a***, **chlorophyll *b***, and **carotenoids**. These pigments are clustered in the thylakoid membrane into units called antenna complexes.

All of the pigments within a unit are able to "gather" light, but they're not able to "excite" electrons. Only one special molecule—located in the **reaction center**—is capable of transforming light energy to chemical energy. In other words, the other pigments, called **antenna pigments**, "gather" light and "bounce" energy to the reaction center.

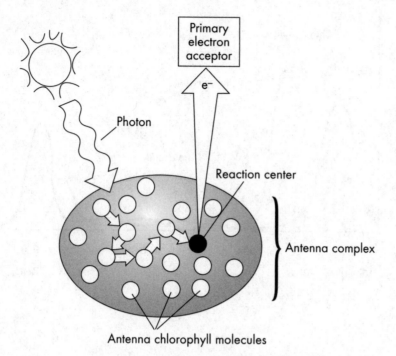

Antenna chlorophyll molecules

There are two types of reaction centers: **photosystem I (PS I)** and **photosystem II (PS II)**. The principal difference between the two is that each reaction center has a specific type of chlorophyll—chlorophyll *a*—that absorbs a particular wavelength of light. For example, **P680**, the reaction center of photosystem II, has a maximum absorption at a wavelength of 680 nanometers. The reaction center for photosystem I —**P700**—best absorbs light at a wavelength of 700 nanometers.

When light energy is used to make ATP, it is called **photophosphorylation**. Autotrophs are using light (that's *photo*) and ADP and phosphates (that's *phosphorylation*) to produce ATP.

An **absorption spectrum** shows how well a certain pigment absorbs electromagnetic radiation. Light absorbed is plotted as a function of radiation wavelength. This spectrum is the opposite of an **emission spectrum**, which gives information on which wavelengths are emitted by a pigment. The absorption spectrum for chlorophyll *a*, chlorophyll *b*, and carotenoids is on the next page. You can see that chlorophyll *a* and chlorophyll *b* absorb blue and red light quite well but do not absorb light in the green part of the spectrum. Light in the green range of wavelengths is reflected, and this is why chlorophyll and many plants are green. Carotenoids absorb light on the blue-green end of the spectrum, but not on the other end. This is why plants rich in carotenoids are yellow, orange, or red.

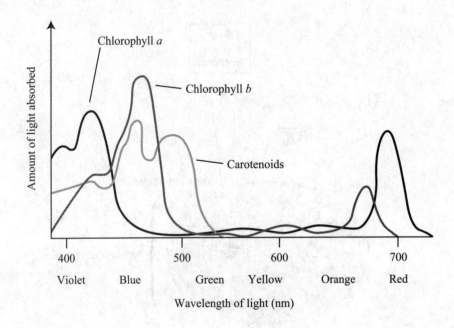

THE LIGHT REACTIONS

When a leaf captures sunlight, the energy is sent to P680, the reaction center for photosystem II. The activated electrons are trapped by P680 and passed to a molecule called the primary acceptor, and then they are passed down to carriers in the electron transport chain.

To replenish the electrons in the thylakoid, water is split into oxygen, hydrogen ions, and electrons. That process is called **photolysis**. The electrons from photolysis replace the missing electrons in photosystem II.

As the energized electrons from photosystem II travel down the electron transport chain, they pump hydrogen ions into the thylakoid lumen. A proton gradient is established. As the hydrogen ions move back into the stroma through ATP synthase, ATP is produced.

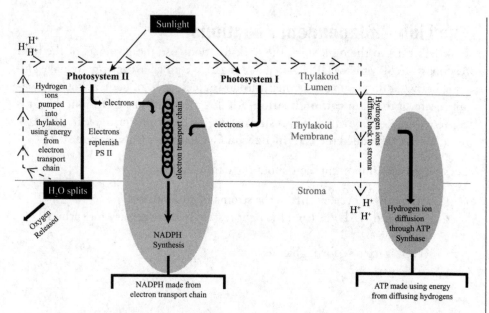

After the electrons leave photosystem II, they go to photosystem I. The electrons are passed through a second electron transport chain until they reach the final electron acceptor NADP⁺ to make **NADPH**.

Photosystem I and photosystem II were numbered in order of their discovery, not the order they are used in photosynthesis.

> The light reactions of photosynthesis occur in the thylakoid membranes.

Here is a summary of the most important points from the first part of photosynthesis:

- P680 in photosystem II captures light and passes excited electrons down an electron transport chain to produce ATP.
- P700 in photosystem I captures light and passes excited electrons down an electron transport chain to produce NADPH.
- A molecule of water is split by sunlight, releasing electrons, hydrogen, and free O_2.
- The light-dependent reactions occur in the grana of chloroplasts, where the thylakoids are found. Remember: the light-absorbing pigments and enzymes for the light-dependent reactions are found within the thylakoids.

Most plants follow this electron flow, which is called linear, or noncyclic. However, some plants (such as C_4 plants discussed below) perform cyclic electron flow instead. Cyclic photophosphorylation is similar to noncyclic, but generates only ATP, and no NADPH. It takes place only at photosystem I: once an electron is displaced from the photosystem, it is passed down electron acceptor molecules and returns to photosystem I, from where it was emitted.

The Light-Independent Reactions

Now let's turn to the dark side. The dark reactions use the products of the light reactions—ATP and NADPH—to make sugar. We now have energy to make glucose, but what do plants use as their carbon source? CO_2, of course. You've probably heard of the term **carbon fixation**. All this means is that CO_2 from the air is converted into carbohydrates. This step occurs in the stroma of the leaf. The dark reactions are also called the **Calvin-Benson Cycle** (or just Calvin cycle).

Let's summarize important facts about the dark reactions:

- The Calvin cycle occurs in the stroma of chloroplasts.
- ATP and NADPH from the light reactions are necessary for carbon fixation.
- CO_2 is fixed to form glucose.

Photosynthesis: A Summary

The following table summarizes the stages of photosynthesis:

Stage of Photosynthesis	Location	Input	Output
Light-Dependent Reactions	Thylakoid membrane of chloroplasts Photosystem II (P680) Photosystem I (P700)	Photons H_2O	NADPH ATP O_2
Light-Independent Reactions (Calvin Cycle)	Stroma	3 CO_2 9 ATP 6 NADPH	sugar

The light-dependent and light-independent reactions are inexorably linked; neither set of reactions alone can produce carbohydrates from CO_2. The light-dependent reactions use water and light to produce ATP, NADPH, and O_2; the light-independent reactions use CO_2, ATP, and NADPH to produce ADP, $NADPH^+$, and carbohydrates.

Stomata are pores on the leaf surface that allow CO_2 to enter the leaf, and O_2 and water to exit. Most plants close their stomata on hot, dry days, to prevent water loss by transpiration (the evaporative loss of water from leaves). While this conserves water, it also limits access to CO_2 and thus reduces photosynthetic yield. With less CO_2 available, O_2 accumulates and plants start performing photorespiration instead of photosynthesis. **Photorespiration** is a wasteful process that uses ATP and O_2, produces more CO_2, and doesn't produce any sugars.

Plants that live in hot climates have evolved two different ways around this: **CAM plants** deal with this problem by temporally separating carbon fixation and the Calvin cycle. They open their stomata at night and incorporate CO_2 into organic acids. During the day, they close their stomata and release CO_2 from the organic acids while the light reactions run. In contrast, **C_4 plants** have slightly different leaf anatomy that allows them to perform CO_2 fixation in a different part of the leaf from the rest of the Calvin cycle. This prevents photorespiration. C_4 plants produce a four-carbon molecule as the first product of carbon fixation and perform cyclic electron flow in the light reactions (discussed above).

CELLULAR RESPIRATION

In the shorthand version, cellular respiration looks something like this.

$$C_6H_{12}O_6 + 6O_2 \rightarrow 6CO_2 + 6H_2O + ATP$$

Notice that we've taken a sugar, perhaps a molecule of glucose, and combined it with oxygen to produce carbon dioxide, water, and energy in the form of our old friend, ATP. However, as you probably already know, the actual picture of what really happens is far more complicated.

Generally speaking, we can break cellular respiration down into two different approaches: aerobic respiration and anaerobic respiration. If ATP is made in the presence of oxygen, we call it **aerobic respiration**. If oxygen isn't present, we call it **anaerobic respiration**. Let's jump right in with aerobic respiration.

Several metabolic pathways, including cell respiration, are conserved across all currently recognized domains of life.

Introduction to Aerobic Respiration

Aerobic respiration consists of four stages:

1. glycolysis
2. formation of acetyl-CoA
3. the Krebs (citric acid) cycle
4. oxidative phosphorylation (the electron transport chain + chemiosmosis)

In the first three stages, glucose is broken down and energy molecules are made. Some of these are ATP, and others are special electron carriers called **NADH** and **FADH$_2$**. In the fourth stage, these electron carriers unload their electrons, and the energy is eventually used to make even more ATP.

Stage 1: Glycolysis

The first stage begins with **glycolysis**, the splitting (*-lysis*) of glucose (*glyco-*). Glucose is a six-carbon molecule that is broken into two three-carbon molecules called **pyruvic acid**. This breakdown of glucose also results in the net production of two molecules of ATP.

$$\text{Glucose} + 2\ \text{ATP} + 2\text{NAD}^+ \rightarrow 2\ \text{Pyruvic acid} + 4\ \text{ATP} + 2\text{NADH}$$

Although we've written glycolysis as if it were a single reaction, this process doesn't occur in one step. In fact, it requires a sequence of enzyme-catalyzed reactions. Let's look at a summary of these reactions:

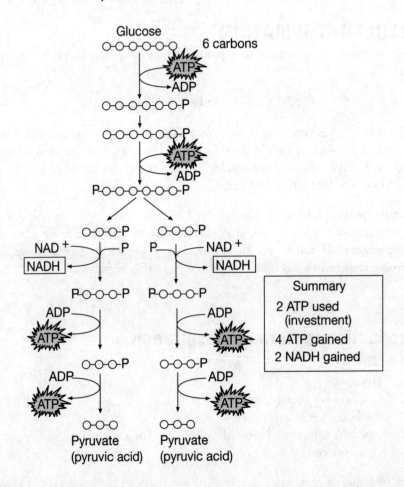

Pyruvic acid is an acid. In your cells, it usually loses a proton, so it is called pyruvate instead.

Glycolysis occurs in the cytoplasm and starts with one glucose and ends with two pyruvates. There is a net production of 2 ATP and 2 NADH.

If you take a good look at the reactions above, you'll see two ATPs are needed to produce four ATP. You've probably heard the expression "You have to spend money to make money." In biology, you have to invest ATP to make ATP: our investment of two ATP yielded four ATP, for a net gain of two. The ATP molecules generated in glycolysis are created from combining ADP and an inorganic phosphate with the help of an enzyme.

Glycolysis also creates two NADH, which result from the transfer of electrons to the carrier NAD⁺, which then becomes NADH. NAD⁺ and NADH are constantly being turned into each other as electrons are being carried and then unloaded.

Electron Carriers

$\text{NAD}^+ + \text{H}^+ + 2\ \text{electrons} = \text{NADH}$

$\text{FAD} + 2\text{H}^+ + 2\ \text{electrons} = \text{FADH}_2$

To carry electrons is to carry energy.

There are four important tidbits to remember regarding glycolysis:

- occurs in the cytoplasm
- net of 2 ATPs produced
- 2 pyruvic acids formed
- 2 NADH produced

Stage 2: Formation of Acetyl-CoA

Pyruvic acid is transported to the mitochondrion. Each pyruvic acid (a three-carbon molecule) is converted to **acetyl-coenzyme A** (a two-carbon molecule, usually just called acetyl-CoA) and CO_2 is released.

$$2 \text{ Pyruvic acid} + 2 \text{ Coenzyme A} + 2NAD^+ \rightarrow 2 \text{ Acetyl-CoA} + 2CO_2 + 2NADH$$

Are you keeping track of our carbons? We've now gone from two three-carbon molecules to two two-carbon molecules. The extra carbons leave the cell in the form of CO_2. Once again, two molecules of NADH are also produced for each glucose you started with. This process of turning pyruvic acid into acetyl-CoA is catalyzed by an enzyme complex called the **pyruvate dehydrogenase complex** (**PDC**).

In the mitochondria, pyruvate is turned into acetyl-CoA and 1 NADH is made; double this if you are counting per glucose.

Stage 3: The Krebs Cycle

The next stage is the **Krebs cycle**, also known as the **citric acid cycle**. Each of the two acetyl-coenzyme A molecules will enter the Krebs cycle, one at a time, and all the carbons will ultimately be converted to CO_2. This stage occurs in the **matrix** of the mitochondria.

The Krebs cycle begins with each molecule of acetyl-CoA produced from the second stage of aerobic respiration combining with **oxaloacetate**, a four-carbon molecule, to form a six-carbon molecule, **citric acid** or citrate (hence its name, the citric acid cycle).

Citrate gets turned into several other things, and because the cycle begins with a four-carbon molecule, oxaloacetate, it eventually gets turned back into oxaloacetate to maintain the cycle by joining with the next acetyl-CoA coming down the pipeline.

The Krebs cycle occurs in the mitochondrial matrix. It begins with acetyl-CoA joining with oxaloacetate to make citric acid and ends with oxaloacetate, 1 ATP, 3 NADH, and 1 $FADH_2$; double this if you are counting per glucose.

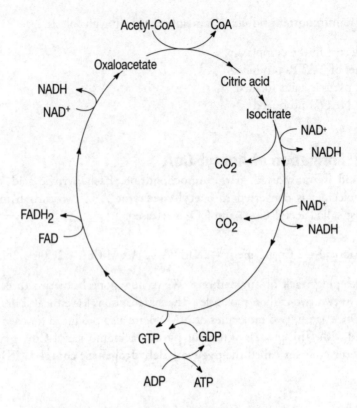

With each turn of the cycle, three types of energy are produced:

- 1 ATP
- 3 NADH
- 1 FADH$_2$

To figure out the total number of products per molecule of glucose, we simply double the number of products—after all, we started off the Krebs cycle with two molecules of acetyl-CoA for each molecule of glucose!

Now we're ready to tally up the number of ATP produced.

So far, we've made only four ATP—two ATP from glycolysis and two ATP from the Krebs cycle.

Although that seems like a lot of work for only four ATP, we have also produced hydrogen carriers in the form of 10 NADH and 2 FADH$_2$. These molecules will, in turn, produce lots of ATP in the next stage of cellular respiration.

Stage 4: Oxidative Phosphorylation

Electron Transport Chain

As electrons (and the hydrogen atoms to which they belong) are removed from a molecule of glucose, they carry with them much of the energy that was originally stored in their chemical bonds. These electrons—and their accompanying energy—are then transferred to readied hydrogen carrier molecules. In the case of cellular respiration, these charged carriers are NADH and $FADH_2$.

Let's see how many "loaded" electron carriers we've produced. We now have:

- 2 NADH molecules from glycolysis
- 2 NADH from the production of acetyl-CoA
- 6 NADH from the Krebs cycle
- 2 $FADH_2$ from the Krebs cycle

That gives us a total of 12 electron or energy carriers altogether.

These electron carriers—NADH and $FADH_2$—"shuttle" electrons to the **electron transport chain**, the resulting NAD^+ and FADH can be recycled to be used as carriers again, and the hydrogen atoms are split into hydrogen ions and electrons.

$$H_2 \rightarrow 2H^+ + 2e^-$$

Then, two interesting things occur. First, the high-energy electrons from NADH and $FADH_2$ are passed down a series of protein carrier molecules that are embedded in the cristae; thus, it is called the electron transport chain. Some of the carrier molecules in the electron transport chain are NADH dehydrogenase and **cytochrome C**.

Each carrier molecule hands down the electrons to the next molecule in the chain. The electrons travel down the electron transport chain until they reach the final electron acceptor, oxygen. Oxygen combines with these electrons (and some hydrogens) to form water. This explains the "aerobic" in aerobic respiration. If oxygen weren't available to accept the electrons, they wouldn't move down the chain at all, thereby shutting down the whole process of electron transport.

Chemiosmosis

At the same time that electrons are being passed down the electron transport chain, another mechanism is at work. Remember those hydrogen ions (also called *protons*) that split off from the original hydrogen atom? The energy released from the electron transport chain is used to pump hydrogen ions across the inner mitochondrial membrane from the matrix into the intermembrane space. The pumping of hydrogen ions into the intermembrane space creates a **pH gradient**, or **proton gradient**. The hydrogen ions really want to diffuse back into the matrix. The potential energy established in this gradient is responsible for the production of ATP. This pumping of ions and diffusion of ions to create ATP is **chemiosmosis**.

The electron transport chain here is similar to that seen in photosynthesis. They differ in their final electron acceptors: $NADP^+$ and O_2.

Oxygen is required as the final stop in the electron transport chain.

$NADP^+$ is the terminal electron acceptor in photosynthesis, while oxygen is the terminal electron acceptor in cellular respiration.

Brown fat, which is found in newborns and hibernating mammals, can also use the hydrogen gradient generated from the electron transport chain to generate heat. This occurs when they decouple the chemiosmosis from oxidative phosphorylation.

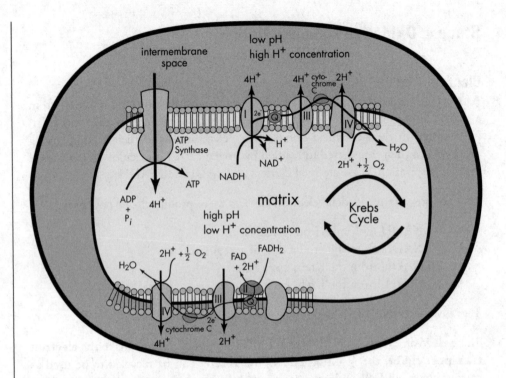

These hydrogen ions can diffuse across the inner membrane only by passing through channels called **ATP synthase**. Meanwhile, ADP and P_i are on the other side of these protein channels. The flow of protons through these channels produces ATP by combining ADP and P_i on the matrix side of the channel. Overall, this process is called **oxidative phosphorylation** because when electrons are given up it is called "oxidation" and then ADP is "phosphorylated" to make ATP.

You're also expected to know the following two things for the AP Biology Exam:

- Every NADH from glycolysis yields 1.5 ATP and all other NADH molecules yield 2.5 ATP.
- Every FADH$_2$ yields 1.5 ATP.

You will also want to make sure you remember the major steps of cell respiration, and the outcome of each step:

STAGES OF AEROBIC RESPIRATION				
Process	**Location**	**Main Input**	**Main Output**	**Energy Output (per glucose)**
Glycolysis	Cytoplasm	1 Glucose	2 pyruvates (2 net)	2 ATP 2 NADH
Formation of Acetyl-CoA	As pyruvate is transported into the mitochondria	2 pyruvates 2 Coenzyme A	2 Acetyl-CoA	2 NADH
Krebs or Citric Acid Cycle	Matrix of mitochondria	2 Acetyl-CoA oxaloacetate	oxaloacetate	6 NADH 2 FADH$_2$ 2 ATP
Oxidative Phosphorylation	Inner mitochondrial membrane	10 NADH 2 FADH$_2$ 2 O$_2$	Oxygen 2 NADH (× 1.5) → 3 ATP 8 NADH (× 2.5) → 20 ATP 2 FADH$_2$ (× 1.5) → 3 ATP 4 ATP	
				Overall Net: 30 ATP

Photosynthesis vs. Cell Respiration

The photosynthesis reactions share some similarities with cell respiration. In both cases, ATP production is driven by a proton gradient, and the proton gradient is created by an electron transport chain. In respiration, protons are pumped from the mitochondrial matrix to the intermembrane space, and they return to the matrix through an ATP synthase down their concentration gradient. In photosynthesis, protons are pumped from the stroma into the thylakoids compartment, and they return to the stroma through an ATP synthase down their concentration gradient.

The Krebs cycle and the Calvin cycle are both series of reactions that ultimately regenerate their starting product. Both cycles have an indirect need for a particular substance; although they do not use it directly, without it they cease to run. For the Krebs cycle, that substance is oxygen; for the Calvin cycle, that "substance" is light.

Furthermore, the goals of the two cycles are opposite; the Krebs cycle seeks to oxidize carbohydrates to CO_2, while the Calvin cycle seeks to reduce CO_2 to carbohydrates.

Anaerobic Respiration

When oxygen is not available, the anaerobic version of respiration occurs. The electron transport chain stops working, and electron carriers have nowhere to drop their electrons. The mitochondrial production of acetyl-CoA and the Krebs cycle cease too.

Glycolysis, however, can continue to run. This means that glucose can be broken down to give net two ATP. Only two instead of 30!

Glycolysis also gives two pyruvates and two NADH. The pyruvate and NADH make a deal with each other, and pyruvate helps NADH get recycled back into NAD$^+$ and takes its electrons. The pyruvate turns into either **lactic acid** (in muscles) or **ethanol** (in yeast). Since these two things are toxic at high concentrations, this process, called **fermentation**, is done only in emergencies. Aerobic respiration is a better option.

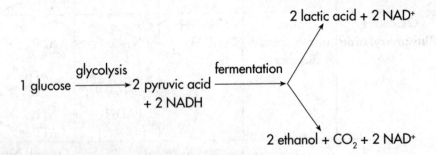

What types of organisms undergo fermentation?

- Yeast cells and some bacteria make ethanol and carbon dioxide.
- Other bacteria produce lactic acid.

Your Muscle Cells Can Ferment

Although human beings are aerobic organisms, they can actually carry out fermentation in their muscle cells. Have you ever had a leg cramp? If so, that cramp was possibly the consequence of anaerobic respiration.

When you exercise, your muscles require a lot of energy. To get this energy, they convert enormous amounts of glucose to ATP. But as you continue to exercise, your body doesn't get enough oxygen to keep up with the demand in your muscles. This creates an oxygen debt. What do your muscle cells do? They switch over to anaerobic respiration. Pyruvic acid produced from glycolysis is converted to lactic acid. As a consequence, lactic acid causes your muscles to ache.

KEY TERMS

bioenergetics

First Law of Thermodynamics

Second Law of Thermodynamics

entropy

exergonic reaction

energy diagram

endergonic reaction

transition state

activation energy

enzymes

enzyme specificity

substrates

active site

enzyme-substrate complex

induced-fit

cofactors

coenzymes

denatured

Q_{10}

allosteric sites

competitive inhibition

allosteric inhibitor

noncompetitive inhibition

cellular respiration

photosynthesis

light reactions

dark reactions

photons

stroma

grana

thylakoids

chlorophyll *a* and *b*

carotenoids

reaction center

antenna pigments

photosystem I (PS I)

photosystem II (PS II)

P680

P700

photophosphorylation

absorption spectrum

emission spectrum

photolysis

NADPH

carbon fixation

Calvin-Benson Cycle

photorespiration

CAM plants

C_4 plants

aerobic respiration

anaerobic respiration

NADH

$FADH_2$

glycolysis

pyruvic acid

acetyl-coenzyme A (acetyl-CoA)

pyruvate dehydrogenase complex

Krebs cycle (citric acid cycle)

matrix

oxaloacetate

citric acid

electron transport chain

cytochrome C

pH gradient (proton gradient)

chemiosmosis

ATP synthase

oxidative phosphorylation

lactic acid

ethanol (ethyl alcohol)

fermentation

Summary

o Energy cannot be created or destroyed. There are endergonic reactions and exergonic reactions. The change in Gibbs free energy indicates whether the reaction will require energy or release energy.

o The study of how cells accomplish biological processes is called bioenergetics.
   • Chemical reactions are catalyzed by enzymes.
   • Enzymes lower the activation energy of chemical reactions by stabilizing the transition state.
   • Enzymes do not change the energy difference between reactants and products, but by lowering the activation energy, they facilitate chemical reactions to occur.

o Enzymes are proteins that are highly specific for their substrate (or reactant, which binds at the active site).

o Enzymes have an optimal, narrow range of temperature and pH in which they have the highest rate of reaction. Outside this range, they may undergo denaturation and, therefore, no longer be active.

o Enzyme activity may also be regulated or altered by allosteric/noncompetitive inhibitors, competitive inhibitors, and activators.

o Enzymes may require coenzymes or cofactors to help catalyze reactions.

o ATP is the universal energy molecule in cells. It is created via cellular respiration.

o Photosynthesis is the process by which plants use energy from sunlight to make sugar.
   • In the light-dependent reactions, sunlight is absorbed by chlorophyll, electrons are passed down a chain, and the energy molecules ATP and NADPH are produced. Water is also split into hydrogen and oxygen.
   • In the light-independent reactions, sugar is made in the Calvin-Benson Cycle (Calvin cycle) by using energy from ATP and NADPH, and the carbon from CO_2.

o Cellular respiration consumes oxygen and produces a lot of ATP energy for the cell, using NADH and $FADH_2$ as electron carriers. There are four stages to cellular respiration:
   • glycolysis
   • formation of acetyl-CoA via the PDC
   • Krebs cycle
   • oxidative phosphorylation

o Anaerobic respiration occurs when cells lack oxygen to act as a final electron acceptor. In this process, the pyruvates generated by glycolysis are broken down by fermentation to produce lactic acid or ethanol and NAD^+, which permits glycolysis to continue and provide the 2 ATP it makes.

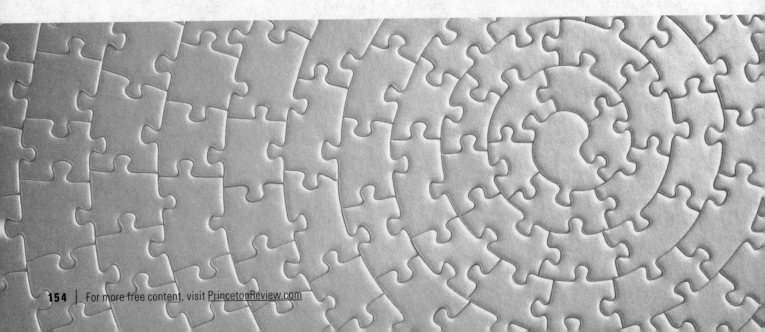

Chapter 6 Drill

Answers and explanations can be found in Chapter 15.

1. The mitochondrion is a critical organelle structure involved in cellular respiration. Below is a simple schematic of the structure of a mitochondrion. Which of the structural components labeled below in the mitochondrion is the primary location of the Krebs cycle?

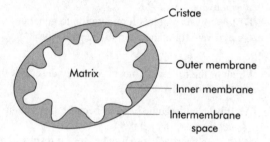

(A) Inner membrane
(B) Matrix
(C) Intermembrane space
(D) Outer membrane

2. Binding of Inhibitor Y as shown below inhibits a key catalytic enzyme by inducing a structural conformation change. Which of the following accurately describes the role of Inhibitor Y?

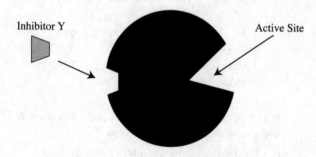

(A) Inhibitor Y competes with substrates for binding in the active site and functions as a competitive inhibitor.
(B) Inhibitor Y binds allosterically and functions as a competitive inhibitor.
(C) Inhibitor Y competes with substrates for binding in the active site and functions as a noncompetitive inhibitor.
(D) Inhibitor Y binds allosterically and functions as a noncompetitive inhibitor.

3. A single-step chemical reaction is catalyzed by the addition of an enzyme. Which of the reaction coordinate diagrams accurately shows the effect of the added enzyme (represented by the dashed line) to the reaction?

(A)

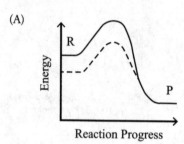

(B)

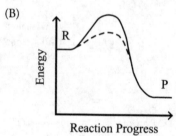

(C)

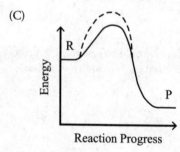

(D)

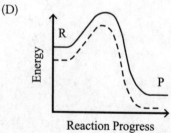

Questions 4–6 refer to the following diagram and paragraph.

Glycolysis (shown below) is a critical metabolic pathway that is utilized by nearly all forms of life. The process of glycolysis occurs in the cytoplasm of the cell and converts 1 molecule of glucose into 2 molecules of pyruvic acid.

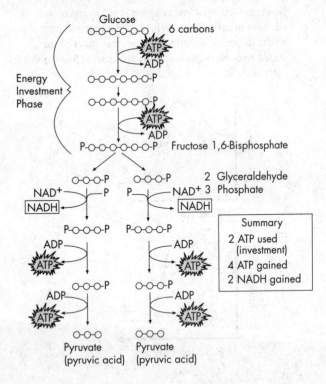

4. How many net ATP would be generated directly from glycolysis from the breakdown of 2 glucose molecules?

(A) 2
(B) 4
(C) 8
(D) 12

5. Glycolysis does not require oxygen to occur in cells. However, under anaerobic conditions, glycolysis normally requires fermentation pathways to occur to continue to produce ATP. Which best describes why glycolysis is dependent on fermentation under anaerobic conditions?

(A) Glycolysis requires fermentation to produce more glucose as a substrate.
(B) Glycolysis requires fermentation to synthesize lactic acid and restore NADH to NAD$^+$.
(C) Glycolysis requires fermentation to generate ATP molecules to complete the early steps of the pathway.
(D) Glycolysis requires fermentation to generate pyruvate for a later step in the pathway.

6. Which of the following most accurately describes the net reaction of glycolysis?

(A) It is an endergonic process because it results in a net increase in energy.
(B) It is an exergonic process because it results in a net increase in energy.
(C) It is an endergonic process because it results in a net decrease in energy.
(D) It is an exergonic process because it results in a net decrease in energy.

7. *Taq* polymerase, a DNA polymerase derived from thermophilic bacteria, is used in polymerase chain reactions (PCR) in the laboratory. During PCR, *Taq* catalyzes DNA polymerization, similar to how it would in bacteria. A normal PCR cycle is as follows:

1. Melting/Denaturing	95°C
2. Primer Annealing	50°C
3. Elongation of DNA	72°C
(repeat 20–30 cycles)	

Which of the following conditions likely describes the living environment of *Taq* bacteria?

(A) Freshwater with acidic pH
(B) Hydrothermal vents reaching temperatures between 70–75°C
(C) Hot springs of 40°C
(D) Tide pools with high salinity

8. The biggest difference between an enzyme-catalyzed reaction and an uncatalyzed reaction is that

   (A) the free energy between the reactants and the products does not change
   (B) the free energy difference between the reactants and the products does not change
   (C) the catalyzed reaction would not occur without the enzyme
   (D) a different amount of energy is required to reach the transition state of the reaction

9. Two groups of cells were grown under identical conditions. Mitochondria from each group were isolated and half of them were placed in a low pH (approximately pH 6.8) and the other half were placed in a neutral pH. Small molecules were allowed to diffuse across the outer membrane via facilitated diffusion. Both samples were exposed to oxygen bubbles through the growth media. What would you expect to see in terms of ATP production in the sample of cells placed in a low pH, with respect to the control population?

   (A) ATP production decreases.
   (B) ATP production increases.
   (C) ATP production stays the same.
   (D) ATP production ceases entirely.

10. The Second Law of Thermodynamics states that entropy, or disorder, is constantly increasing in the universe for spontaneous processes. Therefore, how is it possible that organisms exist in very ordered states?

    (A) The Second Law of Thermodynamics does not apply to biological life.
    (B) Energy can be created or destroyed.
    (C) The catabolic reactions are always equal and opposite of the anabolic reactions.
    (D) Biological life actually creates an increase in entropy through dissemination of heat and waste.

REFLECT

Respond to the following questions:

- Which topics from this chapter do you feel you have mastered?

- Which content topics from this chapter do you feel you need to study more before you can answer multiple-choice questions correctly?

- Which content topics from this chapter do you feel you need to study more before you can effectively compose a free response?

- Was there any content that you need to ask your teacher or another person about?

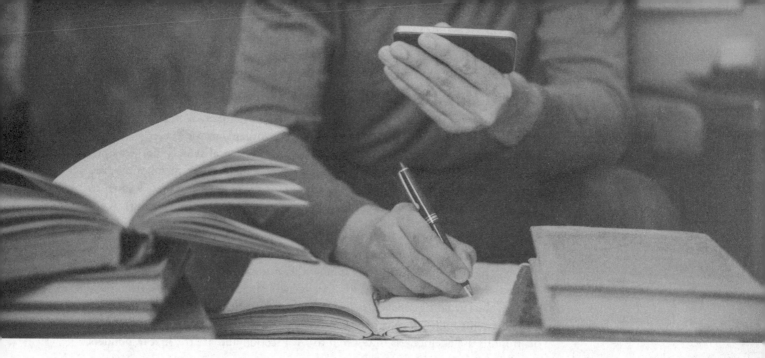

Chapter 7
Cell
Communication
and Cell Cycle

CELL COMMUNICATION

Communication is important for organisms and involves transduction of stimulatory or inhibitory signals from other cells, organisms, or the environment. Unicellular organisms detect and respond to environmental signals. For example, they can use cell communication to locate a suitable mate, adapt metabolism rates in response to changes in nutrient availability, or make their numbers known to other members of their species (a phenomenon known as **quorum sensing**).

Taxis is the movement of an organism in response to a stimulus and can be positive (toward the stimulus) or negative (away from the stimulus). Taxes are innate behavioral responses, or instincts. **Chemotaxis** is movement in response to chemicals. For example, bacteria can control flagella rotation to direct their motion, thus avoiding repellents (such as poisons), or helping them find favorable locations with high concentrations of attractants (such as food). In our bodies, neutrophils use chemotaxis to respond to an infection and are the first responders to inflammation.

The cells of multicelled organisms must communicate with one another to coordinate the activities of the organism as a whole. Signaling can be short-range (affecting only nearby cells) or long-range (affecting cells throughout the organism). It can be done by cell junctions or signaling molecules called **ligands** that bind to **receptors** and trigger a response. Hormones and neurotransmitters are common signaling molecules, which will be covered in Chapter 9.

Signal transduction is the process by which an external signal is transmitted to the inside of a cell. It usually involves the following three steps:

1. a signaling molecule binding to a specific receptor
2. activation of a signal transduction pathway
3. production of a cellular response

Hydrophobic signaling molecules, like cholesterol-based steroid hormones, can diffuse across the plasma membrane. These molecules can bind receptors inside the cell and often travel to the nucleus where they regulate gene expression.

For signaling molecules that cannot enter the cell, a plasma membrane receptor is required. Plasma membrane receptors form an important class of integral membrane proteins that transmit signals from the extracellular space into the cytoplasm. Each receptor binds a particular molecule in a highly specific way. There are three classes of membrane receptors.

1. **Ligand-gated ion channels** in the plasma membrane open an ion channel upon binding a particular ligand. An example is the ligand-gated sodium channel on the surface of a skeletal muscle cell at the neuromuscular junction. This channel opens in response to acetylcholine, and a massive influx of sodium depolarizes the muscle cell and causes it to contract.

2. **Catalytic (enzyme-linked) receptors** have an enzymatic active site on the cytoplasmic side of the membrane. Enzyme activity is initiated

by ligand binding at the extracellular surface. The insulin receptor is an example of an enzyme-linked receptor. After binding insulin, enzymatic activity initiates a complex signaling pathway that allows the cell to grow, synthesize lipids, and import glucose.

3. A **G-protein-linked receptor** does not act as an enzyme, but instead will bind a different version of a G-protein (often GTP or GDP) on the intracellular side when a ligand is bound extracellularly. This causes activation of secondary messengers within the cell. One important second messenger is cyclic AMP (cAMP). It is known as a "universal hunger signal" because it is the second messenger of the hormones epinephrine and glucagon, which cause energy mobilization. Secondary messengers are usually small molecules that can move quickly through the cell. They can be made and destroyed quickly and help the signal amplify throughout the cell.

There are several other types of second messengers, such as cyclic GMP, calcium ions, and inositol triphosphate.

Signal transduction in eukaryotic cells usually involves many steps and complex regulation. Many signal transduction pathways include protein modifications and phosphorylation cascades, in which a series of protein kinases add a phosphate group to the next protein in the cascade sequence. In contrast, bacterial cells usually use a simpler two-component regulatory system in transduction pathways.

Signal transduction cascades are helpful to amplify a signal. They are also helpful in quickly turning a response on or off, and can lead to important changes in gene expression and cell function.

Mutations in the receptor or the ligand can alter signaling. Drugs that inhibit or excite part of the pathway can also influence the gene expression within the cell.

Homeostasis

All living things have a "steady state" that they try to maintain. This is helpful because the structures and processes within living things are sensitive to things like temperature, pH, pressure, salinity, osmotic pressure, and many other things. The set of conditions under which living things can successfully survive is called **homeostasis**. The body is constantly working to maintain this state by taking measurements and then responding appropriately. That might mean shivering if it is cold or releasing the hormone insulin if blood glucose is high. Depending on the condition, the body will respond accordingly.

Your blood glucose levels are regulated by insulin and glucagon, two hormones released from your pancreas. For example, when you've just eaten, blood sugar levels are high, and this stimulates insulin release. The cells of your body take up glucose and hepatocytes in your liver store extra glucose, as glycogen. Blood sugar levels go down, your body reestablishes homeostasis, and insulin release is shut off via negative feedback.

Many of these responses are controlled by negative or positive feedback pathways. A **negative feedback pathway** (also called **feedback inhibition**) works by turning itself off using the end product of the pathway. The end product inhibits the process from beginning, thus shutting down the pathway. This is a common strategy to conserve energy. If plenty of X is made, then the pathway to make X can be turned off. Sometimes, a very distant end product will turn off a process occurring several steps back. For example, ATP turns off the beginning stages of glycolysis, although ATP is not made until much later in cellular respiration.

A **positive feedback pathway** also involves an end product playing a role, but instead of inhibiting the pathway, it further stimulates it. For example, once X is made, it tells the pathway to make even more X. This is less common, but it occurs during fruit ripening and labor and delivery, which ramps up and up as it proceeds.

Cell Communication in Plants

Plants do not have a nervous system but can produce several proteins found in animal nervous systems, such as certain neurotransmitter receptors. Plants can also generate electrical signals in response to environmental stimuli, and this can affect flowering, respiration, photosynthesis, and wound healing. Light receptors are common in plants and help link environmental cues to biological processes such as seed germination, the timing of flowering, and chlorophyll production. Some plants can also use chemicals to communicate with nearby plants. For example, wounded tomatoes produce a volatile chemical as an alarm signal. This warns nearby plants and allows them to prepare a defense or immune response.

CELL DIVISION

Every second, thousands of cells are dying throughout our bodies. Fortunately, the body replaces them at an amazing rate. In fact, epidermal cells, or skin cells, die off and are replaced so quickly that the average 18-year-old grows an entirely new skin every few weeks. The body keeps up this unbelievable rate thanks to the mechanisms of **cell division**.

The remainder of this chapter takes a closer look at how cells divide. But remember, cell division is only a small part of the life cycle of a cell. Most of the time, cells are busy carrying out their regular activities. There are also some types of cells that are nondividing. Often, these are highly specialized cells that are created from a population of less specialized cells. The body continually makes them as needed, but they do not directly replicate themselves. Red blood cells are an example of nondividing cells. Some other cells are just temporarily nondividing. They enter a phase called G_0, where they hang out until they get a signal to reenter the normal cell cycle.

THE CELL CYCLE

Every cell has a life cycle—the period from the beginning of one division to the beginning of the next. The cell's life cycle is known as the **cell cycle**. The cell cycle is divided into two periods: **interphase** and **mitosis**. Take a look at the cell cycle of a typical cell.

Special gametes split their time between interphase and meiosis.

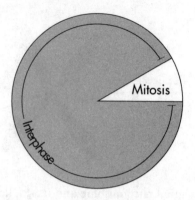

Notice that most of the life of a cell is spent in interphase.

INTERPHASE: THE GROWING PHASE

Interphase is the time span from one cell division to another. We call this stage interphase (*inter-* means between) because the cell has not yet started to divide. Although biologists sometimes refer to interphase as the "resting stage," the cell is definitely not inactive. During this phase, the cell carries out its regular activities. All the proteins and enzymes it needs to grow are produced during interphase.

The Three Stages of Interphase

Interphase can be divided into three stages: G_1, S, G_2.

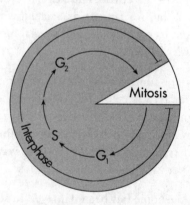

Sometimes certain cells can enter a stage called G_0 during which it is in a non-dividing state.

The most important phase is the S phase. That's when the cell replicates its genetic material. The first thing a cell has to do before undergoing mitosis is to duplicate all of its chromosomes, which contain the organism's DNA "blueprint." During interphase, every single chromosome in the nucleus is duplicated.

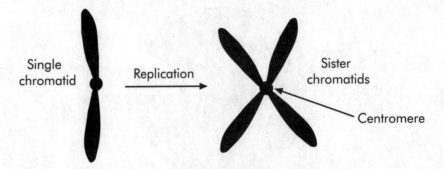

Single chromatid → Replication → Sister chromatids / Centromere

You'll notice that the original chromosome and its duplicate are still linked, like conjoined twins. These identical strands of DNA are now called **sister chromatids**. The chromatids are held together by a structure called the **centromere**. You can think of each chromatid as a chromosome, but because they remain attached, they are called chromatids instead. To be called a chromosome, each needs to have its own centromere. Once the chromatids separate, they will be full-fledged chromosomes.

Cell Cycle Regulation

We've already said that replication occurs during the S phase of interphase, so what happens during G_1 and G_2? During these stages, the cell performs metabolic reactions and produces organelles, proteins, and enzymes. For example, during G_1, the cell produces all of the enzymes required for DNA replication (DNA helicase, DNA polymerase, and DNA ligase, which are explained later in this chapter). By the way, *G* stands for "gap," but we can also associate it with "growth." These three phases are highly regulated by checkpoints and special proteins called **cyclins** and **cyclin-dependent kinases** (CDKs).

Cell cycle checkpoints are control mechanisms that make sure cell division is happening properly in eukaryotic cells. They monitor the cell to make sure it is ready to progress through the cell cycle and stop progression if it is not. In eukaryotes, checkpoint pathways function mainly at phase boundaries (such as the G_1/S transition and the G_2/M transition). Checkpoint pathways are an example of an important cell signaling pathway inside cells.

For example, if the DNA genome has been damaged in some way, the cell should not divide. If it does, damaged DNA will be passed onto daughter cells, which can be disastrous. When damaged DNA is found, checkpoints are activated and cell cycle progression stops. The cell uses the extra time to repair damage in DNA. If the DNA damage is so extensive that it cannot be repaired, the cell can undergo **apoptosis**, or programmed cell death. Apoptosis cannot stop once it has begun, so it is a highly regulated process. It is an important part of normal cell turnover in multicellular organisms.

Cell cycle checkpoints control cell cycle progression by regulating two families of proteins: cyclin-dependent kinases (CDKs) and cyclins. To induce cell cycle progression, an inactive CDK binds a regulatory cyclin. Once together, the complex is activated, can affect many proteins in the cell, and causes the cell cycle to continue. To inhibit cell cycle progression, CDKs and cyclins are kept separate. CDKs and cyclins were first studied in yeast, unicellular eukaryotic fungi. Budding yeast and fission yeast were good model organisms because they have only one CDK protein each and a few different cyclins. This relatively simple system allowed biologists to figure out how cell cycle progression is controlled. Their findings were then used to figure out how more complex eukaryotes (such as mammals) control the cell cycle.

Cancer

A cell losing control of the cell cycle can have disastrous consequences. For example, a mutation in a protein that normally controls progression through the cell cycle can result in unregulated cell division and cancer. **Cancer** means "crab," as in the zodiac sign. The name comes from the observation that malignant tumors grow into surrounding tissue, embedding themselves like clawed crabs. Cancer occurs when normal cells start behaving and growing very abnormally and spread to other parts of the body.

Mutated genes that induce cancer are called **oncogenes** ("onco-" is a prefix denoting cancer). Normally, these genes are required for proper growth of the cell and regulation of the cell cycle. Oncogenes, then, are genes that can convert normal cells into cancerous cells. Often these are abnormal or mutated versions of normal genes. The normal healthy version is called a proto-oncogene.

Tumor suppressor genes produce proteins that prevent the conversion of normal cells into cancer cells. They can detect damage to the cell and work with CDK/cyclin complexes to stop cell growth until the damage can be repaired. They can also trigger apoptosis if the damage is too severe to be repaired.

Cell cycle checkpoints and apoptosis have been intimately linked with cancer. In order for a normal cell to become a cancer cell, it must override cell cycle checkpoints and grow in an unregulated way. It must also avoid cell death. Tumors often accumulate DNA damage. Cancer treatments target these changes, because they are what make a cancer cell different from a normal cell.

Let's recap:

- The cell cycle consists of two things: interphase and mitosis.
- During the S phase of interphase, the chromosomes replicate via DNA replication.
- Growth and preparation for mitosis occur during the G_1 and G_2 stages of interphase.
- Cell cycle checkpoints make sure the cell is ready to continue through the cell cycle.
- CDK and cyclin proteins work together to promote cell cycle progression.
- Oncogenes promote cell growth and tumor suppressor genes inhibit cell growth.

Both oncogenes and tumor suppressors can cause cancer, but oncogenes do it by stepping on the cell cycle gas pedal, while tumor suppressors do it by removing the brake pedal.

MITOSIS: THE DANCE OF THE CHROMOSOMES

Once the chromosomes have replicated, the cell is ready to begin mitosis—the period during which the cell divides. Mitosis consists of a sequence of four stages: **prophase**, **metaphase**, **anaphase**, and **telophase**. It is not important to memorize the name of each phase, but it is important that you know the basic order of operations for mitosis.

Stage 1: Prophase: Prep (the cell prepares to divide)

One of the first signs of prophase is the disappearance of the nucleolus (ribosome-making area) and the nuclear envelope. In addition, the genome becomes visible as individual chromosomes; diffuse **chromatin** (which looks like a pile of scrunched up dental floss in the nucleus) condenses into densely packed chromosomes. To accomplish this, chromatin forms coils upon coils, and thickens. If you look at a human cell under the light microscope at the beginning of prophase, you would see 46 differently sized chromosomes, and each one would be the shape of an X. During interphase, individual chromosomes are not visible.

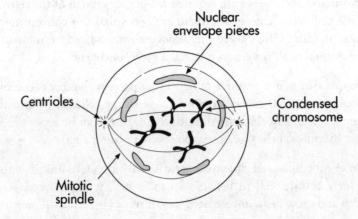

Now the cell has plenty of room to "sort out" the chromosomes. Remember centrioles? During prophase, these cylindrical bodies found within microtubule organizing centers (MTOCs) start to move away from each other, toward opposite ends of the cell. The centrioles will spin out a system of microtubules known as the **spindle fibers**. These spindle fibers will attach to a **kinetochore**, a structure on each chromatid. The kinetochores are part of the centromere.

Stage 2: Metaphase: Middle (chromosomes align in the middle)

The next stage is called metaphase. The chromosomes now begin to line up along the equatorial plane, or the **metaphase plate**, of the cell. This happens because spindle fibers are attached to the kinetochore of each chromatid.

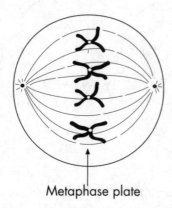

Metaphase plate

Stage 3: Anaphase: Apart (chromatids are pulled apart)

During anaphase, sister chromatids of each chromosome separate at the centromere and migrate to opposite poles. Chromatids are pulled apart by the microtubules, which begin to shorten. Each half of a pair of sister chromatids now moves to opposite poles of the cell. Non-kinetochore microtubules elongate the cell.

Anaphase

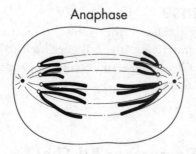

Stage 4: Telophase: Two (the cell completes splitting in two)

The final phase of mitosis is telophase. A nuclear membrane forms around each set of chromosomes and the nucleoli reappear.

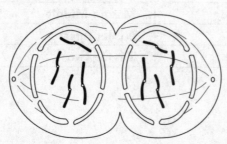

The nuclear membrane is ready to divide. Now it's time to split the cytoplasm in a process known as **cytokinesis**. Look at the figure below and you'll notice that the cell has begun to split along a **cleavage furrow** (which is produced by actin microfilaments).

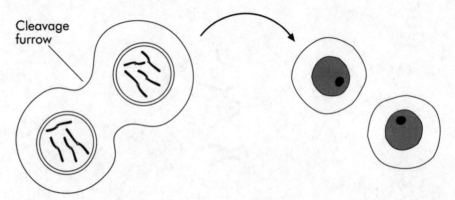

Cleavage furrow

A cell membrane forms around each cell, and the cell splits into two distinct daughter cells. The division of the cytoplasm yields two daughter cells.

Here's one thing to remember: cytokinesis occurs differently in plant cells. The cell doesn't form a cleavage furrow. Instead, a partition called a **cell plate** forms down the middle region.

Stage 5: Interphase

Once daughter cells are produced, they reenter the initial phase—interphase—and the whole process starts over. The cell goes back to its original state. Once again, the chromosomes decondense and become invisible, and the genetic material is called chromatin again.

But How Will I Remember All That?

For mitosis, you may already have your own mnemonic. If not, here's a table with a mnemonic we created for you.

IPMAT	
Interphase	**I** is for **Interlude**
Prophase	**P** is for **Prepare**
Metaphase	**M** is for **Middle**
Anaphase	**A** is for **Apart**
Telophase	**T** is for **Two**

Purpose of Mitosis

Mitosis achieves two things:

- the production of daughter cells that are identical copies of the parent cell
- maintaining the proper number of chromosomes from generation to generation

The impetus to divide occurs because an organism needs to grow, a tissue needs repair, or asexual reproduction must take place. For our purposes, we can say that mitosis occurs in just about every cell except sex cells. When you think of mitosis, remember: "Like begets like." Hair cells "beget" other hair cells, skin cells "beget" other skin cells, and so on. Mitosis is involved in growth, repair, and asexual reproduction.

KEY TERMS

quorum sensing
taxis
chemotaxis
ligands
receptors
signal transduction
ligand-gated ion channel
catalytic (or enzyme-linked) receptor
G-protein-linked receptor
homeostasis
negative feedback pathway (feedback inhibition)
positive feedback pathway
cell division
G_0 phase
cell cycle
interphase
mitosis
G_1 phase
S phase
G_2 phase

sister chromatids
centromere
cyclins
cyclin-dependent kinases (CDKs)
cell cycle checkpoints
apoptosis
cancer
oncogene
tumor suppressor gene
prophase
metaphase
anaphase
telophase
chromatin
spindle fibers
kinetochores
metaphase plate
cytokinesis
cleavage furrow
cell plate

Summary

○   Sometimes cells do not need to pass any molecules through the membrane, but instead they just pass a message. This is signal transduction. A ligand binds to a receptor on the outside of the cell and causes changes to the inside of the cell. Ligand-gated ion channels, catalytic receptors, and G-protein-linked receptors are common examples.

○   The cell cycle is divided into interphase and mitosis, or cellular division.

○   The three stages of interphase are G_1, G_2, and S phase.
  •   S phase is the "synthesis" phase, when chromosomes replicate.
  •   Growth and preparation for mitosis occur in G_1 and G_2.

○   Cell cycle progression is controlled by checkpoint pathways and CDK/cyclin complexes.

○   Cancer occurs when cells grow abnormally and spread to other parts of the body. Tumor-suppressor genes are genes that prevent the cell from dividing when it shouldn't. Proto-oncogenes are genes that help the cell divide. If either type is mutated (a mutated proto-oncogene is called an oncogene), it can lead to cell growth that is out of control.

○   Mitosis, or cellular division, occurs in four stages: prophase, metaphase, anaphase, and telophase.
  •   During prophase, the nuclear envelope disappears and chromosomes condense.
  •   Next is metaphase, when chromosomes align at the metaphase plate and mitotic spindles attach to kinetochores.
  •   In anaphase, chromosomes are pulled away from the center.
  •   Telophase terminates mitosis, and the two new nuclei form.
  •   The process of cytokinesis, which occurs during telophase, ends mitosis, as the cytoplasm and plasma membranes pinch to form two distinct, identical daughter cells.

Chapter 7 Drill

Answers and explanations can be found in Chapter 15.

1. A scientist is testing new chemicals designed to stop the cell cycle at various stages of mitosis. Upon applying one of the chemicals, she notices that all of the cells appear as shown below. Which of the following best explains how the chemical is likely acting on the cells?

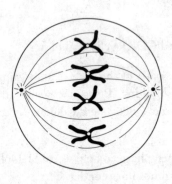

(A) The chemical has arrested the cells in prophase and has prevented attachment of the spindle fibers to the kinetochore.
(B) The chemical has arrested the cells in metaphase and has prevented dissociation of the spindle fibers from the centromere.
(C) The chemical has arrested the cells in metaphase and is preventing the shortening of the spindle fibers.
(D) The chemical has arrested the cells in anaphase and is preventing the formation of a cleavage furrow.

Questions 2 and 3 refer to the following graph and paragraph.

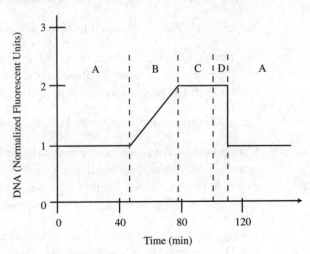

An experiment is performed to evaluate the amount of DNA present during a complete cell cycle. All of the cells were synced prior to the start of the experiment. During the experiment, a fluorescent chemical was applied to cells, which would fluoresce only when bound to DNA. The results of the experiment are shown above. Differences in cell appearance by microscopy or changes in detected DNA were determined to be phases of the cell cycle and are labeled with the letters A–D.

2. Approximately how long does S phase take to occur in these cells?

(A) 15 min
(B) 20 min
(C) 30 min
(D) 40 min

3. During which of the labeled phases of the experiment would the cell undergo anaphase?

(A) Phase A
(B) Phase B
(C) Phase C
(D) Phase D

4. Trisomy 21, which results in Down syndrome, results from nondisjunction of chromosome 21 in humans. Nondisjunction occurs when two homologous chromosomes, or two sister chromatids, do not separate. Which of the following describes the mechanism of this defect?

   (A) During DNA replication in S phase of the cell cycle, the two new strands do not separate.
   (B) During mitosis, at the metaphase plate, non-sister chromatids do not separate.
   (C) The mitotic spindle attaches to chiasmata rather than kinetochores.
   (D) The same microtubule in the spindle attaches to both sister chromatids during meiosis II.

5. A researcher isolates DNA from different types of cells and determines the amount of DNA for each type of cell. The samples may contain cells in various stages of the cell cycle. In order of increasing DNA content, which of the following would have the least amount of DNA to the greatest amount of DNA?

   (A) Sperm, neuron, liver
   (B) Ovum, muscle, taste bud
   (C) Liver, heart, intestine
   (D) Skin, neuron, ovum

6. Which of the following mutations would be least likely to have any discernable phenotype on the individual?

   (A) Translocation of the last 1,000 base pairs of chromosome 1 onto chromosome 2
   (B) Nondisjunction of chromosome 7 to produce trisomy-7
   (C) Nondisjunction of chromosome 8 to produce monosomy-8
   (D) Deletion of 400 base pairs, resulting in the loss of an enhancer

REFLECT

Respond to the following questions:

- Which topics from this chapter do you feel you have mastered?

- Which content topics from this chapter do you feel you need to study more before you can answer multiple-choice questions correctly?

- Which content topics from this chapter do you feel you need to study more before you can effectively compose a free response?

- Was there any content that you need to ask your teacher or another person about?

Chapter 8
Heredity

HAPLOIDS VERSUS DIPLOIDS

Every organism has a certain number of unique chromosomes. For example, fruit flies have 4 chromosomes, humans have 23 chromosomes, and dogs have 39 chromosomes. It turns out that most eukaryotic cells in fact have two full sets of chromosomes—one set from each parent. Humans, for example, have two sets of 23 chromosomes, giving us our grand total of 46.

A cell that has two sets of chromosomes is a **diploid cell**, and the zygotic chromosome number is given as "$2n$." That means we have two copies of each chromosome.

If a cell has only one set of chromosomes, we call it a **haploid cell**. This kind of cell is given the symbol n. For example, we would say that the haploid number of chromosomes for humans is 23.

Remember:

- *Diploid* refers to any cell that has two sets of chromosomes.
- *Haploid* refers to any cell that has one set of chromosomes.

Why do we need to know the terms *haploid* and *diploid*? Because they are extremely important when it comes to sexual reproduction. As we've seen, 46 is the normal diploid number for human beings, but there are only 23 different chromosomes. The duplicate versions of each chromosome are called **homologous chromosomes**. The homologous chromosomes that make up each pair are similar in size and shape and contain the same genes in the same locations (although they might have different versions of those genes). This is the case in all sexually reproducing organisms. In fact, this is the essence of sexual reproduction: each parent donates half its chromosomes to its offspring.

Gametes

Although most cells in the human body are diploid, there are special cells that are haploid. These haploid cells are called **sex cells**, or **gametes**. Why do we have haploid cells?

As we've said, an offspring has one set of chromosomes from each of its parents. A parent, therefore, contributes a gamete with one set that will be paired with the set from the other parent to produce a new diploid cell, or zygote. This produces offspring that are a combination of their parents.

Remember that the *n* stands for the number of unique chromosomes.

GREGOR MENDEL: THE FATHER OF GENETICS

What is genetics? In its simplest form, genetics is the study of heredity. It explains how certain characteristics are passed on from parents to children. Much of what we know about genetics was discovered by the monk **Gregor Mendel** in the 19th century. Since then, the field of genetics has vastly expanded.

Let's begin then with some of the fundamental points of genetics:

- **Traits**—or expressed characteristics—are influenced by one or more of your genes. Remember, a **gene** is simply a chunk of DNA that codes for a particular recipe. Different recipes affect different traits. DNA is passed from generation to generation, and from this process the genes and the traits associated with those genes are inherited. Within a chromosome, there are many genes, and each can contribute to a different trait. For example, in pea plants, there's a gene on a chromosome that codes for seed coat characteristics. The position of a gene on a chromosome is called a **locus**.

- Diploid organisms (organisms that have two sets of chromosomes) usually have two copies of each gene, one on each homologous chromosome. Homologous chromosomes are two copies or versions of the same chromosome in a diploid cell or organism. Humans have 23 pairs of homologous chromosomes.

  The two gene copies may be different from one another—that is, they may be alternate versions, or **alleles**, of the same gene. For example, if we're talking about the height of a pea plant, there's an allele for tall and an allele for short. In other words, both alleles are alternate forms of the gene for height.

- When an organism has two identical alleles for a given trait, the organism is **homozygous**. If an organism has two *different* alleles for a given trait, the organism is **heterozygous**.

- When discussing the physical appearance of an organism, we refer to its **phenotype**. The phenotype tells us what the organism looks like. When talking about the genetic makeup of an organism, we refer to its **genotype**. The genotype tells us which alleles the organism possesses.

- An allele can be **dominant** or **recessive**. This is determined by which allele "wins out" over the other in a heterozygote. The convention is to assign one of two letters for the two different alleles. The dominant allele receives a capital letter and the recessive allele receives a lowercase of the same letter. For instance, we might give the dominant allele for height in pea plants a *T* for tall. This means that the recessive allele would be *t* for short.

Overall then, homologous chromosomes are the same size and shape, and contain the same genes. However, they can contain different versions (or alleles) of those genes, and thus have different genetic sequences.

Name	Genotype	Phenotype
homozygous dominant	*TT*	Tall
homozygous recessive	*tt*	Short
heterozygous	*Tt*	Tall

One of the major ways the testing board likes to test genetics is by having you do crosses. Crosses involve mating hypothetical organisms with specific phenotypes and genotypes. We'll look at some examples in a moment, but for now, keep these test-taking tips in mind:

- Label each generation in the cross. The first generation in an experiment is always called the **parent**, or **P generation**. The offspring of the P generation are called the first **filial**, or **F_1 generation**. Members of the next generation, the grandchildren, are called the **F_2 generation**.
- Always write down the symbols you're using for alleles, along with a key to remind yourself what the symbols refer to. Use uppercase for dominant alleles and lowercase for recessive alleles.

Now let's look at some basic genetic principles.

MENDELIAN GENETICS

One of Mendel's hobbies was to study the effects of cross-breeding on different strains of pea plants. Mendel worked exclusively with true-breeder pea plants. This means the plants he used were genetically pure and consistently produced the same traits. For example, tall plants always produced tall plants; short plants always produced short plants. Through his work, he came up with three principles of genetics: the **Law of Dominance**, the **Law of Segregation**, and the **Law of Independent Assortment**.

The Law of Dominance

Mendel crossed two true-breeding plants with contrasting traits: tall pea plants and short pea plants.

To his surprise, when Mendel mated these plants, the characteristics didn't blend to produce plants of average height. Instead, all the offspring were tall.

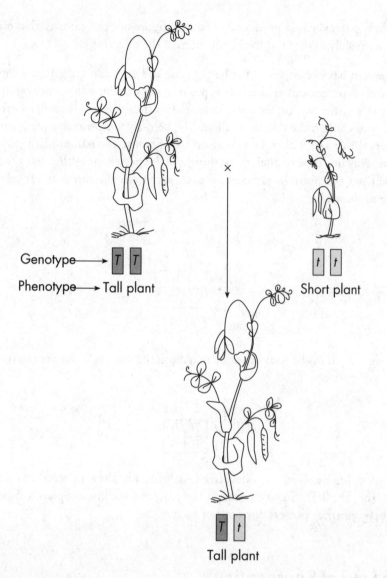

Genotype → T T
Phenotype → Tall plant

×

Short plant — t t

Tall plant — T t

Mendel recognized that one trait must be masking the effect of the other trait. This is called the Law of Dominance. The dominant tall allele, *T*, somehow masked the presence of the recessive short allele, *t*. Consequently, all a plant needs is one tall allele to make it tall.

Monohybrid Cross

A **monohybrid cross** occurs when two individuals are crossed and one gene is being studied. A simple way to represent a monohybrid cross is to set up a **Punnett square**. Punnett squares are used to predict the results of a cross. Let's construct a Punnett square for the cross between Mendel's true-breeding tall and short pea plants. Let's first designate the alleles for each plant. As we saw earlier, we can use the letter *T* for the tall dominant allele and *t* for the recessive short allele.

Since one parent was a pure, tall pea plant, we'll give it two dominant alleles (*TT* homozygous dominant). The other parent was a pure, short pea plant, so we'll give it two recessive alleles (*tt* homozygous recessive). To look at the types of offspring

Know Your Way Around a Square

A Punnett square is just a visual way to see all the possible offspring combos that could result from two parents mating. The possible gametes for each parent go on the sides and then they get matched up with each other to show the possible offspring genotypes.

Remember, these squares are looking only at one (or sometimes two) genes. So, they show possible offspring genotypes for that one gene only.

that these parents could produce, we have to figure out the gametes that each parent can possibly make (and thus contribute to the offspring).

Each parent has two copies of the height gene and each gamete gets one copy. This means each parent can make two types of gametes. For the homozygous dominant parent, the two copies are *T* and *T*. In other words, all gametes from this parent will contain the tall, or *T*, allele. The homozygous recessive plant can make gametes with *t* and with *t*. Notice again that because the parent plant was homozygous, only one type of allele (*t*, or short) will be passed on to the next generation. We will put the possible gametes for each parent on the top and left side of the square as shown below.

Keep This In Mind

When a parent is homozygous, both of their gametes will be identical, but it is still helpful to list each gamete individually on the square.

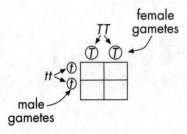

Now we can fill in the four boxes by matching the letters. What are the results for the F₁ generation?

	T	T
t	Tt	Tt
t	Tt	Tt

Each offspring received one allele from each parent. They all received one *T* and one *t*. They're all *Tt*! So, even though the parents were homozygous, offspring are heterozygous: they possess one copy of each allele.

The Law of Segregation

Next, Mendel took the offspring and self-pollinated them. Let's use a Punnett square to spell out the results. This time we're mating the offspring of the first generation—F₁; we are crossing those heterozygotes. Take a look at the results:

	T	t
T	TT	Tt
t	Tt	tt

F₂ Generation

Check out the lower right box. One of the offspring could be a short pea plant!

Although all of the F₁ plants were tall, they were heterozygotes that could make gametes containing the short *t* allele. In other words, their *T* and *t* alleles were segregated (separated) during gamete formation. This is an example of the Law of Segregation. It just means that each gamete gets only one of two copies of a gene. Gametes are haploid.

What about the genotype and phenotype for this cross? Remember, genotype refers to the genetic makeup of an organism, whereas phenotype refers to the appearance of the organism. According to the results of our Punnett square, what is the ratio of phenotypes and genotypes in the offspring?

Let's sum up the results. We have four offspring with two different phenotypes: three of the offspring are tall, whereas one of them is short. On the other hand, we have three genotypes: 1 *TT*, 2 *Tt*, and 1 *tt*.

Here's a summary of the results:

- The ratio of phenotypes is 3:1 (three tall:one short).
- The ratio of genotypes is 1:2:1 (one *TT*:two *Tt*:one *tt*).

The Law of Independent Assortment

So far, we have looked at only one trait: tall versus short. What happens when we study two traits at the same time? Each allele of the two traits will get segregated into two gametes, but how one trait gets split up into gametes has no bearing on how the other trait gets split up. In other words, the two alleles for each trait are sorted independently of the two alleles for other traits. For example, let's look at two traits in pea plants: height and color. When it comes to height, a pea plant can be either tall (*T*) or short (*t*). As for color, the plant can be either green (*G*) or yellow (*g*), with green being dominant. Each gene comes in two alleles, so this gives us four alleles total. By the Law of Independent Assortment, these four alleles can combine to give us four different gametes.

TG Tg tG tg

In other words, each allele of the height gene can be paired with either allele of the color gene. The reason alleles segregate independently is because chromosomes segregate independently. During meiosis I, each pair of homologous chromosomes is split, and which one aligns left or right during metaphase I is different for each pair.

Dihybrid Cross

Now that we know that different genes assort independently into gametes, let's look at a cross of two traits using those four gametes that we created above. A **dihybrid cross** is just like the monohybrid, but studies how two genes passed on to offspring

Here is the Punnett Square for a cross between two double heterozygotes (*Tt Gg*).

	TG	Tg	tG	tg
TG	TTGG	TTGg	TtGG	TtGg
Tg	TTGg	TTgg	TtGg	Ttgg
tG	TtGG	TtGg	ttGG	ttGg
tg	TtGg	Ttgg	ttGg	ttgg

This is an example of the Law of Independent Assortment. Each of the traits segregated independently. Don't worry about the different combinations in the cross—you'll make yourself dizzy with all those letters. Simply memorize the phenotype ratio of the pea plants. For the 16 offspring there are:

- 9 tall and green
- 3 tall and yellow
- 3 short and green
- 1 short and yellow

That's 9:3:3:1, but don't forget the circumstances to which this applies. It is only when two heterozygotes for two genes are crossed. It is not just a magic ratio that always works.

Rules of Probability

The Punnett square method works well for monohybrid crosses and helps us visualize possible combinations. However, a better method for predicting the likelihood of certain results from a dihybrid cross is to apply the Rules of Probability. To determine the probability that two or more independent events will occur simultaneously, one can simply calculate the product of the probability that each will occur independently. This is called the **Product Rule**. To determine the likelihood that EITHER event occurs, but not both, use another rule called the **Sum Rule**. To illustrate the product rule, let's consider again the cross between two identical dihybrid tall, green plants with the genotype *TtGg*.

> Product Rule: If A and B are independent, then:
> P(A and B) = P(A) times P(B)
>
> Sum Rule: If A and B are mutually exclusive, then:
> P(A or B) = P(A) + P(B)

Probability can be expressed as a fraction, percentage, or a decimal. Remember that this rule works only if the results of one cross are not affected by the results of another cross.

To find the probability of having a tall, yellow plant, simply multiply the probabilities of each event. If the probability of being tall is $\frac{3}{4}$ and the probability of being yellow is $\frac{1}{4}$, then the probability of being tall and yellow is $\frac{3}{4} \times \frac{1}{4} = \frac{3}{16}$.

Let's summarize Mendel's three laws.

SUMMARY OF MENDEL'S LAWS	
Laws	**Definition**
Law of Dominance	One trait masks the effects of another trait.
Law of Segregation	Each gamete gets only one of the copies of each gene.
Law of Independent Assortment	Each pair of homologous chromosomes splits independently, so the alleles of different genes can mix and match.

Test Cross

Suppose we want to know if a tall plant is homozygous (TT) or heterozygous (Tt). Its physical appearance doesn't necessarily tell us about its genetic makeup. The only way to determine its genotype is to cross the plant with a recessive, short plant, tt. This is known as a **test cross**. When the recessive plant is used, there will be only two possible outcomes in the daughter plants. Our mystery tall plant could be TT or Tt so when we mate it to a short plant, these are the possible situations: (1) $TT \times tt$ or (2) $Tt \times tt$. Let's take a look.

If none of the offspring is short, our original plant must have been homozygous, TT. If, however, even one short plant appears in the bunch, we know that our original pea plant was heterozygous, Tt. In other words, it wasn't a pure-breeding plant. Overall then, a test cross uses a recessive organism to determine the genotype of an organism of unknown genotype.

Linked Genes

Sometimes genes on the *same* chromosome stay together during assortment and move as a group. The group of genes is considered linked and tends to be inherited together. For example, the genes for flower color and pollen shape are linked on the same chromosomes and show up together.

Since **linked genes** are found on the same chromosome, they cannot segregate independently. This violates the Law of Independent Assortment.

Let's pretend that the height and color genes we looked at in the dihybrid cross are linked. A heterozygote for both traits still has two alleles for height (T and t) and two alleles for color (G and g). However, because height and color are located on the same chromosome, one allele for height and one allele for color are physically linked. For example, maybe the heterozygote has one chromsome with Tg and one chromosome with tG. When gametes are formed, the T and g will travel together, and t and G will travel together and be packaged into a gamete together. So, in the unlinked dihybrid shown earlier, there were four possible gamete combinations (*TG*, *Tg*, *tG*, *tg*), but now there are only two (*Tg* and *tG*). The only way to physically separate linked alleles is by crossing-over. If a crossover event occurs between linked genes, then recombinant gametes can occur. However, because linked genes are physically close together, crossing-over is not likely to occur between them.

If the genes were unlinked, then four gametes (*TG*, *Tg*, *tG*, *tg*) would be equally likely (as we saw when we talked about independent assortment). However, if certain combinations of alleles are found more often in offspring than they should (based on Mendelian ratios), this is a red flag that the two genes are close together and linked.

Look at the example below from a cross of *TtGg* and *ttgg*. If you draw out the Punnett square, you should get equal numbers of each phenotype occurring if the genes are assumed to be unlinked. Yet, what if the numbers below were the actual numbers of offspring of each phenotype?

Tall and Green: 6 offspring

short and Green: 37 offspring

Tall and yellow: 45 offspring

short and yellow: 3 offspring

There are more Green/short and yellow/Tall offspring. This is a sign that these alleles are linked on a chromosome because they occur more often than they statistically should. This shows that most of the gametes of the *TtGg* parent had *tG* together and *Tg* together since those gamete combinations would result in Green/short and yellow/Tall offspring.

So, where did the Green/Tall and yellow/short plants come from? They are the result of gametes from the *TtGg* parent that are *TG* and *tg*, respectively. If the *Tg* and *tG* alleles were linked in that parent, how could they produce a few *TG* and *tg* gametes? The answer is through recombination! Even though those alleles were not found on the same chromosome, recombination must have occurred during meiosis to swap them. This allowed gametes to be formed with *TG* and *tg*, and offspring to occur that were Tall/Green and short/yellow. The offspring formed from reccombination events are called **recombinants**. The **percentage of recombination (recombination frequency)** can be determined by adding up the recombinants and dividing by the total number of offspring.

$$9/91 = 9.9\%$$

This percentage can also be used as a measure of how far apart the genes are. The distance on a chromosome is measured in **map units**, or centimorgans.

The frequency of crossing-over between any two linked alleles is proportional to the distance between them. That is, the farther apart two linked alleles are on a chromosome, the more often the chromosome will cross over between them. This finding led to recombination mapping—mapping of linkage groups with each map unit being equal to 1 percent recombination. For example, if two linked genes, *A* and *B*, recombine with a frequency of 15 percent, and *B* and *C* recombine with a frequency of 9 percent, and *A* and *C* recombine with a frequency of 24 percent, what is the sequence and the distance between them?

The sequence and the distance of *A-B-C* is

$$15 \text{ units} \qquad 9 \text{ units}$$
$$A_____B_____C$$

If the recombination frequency between *A* and *C* had been 6 percent instead of 24 percent, the sequence and distance of *A-B-C* would instead be

$$6 \text{ units} \qquad 9 \text{ units}$$
$$A_____C_____B$$

SEX-LINKED TRAITS

We already know that humans contain 23 pairs of chromosomes. Twenty-two of the pairs of chromosomes are called **autosomes**. They code for many different traits. The other pair contains the **sex chromosomes**. This pair determines the sex of an individual. A female has two X chromosomes. A male has one X and one Y chromosome—an X from his mother and a Y from his father. Some traits, such as **color blindness** and **hemophilia**, are carried on sex chromosomes. These are called **sex-linked traits**. Most sex-linked traits are found on the X chromosome and are more properly referred to as "X-linked."

Since males have one X and one Y chromosome, what happens if a male has an X-chromosome with the color blindness allele? Unfortunately, he'll express the sex-linked trait, even if it is recessive. Why? Because his one and only X chromosome is color blind-X. He doesn't have another X to mask the effect of the color blind-X. However, if a female has only one color blind-X chromosome, she won't express a recessive sex-linked trait. For her to express the trait, she has to inherit two color blind-X chromosomes. A female with one color blind-X is called a **carrier**. Although she does not exhibit the trait, she can still pass it on to her children.

You can also use the Punnett square to figure out the results of sex-linked traits. Here's a classic example: a male who has normal color vision and a woman who is a carrier for color blindness have children. How many of the children will be color blind? To figure out the answer, let's set up a Punnett square.

Both red/green color blindness and hemophilia are X-linked recessive traits. Other X-linked traits are dominant, meaning only one copy of the mutant allele is required for expression of the trait in females.

$\bar{X}$ = color blind-*X*

Notice that we placed a bar above any color blind-X to indicate the presence of a color blindness allele. And now for the results. The couple could have a son who is color-blind, a son who sees color normally, a daughter who is a carrier, or a daughter who sees color normally.

Other Sex-Linked Diseases

There are a few X-linked conditions you should be familiar with for test day. Some are caused by a dominant allele, and others are inherited by a recessive allele.

Examples of X-Linked Dominant Traits	
Trait	**Phenotype**
Rett Syndrome	Postnatal neurological disorder of the grey matter of the brain
Vitamin D Resistant Rickets	Bone deformity including short stature and bow leggedness
Fragile X Syndrome	Developmental problems including learning disabilities and cognitive impairment

Examples of X-Linked Recessive Traits	
Trait	**Phenotype**
Red/Green Color Blindness	Difficulty discriminating red and green hues
Hemophilia	Impaired blood clotting, bruising easily
Duchenne Muscular Dystrophy	Muscle weakness and loss
Hunter Syndrome	Decreased ability to break down and recycle mucopolysaccharides
Juvenile Gout	Uric acid build-up, swollen joints, kidney problems
Hereditary Nephritis	Kidney disease and hearing loss

Barr Bodies

A look at the cell nucleus of normal females will reveal a dark-staining body known as a **Barr body**. A Barr body is an X chromosome that is condensed and visible. In every female cell, one X chromosome is activated and the other X chromosome is deactivated during embryonic development. Surprisingly, the X chromosome destined to be inactivated is randomly chosen in each cell. Therefore, in every tissue in the adult female, one X chromosome remains condensed and inactive. However, this X chromosome is replicated and passed on to a daughter cell. X-inactivation is the reason it is okay that females have two X chromosomes and males have only one. After X-inactivation, it is like everyone has one copy.

Other Inheritance Patterns

Not all patterns of inheritance obey the principles of Mendelian genetics. In fact, many traits we observe are due to a combined expression of alleles. Here are a couple of examples of non-Mendelian forms of inheritance:

- **Incomplete dominance (blending inheritance):** In some cases, the traits will blend. For example, if you cross a white snapdragon plant (genotype *WW*) with a red snapdragon plant (*RR*), the resulting progeny will be pink (*RW*). In other words, neither color is dominant over the other.

- **Codominance:** Sometimes you'll see an equal expression of both alleles. For example, an individual can have an AB blood type. In this case, each allele is equally expressed. That is, both the A allele and the B allele are expressed ($I^A I^B$). That's why the person is said to have AB blood. The expression of one allele does not prevent the expression of the other.

- **Polygenic inheritance:** In some cases, a trait results from the interaction of *many* genes. Each gene will have a small effect on a particular trait. Height, skin color, and weight are all examples of polygenic traits.

- **Non-nuclear inheritance**: Apart from the genetic material held in the nucleus, there is also genetic material in the mitochondria. The mitochondria are always provided by the egg during sexual reproduction, so mitochondrial inheritance is always through the maternal line, not the male line. In plants the mitochondria are provided by the ovule and are maternally inherited. Chloroplasts also contain DNA, in the form of one circular double-stranded molecule. Both chloroplasts and mitochondria assort randomly into daughter cells and gametes.

Pedigrees

One way to study genetic inheritance is by looking at a special family tree called a **pedigree**. A pedigree shows which family members have a particular trait and it can help determine if a trait is recessive or dominant and if it is sex-linked. Traits that skip generations are usually recessive. Traits that appear more in one sex than the other are usually sex-linked. In a pedigree chart, the males are squares and the females are circles. Matings are drawn with horizontal lines and the offspring are shown below the mated couple.

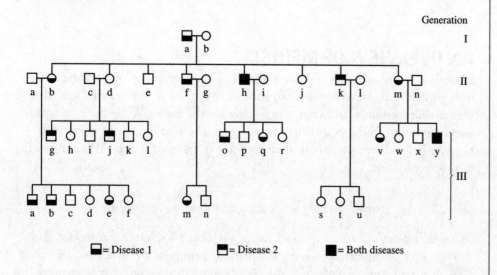

= Disease 1   = Disease 2   = Both diseases

ENVIRONMENTAL EFFECTS ON TRAITS

Changes in genotypes can result in changes in phenotype, but environmental factors also influence many traits, both directly and indirectly. In addition, an organism's adaptation to the local environment reflects a flexible response of its genome. For example, gene expression and protein function can be affected by temperature. **Phenotypic plasticity** occurs if two individuals with the same genotype have different phenotypes because they are in different environments. Here are some examples of how the environment can influence traits both directly and indirectly:

Height and Weight in Humans
- Controlled by both genetic and environmental factors
- **Examples:** nutrition, disease, physical activity, medication use

Soil pH Affects Flower Color
- Soil pH affects nutrient accessibility, which can change flower color.
- In acidic soil (pH<5.5), aluminum is available to the roots of hydrangeas, resulting in blue flowers.
- When no aluminum is available, flowers are white.

Seasonal Fur Color in Arctic Animals
- Arctic hare is brown or gray in the summer, when their habitat is full of greens and browns.
- Arctic hare is white in winter because everything is covered in snow.
- Color change is linked to photoperiod (or how much light is received during the day).

Sex Determination in Reptiles
- Temperatures experienced during embryonic/larval development determine the sex of offspring.
- In lizards, male embryos develop at intermediate temperatures and female embryos are generated at both extremes.

AN OVERVIEW OF MEIOSIS

Meiosis is the production of gametes. Since sexually reproducing organisms need only haploid cells for reproduction, meiosis is limited to sex cells in special sex organs called **gonads**. In males, the gonads are the **testes**, while in females they are the **ovaries**. The special cells in these organs—also known as **germ cells**—produce haploid cells (n), which then combine to restore the diploid ($2n$) number during fertilization.

$$\text{female gamete } (n) + \text{male gamete } (n) = \text{zygote } (2n)$$

When it comes to genetic variation, meiosis is a big plus. Variation, in fact, is the driving force of evolution. The more variation there is in a population, the more likely it is that some members of the population will survive extreme changes in the environment. Meiosis is far more likely to produce these sorts of variations than is mitosis, which therefore confers selective advantage on sexually reproducing organisms. We'll come back to this theme in Chapter 10.

A Closer Look at Meiosis

Meiosis actually involves two rounds of cell division: **meiosis I** and **meiosis II**.

Before meiosis begins, the diploid cell goes through interphase. Just as in mitosis, double-stranded chromosomes are formed during S phase.

Meiosis I

Meiosis I consists of four stages: prophase I, metaphase I, anaphase I, and telophase I.

Prophase I

Prophase I is a little more complicated than regular prophase. As in mitosis, the nuclear membrane disappears, the chromosomes become visible, and the centrioles move to opposite poles of the nucleus. But that's where the similarity ends.

The major difference involves the movement of the chromosomes. In meiosis, the chromosomes line up side-by-side with their counterparts (homologs). This event is known as **synapsis**.

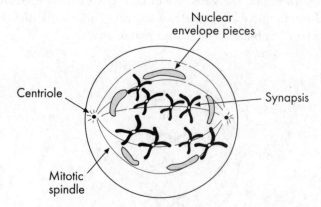

Synapsis involves two sets of chromosomes that come together to form a **tetrad** (a **bivalent**). A tetrad consists of four chromatids. Synapsis is followed by **crossing-over**, the exchange of segments between homologous chromosomes.

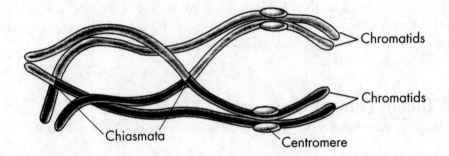

What's unique in prophase I is that pieces of chromosomes are exchanged between homologous partners. This is one of the ways organisms produce genetic variation. By the end of prophase I, chromosomes will have exchanged regions containing several **alleles**, or different forms of the same gene. This means that each chromosome will have a different mix of alleles than it began with. Remember, the two homologous chromosomes are versions that came from the person's mother and father. So, a chromosome that has experienced crossing-over will now be a mix of alleles from the mother and father versions.

Metaphase I

As in mitosis, the chromosome pairs—now called tetrads—line up at the metaphase plate. By contrast, you'll recall that in regular metaphase, the chromosomes line up individually. One important concept to note is that the alignment during metaphase is random, so the copy of each chromosome that ends up in a daughter cell is random. Therefore, the gamete will be created with a mixture of genetic information from the person's father and mother. This means that the offspring created would be a combination of all four grandparents.

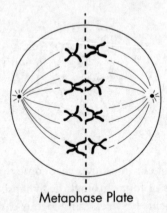

Metaphase Plate

Anaphase I

During anaphase I, each pair of chromatids within a tetrad moves to opposite poles. Notice that the chromatids do not separate at the centromere. The homologs separate with their centromeres intact.

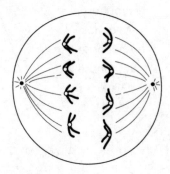

The chromosomes now move to their respective poles.

Telophase I

During telophase I, the nuclear membrane forms around each set of chromosomes.

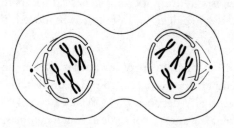

Finally, the cells undergo cytokinesis, leaving us with two daughter cells. Notice that at this point the nucleus contains the haploid number of chromosomes, but each chromosome is a duplicated chromosome, which consists of two chromatids.

Meiosis II

The purpose of the second meiotic division is to separate sister chromatids, and the process is virtually identical to mitosis. Let's run through the steps of meiosis II.

During prophase II, chromosomes once again condense and become visible. In metaphase II, chromosomes move toward the metaphase plate. This time they line up single file, not as pairs. During anaphase II, chromatids of each chromosome split at the centromere, and each chromatid is pulled to opposite ends of the cell. At telophase II, a nuclear membrane forms around each set of chromosomes and a total of *four* haploid cells are produced.

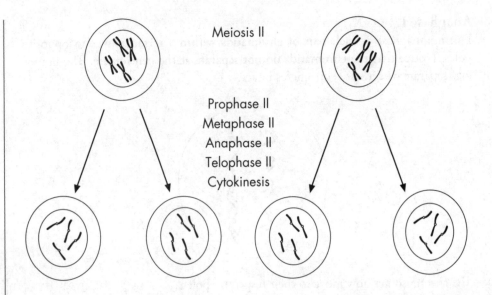

Meiosis II

Prophase II
Metaphase II
Anaphase II
Telophase II
Cytokinesis

Gametogenesis

Meiosis is also known as **gametogenesis.** If sperm cells are produced, then meiosis is called **spermatogenesis**. During spermatogenesis, four sperm cells are produced for each diploid cell. If an egg cell or an ovum is produced, this process is called **oogenesis**.

Oogenesis is a little different from spermatogenesis. Oogenesis produces only one ovum, not four. The other three cells, called **polar bodies,** get only a tiny amount of cytoplasm and eventually degenerate. Why does oogenesis produce only one ovum? Because the female wants to conserve as much cytoplasm as possible for the surviving gamete, the **ovum**.

Here's a summary of the major differences between mitosis and meiosis.

MITOSIS	MEIOSIS
• Occurs in somatic (body) cells	• Occurs in germ (sex) cells
• Produces identical cells	• Produces gametes
• Diploid cell → diploid cells	• Diploid cell → haploid cells
• 1 cell becomes 2 cells	• 1 cell becomes 4 cells
• Number of divisions: 1	• Number of divisions: 2

Meiotic Errors

Sometimes, a set of chromosomes has an extra or a missing chromosome. This occurs because of **nondisjunction**—chromosomes failed to separate properly during meiosis. This error, which produces the wrong number of chromosomes in a cell, usually results in miscarriage or significant genetic defects. For example, humans typically have two copies of each chromosome, but individuals with **Down syndrome** have three—instead of two—copies of the 21st chromosome.

Down syndrome is also known as trisomy 21. Two other conditions you should be familiar with are Turner Syndrome (45XO), which occurs when a female is missing part of or a whole X chromosome, and Klinefelter's Syndrome (47XXY), which occurs when a male has an extra X chromosome.

Nondisjunction can occur in anaphase I (meaning chromosomes don't separate when they should), or in anaphase II (meaning chromatids don't separate). Either one can lead to **aneuploidy**, or the presence of an abnormal number of chromosomes in a cell.

Chromosomal abnormalities also occur if one or more segments of a chromosome break and are either lost or reattach to another chromosome. The most common example is **translocation**, which occurs when a segment of a chromosome moves to another nonhomologous chromosome.

Here's an example of a translocation.

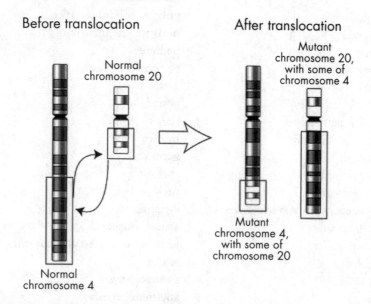

Before translocation

Normal
chromosome 20

Normal
chromosome 4

After translocation

Mutant
chromosome 20,
with some of
chromosome 4

Mutant
chromosome 4,
with some of
chromosome 20

Translocations can occur when recombination (crossing-over) occurs incorrectly. Fortunately, in most cases, damaged DNA can usually be repaired by special repair enzymes.

Many ethical, social, and medical issues surround human genetic disorders. Individuals who are carriers for genetic diseases are at increased risk of having a child with a genetic disease. A prenatal diagnosis can identify some genetic diseases before birth, and parents can then prepare to have a child with special needs or have the option of terminating the pregnancy. In addition, individuals who carry genetic diseases sometimes use an egg or sperm donor instead of using their own gametes.

Civic issues are also important to consider. Genetic discrimination occurs when people are treated unfairly because of actual or perceived differences in their genetic information that may cause or increase the risk of developing a disorder or disease. For example, a health insurer might refuse to give coverage to an individual who has a genetic difference that raises their odds of getting cancer. Many G7 countries protect genetic test information to help eliminate genetic discrimination and safeguard individuals' genetic privacy.

KEY TERMS

diploid cell
haploid cell
homologous chromosomes
sex cells (or gametes)
Gregor Mendel
trait
genes
locus
alleles
homozygous
heterozygous
phenotype
genotype
dominant
recessive
parent, or P generation
filial, or F_1 generation
F_2 generation
Law of Dominance
Law of Segregation
Law of Independent Assortment
monohybrid cross
Punnett square
dihybrid cross
Product Rule
Sum Rule
test cross
linked genes
recombinants
percentage of recombination
   (recombination frequency)
map units (centimorgans)
autosomes

sex chromosomes
color blindness
hemophilia
sex-linked traits
carrier
Barr body
incomplete dominance
   (blending inheritance)
codominance
polygenetic inheritance
non-nuclear inheritance
pedigree
phenotypic plasticity
meiosis
gonads
testes
ovaries
germ cells
meiosis I
meiosis II
synapsis
tetrad (bivalent)
crossing-over (recombination)
alleles
gametogenesis
spermatogenesis
oogenesis
polar bodies
ovum
nondisjunction
Down syndrome
aneuploidy
translocation

Summary

- Genetics is the study of heredity. The foundation of genetics is that traits are produced by genes, which are found on chromosomes.

- Diploid organisms carry two copies of every gene, one on each of two homologous chromosomes, one that came from their mother and one that came from their father.

- When two alleles are the same, the organism is homozygous for that trait. When the two alleles are different, the organism is heterozygous for the trait.

- Genes are described as the genotype of the organism. The physical expression of the traits, or what you see in the organism, is the phenotype.

- Homozygous individuals have only one allele that will determine the phenotype. However, heterozygous individuals have two different alleles that could contribute to the phenotype. The allele that "wins out" and directs the phenotype is the dominant allele. The one that doesn't show up in the phenotype is the recessive allele.

- Simple genetics with dominants and recessives was illuminated by Gregor Mendel. In Mendelian genetics, there are three important laws: the Law of Dominance, the Law of Segregation, and the Law of Independent Assortment.

- If two genes are on the same chromosome and close together, they are linked. Completely linked alleles on the same chromosome are locked together and travel together during gamete formation. Recombination frequency can be used to measure how far apart two genes are on a chromosome.

- Humans have 23 pairs of homologous chromosomes for a total of 46. Of these, 22 pairs are autosomes and 2 are designated as sex chromosomes:
  - Females have two X chromosomes; males have one X and one Y chromosome.
  - Because females have two X chromosomes, one is inactivated in each cell and condenses into a Barr body.

- Non-Mendelian genetics refers to a situation in which traits do not follow the Mendelian laws. Examples include:
  - incomplete dominance
  - codominance
  - polygenic inheritance

o Meiosis produces four genetically distinct haploid gametes (sperm or egg cells). Meiosis involves two rounds of cell division: meiosis I and meiosis II.

- During meiosis I, the homologous chromosome pairs separate, but before they can, they undergo crossing-over and swap some DNA.
- During meiosis II, sister chromatids separate.
- Mutations, such as nondisjunction events or whole chromosome translocations, can lead to genetic diseases.

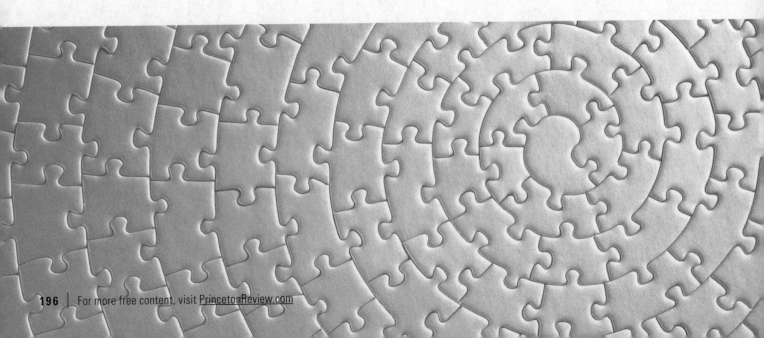

Chapter 8 Drill

Answers and explanations can be found in Chapter 15.

1. Two expecting parents wish to determine all possible blood types of their unborn child. If both parents have an AB blood type, which of the following blood types will their child NOT possess?

    (A) A
    (B) B
    (C) AB
    (D) O

2. A new species of tulip was recently discovered. A population of pure red tulips was crossed with a population of pure blue tulips. The resulting F_1 generation was all purple. This result is an example of which of the following?

    (A) Complete dominance
    (B) Incomplete dominance
    (C) Codominance
    (D) Linkage

3. In pea plants, tall (T) is dominant over short (t) and green (G) is dominant over yellow (g). If a pea plant that is heterozygous for both traits is crossed with a plant that is recessive for both traits, approximately what percentage of the progeny plants will be tall and yellow?

    (A) 0%
    (B) 25%
    (C) 66%
    (D) 75%

Questions 4-6 refer to the following pedigree tree and paragraph.

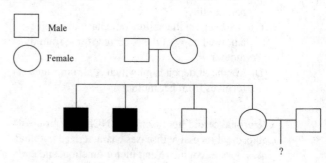

Hemophilia is an X-linked disease associated with the inability to produce specific proteins in the blood-clotting pathway. Shown above is a family pedigree tree in which family members afflicted with the disease are shown with filled-in squares (male) or circles (females). A couple is trying to determine the likelihood of passing on the disease to their future children (represented by the *?* symbol above) because the hemophilia runs in the woman's family.

4. Assuming that the woman in the couple is a carrier, what is the probability that the couple's first son will have hemophilia?

    (A) 0%
    (B) 25%
    (C) 50%
    (D) 100%

5. Why were both women in the family tree above free of the disease?

    (A) They were lucky because they didn't receive an X chromosome with the diseased allele.
    (B) They cannot get hemophilia because it is associated only with a diseased Y chromosome.
    (C) They are carriers and will get the disease only if they have children.
    (D) They have at least one normal (unaffected) copy of the X chromosome.

6. Turner syndrome is a disease in which an individual is born with only a single X chromosome. Suppose the woman in the couple is a carrier for hemophilia and has a child with Turner syndrome. Would this child have the disease?

   (A) Yes, because the child would have only one copy of the X chromosome and it would be affected.
   (B) No, because women cannot be affected by hemophilia.
   (C) No, because the child would have to receive a normal copy of the X chromosome from its mother.
   (D) Maybe, it depends on which X chromosome she receives from her mother.

7. Neuronal ceroid lipofuscinosis (NCL) is a group of autosomal recessive diseases characterized by blindness, loss of cognitive and motor function, and early death. One of the genes that is mutated in this disease is CLN3. When functional CLN3 protein is absent, neurons die because of increased storage material in the cells, presumably because the lysosomes aren't working properly. What is a valid explanation as to why parents of children with this devastating disease can be completely normal?

   (A) Parents of the autosomal recessive disease must both be carriers of the mutation on the CLN3 gene.
   (B) Carriers of the CLN3 mutation genotype do not show the phenotype because one normal CLN3 allele is present to provide a functioning CLN3 protein.
   (C) Redundant proteins take over the function of the mutant CLN3 protein in the parents.
   (D) X inactivation prevents expression of the CLN3 mutated protein in the parents.

8. A breeder of black Labrador puppies notices that in his line a gene predisposing his dogs for white spots has arisen. He believes the white spot allele is autosomal recessive, and he wants to prevent it from continuing in his dogs. He sees that one of his female dogs, Speckle, has white spots, and he therefore assumes she is homozygous for the gene. But he must determine which males in his pack are heterozygous to avoid breeding them in the future. What is a reasonable plan?

   (A) Stop breeding Speckle.
   (B) Pair Speckle with a male with white spots.
   (C) Perform a testcross of Speckle with a black Labrador male. If any pups are born with white spots, stop breeding that male because he is heterozygous.
   (D) Breed Speckle with a chocolate Labrador.

9. Which of the following is a reason why certain traits do not follow Mendel's Law of Independent Assortment?

   (A) The genes are linked on the same chromosome.
   (B) It applies only to eukaryotes.
   (C) Certain traits are not completely dominant.
   (D) Heterozygotes have both alleles.

10. The gene for coat color in some breeds of cats is found on the X chromosome. Calico cats are mottled with orange and black colorings. What is a possible explanation for the fact that true calico cats are only female?

    (A) The allele for coat color is randomly chosen by X-inactivation.
    (B) The coat color is linked with genes that are lethal to male cats, so male calico cats never live past birth.
    (C) Coat orange and black color is expressed only in female cats; males have only one color.
    (D) Male cats have Y chromosomes.

REFLECT

Respond to the following questions:

- Which topics from this chapter do you feel you have mastered?

- Which content topics from this chapter do you feel you need to study more before you can answer multiple-choice questions correctly?

- Which content topics from this chapter do you feel you need to study more before you can effectively compose a free response?

- Was there any content that you need to ask your teacher or another person about?

Chapter 9
Gene Expression
and Regulation

DNA: THE BLUEPRINT OF LIFE

All living things possess an astonishing degree of organization. From the simplest single-celled organism to the largest mammal, millions of reactions and events must be coordinated precisely for life to exist. This coordination is directed by DNA, which is the hereditary blueprint of the cell.

THE MOLECULAR STRUCTURE OF DNA

DNA is made up of repeated subunits of **nucleotides**. Each nucleotide has a **five-carbon sugar**, a **phosphate**, and a **nitrogenous base**. Take a look at the nucleotide below.

The name of the pentagon-shaped sugar in DNA is **deoxyribose**. Hence, the name *deoxyribo*nucleic acid. Notice that the sugar is linked to two things: a phosphate and a nitrogenous base. A nucleotide can have one of four different nitrogenous bases:

- **adenine**—a purine (double-ringed nitrogenous base)
- **guanine**—a purine (double-ringed nitrogenous base)
- **cytosine**—a pyrimidine (single-ringed nitrogenous base)
- **thymine**—a pyrimidine (single-ringed nitrogenous base)

Any of these four bases can attach to the sugar. As we'll soon see, this is extremely important when it comes to the message of the genetic code in DNA.

The nucleotides can link up in a long chain to form a single strand of DNA. Here's a small section of a DNA strand.

The nucleotides themselves are linked together by **phosphodiester bonds** between the sugars and the phosphates. This is called the sugar-phosphate backbone of DNA and it serves as a scaffold for the bases.

Two DNA Strands

Each DNA molecule consists of two strands that wrap around each other to form a long, twisted ladder called a **double helix**. The structure of DNA was brilliantly deduced in 1953 by three scientists: **Watson, Crick,** and **Franklin.**

Now let's look at the way two DNA strands get together. Again, think of DNA as a ladder. The sides of the ladder consist of alternating sugar and phosphate groups, while the rungs of the ladder consist of pairs of nitrogenous bases.

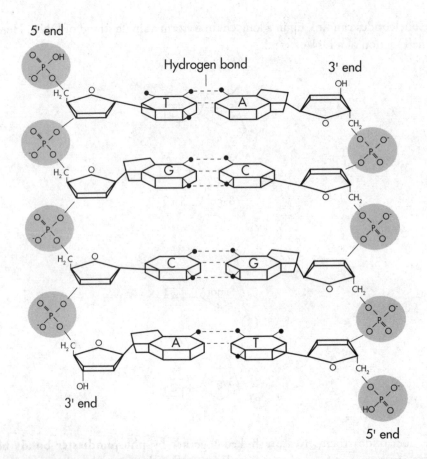

The nitrogenous bases pair up in a particular way. Adenine in one strand always binds to thymine (A–T or T–A) in the other strand. Similarly, guanine always binds to cytosine (G–C or C–G). This predictable matching of the bases is known as **base pairing**.

The two strands are said to be **complementary**. This means that if you know the sequence of bases in one strand, you'll know the sequence of bases in the other strand. For example, if the base sequence in one DNA strand is A–T–C, the base sequence in the complementary strand will be T–A–G.

The two DNA strands run in opposite directions. You'll notice from the figure above that each DNA strand has a 5' end and a 3' end, so-called for the carbon that ends the strand (i.e., the fifth carbon in the sugar ring is at the 5' end, while the third carbon is at the 3' end). The 5' end has a phosphate group, and the 3' end has an OH, or "hydroxyl," group. The 5' end of one strand is always opposite to the 3' end of the other strand. The strands are therefore said to be **antiparallel**.

The DNA strands are linked by **hydrogen bonds**. Two hydrogen bonds hold adenine and thymine together, and three hydrogen bonds hold cytosine and guanine together.

Before we go any further, let's review the base pairing in DNA:

- Adenine pairs up with thymine (A–T or T–A) by forming two hydrogen bonds.
- Cytosine pairs up with guanine (G–C or C–G) by forming three hydrogen bonds.

GENOME STRUCTURE

The order of the four base pairs in a DNA strand is the genetic code. Like a special alphabet in our cells, these four nucleotides spell out thousands of recipes. Each recipe is called a **gene**. The human genome has around 20,000 genes.

The recipes of the genes are spread out among the millions of nucleotides of DNA, and all of the DNA for a species is called its **genome**. Each separate chunk of DNA in a genome is called a **chromosome**.

Prokaryotes have one circular chromosome, and eukaryotes have linear chromosomes. In eukaryotes, the DNA is further structured, likely since linear chromosomes are more likely to get tangled. To keep it organized, the DNA is wrapped around proteins called **histones**, and then the histones are bunched together in groups called a **nucleosome**. Not all DNA is wound up equally, because it must be unwound in order to be read. How tightly DNA is packaged depends on the section of DNA and also what is going on in the cell at that time. When the genetic material is in a loose form in the nucleus, it is called **euchromatin,** and its genes are active, or available for transcription. When the genetic material is fully condensed into coils, it is called **heterochromatin,** and its genes are generally inactive. Situated in the nucleus, chromosomes contain the recipe for all the processes necessary for life, including passing themselves and their information on to future generations. In this chapter, we'll look at precisely how they accomplish this.

Humans have 23 different chromosomes and they have two copies of each, so each human cell has 46 linear chromosome segments in the nucleus.

DNA REPLICATION

We said in the beginning of this chapter that DNA is the hereditary blueprint of the cell. By directing the manufacture of proteins, DNA serves as the cell's blueprint. But how is DNA inherited? For the information in DNA to be passed on, it must first be copied. This copying of DNA is known as **DNA replication**.

Because the DNA molecule is twisted over on itself, the first step in replication is to unwind the double helix by breaking the hydrogen bonds. This is accomplished by an enzyme called **helicase**. The exposed DNA strands now form a *y*-shaped **replication fork.**

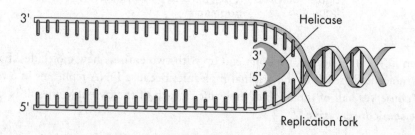

Now each strand can serve as a template for the synthesis of another strand. DNA replication begins at specific sites called **origins of replication**. Because the DNA helix twists and rotates during DNA replication, another class of enzymes, called

Neither DNA strand is directly copied; however, by building a new partner strand for each of them, a two-stranded molecule is created that is identical to the original.

DNA **topoisomerases**, cuts, and rejoins the helix to prevent tangling. The enzyme that performs the actual addition of nucleotides to the freshly built strand is **DNA polymerase**. But DNA polymerase, oddly enough, can add nucleotides only to the 3' end of an existing strand. Therefore, to start off replication, an enzyme called **RNA primase** adds a short strand of RNA nucleotides called an **RNA primer**. After replication, the primer is degraded by enzymes and replaced with DNA so that the final strand contains only DNA.

During DNA replication, one DNA strand is called the **leading strand**, and it is made continuously. That is, the nucleotides are steadily added one after the other by DNA polymerase. The other strand—the **lagging strand**—is made discontinuously. Unlike the leading strand, the lagging strand is made in pieces of nucleotides known as **Okazaki fragments**. Why is the lagging strand made in small pieces?

Nucleotides are added only in the 5' to 3' direction since nucleotides can be added only to the 3' end of the growing chain. However, when the double-helix is "unzipped," one of the two strands is oriented in the opposite direction—3' to 5'. Because DNA polymerase doesn't work in this direction, the strand needs to be built in pieces. You can see in the figure below that the leading strand is being created toward the opening of the helix and the helix continually opens ahead of it to accommodate it. The lagging strand is built in the opposite direction of the way the helix is opening, so it can build only until it hits a previously built stretch. Once the helix unwinds a bit more, it can build another Okazaki fragment, and so on. These fragments are eventually linked together by the enzyme **DNA ligase** to produce a continuous strand. Finally, hydrogen bonds form between the new base pairs, leaving two identical copies of the original DNA molecule.

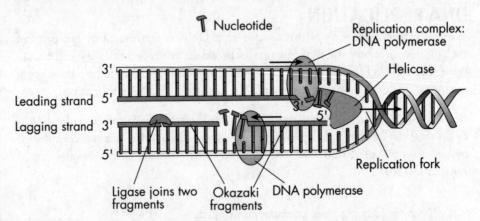

When DNA is replicated, we don't end up with two entirely new molecules. Each new molecule has half of the original molecule. Because DNA replicates in a way that conserves half of the original molecule in each of the two new ones, it is said to be **semiconservative**.

An interesting problem is that a few bases at the very end of a DNA molecule cannot be replicated because DNA polymerase needs space to bind template DNA. This means that every time replication occurs, the chromosome loses a few base pairs at the end. The genome has compensated for this over time, by putting bits of unimportant (or at least less important) DNA at the ends of a molecule. These ends are called **telomeres**. They get shorter and shorter over time.

Many enzymes and proteins are involved in DNA replication. The ones you'll need to know for the AP Biology Exam are DNA helicase, DNA polymerase, DNA ligase, topoisomerase, and RNA primase:

- Helicase unwinds our double helix into two strands.
- DNA Polymerase adds nucleotides to an existing strand.
- Ligase brings together the Okazaki fragments.
- Topoisomerase cuts and rejoins the helix.
- RNA primase catalyzes the synthesis of RNA primers.

Key History

The reason that we know that the DNA is the inheritable material is because of several key experiments. The first turning point was achieved by **Avery, MacLeod, and McCarty.** They isolated various cellular components from a dead virulent strain of bacteria. Then they followed up on a previous experiment by Griffiths and added each of these cellular components to a stain of living non-virulent bacteria. Only the DNA component of the deadly bacteria was able to change the second bacteria into a deadly strain capable of reproducing. This means that DNA must be responsible for passing traits and is inheritable.

The next important experiment was done by **Hershey-Chase**. They used bacteriophages, which are a special type of virus. They labeled the protein parts of some viruses with radiolabeled sulfur and they labeled the DNA parts of other viruses with radiolabeled phosphorus. When the viruses infected bacteria, only the labeled DNA was inside, but they were still able to replicate and make progeny viruses. Thus they proved that only DNA is required to pass on information. The protein was not required.

Central Dogma

DNA's main role is directing the manufacture of molecules that actually do the work in the body. DNA is just the recipe book, not the chef or the meal. When a DNA recipe is used, scientists say that that DNA is expressed.

The first step of DNA expression is to turn it into RNA. The RNA is then sent out into the cell and often gets turned into a protein. Proteins are the biggest group of "worker-molecules," and most expressed DNA turns into proteins. These proteins, in turn, regulate almost everything that occurs in the cell.

The process of making an RNA from DNA is called **transcription,** and the process of making a protein from an RNA is called **translation**. In eukaryotic cells, transcription takes place in the nucleus (where the DNA is kept), and translation takes place in the cytoplasm.

Obviously, in prokaryotes, both occur in the cytoplasm because prokaryotes don't contain a nucleus; this allows transcription and translation to occur at the same time.

The flow of genetic information, the **Central Dogma of Biology**, is therefore:

$$DNA \xrightarrow[\text{in the nucleus}]{\text{transcription}} RNA \xrightarrow[\text{in the cytoplasm}]{\text{translation}} Proteins$$

RNA

Before we discuss transcription of RNA, let's talk about its structure. Although RNA is also made up of nucleotides, it differs from DNA in three ways.

1. RNA is single-stranded, not double-stranded.
2. The five-carbon sugar in RNA is **ribose** instead of deoxyribose.
3. The RNA nitrogenous bases are adenine, guanine, cytosine, and a different base called **uracil**. Uracil replaces thymine as adenine's partner.

Here's a table to compare DNA and RNA. Keep these differences in mind—the testing board loves to test you on them.

DIFFERENCES BETWEEN DNA AND RNA		
	DNA (double-stranded)	**RNA (single-stranded)**
Sugar	deoxyribose	ribose
Bases	adenine	adenine
	guanine	guanine
	cytosine	cytosine
	thymine	uracil

There are three main types of RNA: messenger RNA (mRNA), ribosomal RNA (rRNA), and transfer RNA (tRNA). All three types of RNA are key players in the synthesis of proteins:

- **Messenger RNA (mRNA)** is a temporary RNA version of a DNA recipe that gets sent to the ribosome.
- **Ribosomal RNA (rRNA)**, which is produced in the nucleolus, makes up part of the ribosomes. You'll recall from our discussion of the cell in Chapter 5 that the ribosomes are the sites of protein synthesis. We'll see how they function a little later on.
- **Transfer RNA (tRNA)** shuttles amino acids to the ribosomes. It is responsible for bringing the appropriate amino acids into place at the appropriate time. It does this by reading the message carried by the mRNA.

There is also another class of RNA, called interfering RNAs, or **RNAi.** These are small snippets of RNA that are naturally made in the body or intentionally created by humans. These interfering RNAs, called siRNA and miRNA, can bind to specific sequences of RNA and mark them for destruction. This will be discussed more later in this chapter. Now that we know about the different types of RNA, let's see how they direct the synthesis of proteins.

Transcription

Transcription involves making an RNA copy of a bit of DNA code. The initial steps in transcription are similar to the initial steps in DNA replication. The obvious difference is that, whereas in replication we end up with a complete copy of the cell's DNA, in transcription we end up with only a tiny specific section copied into an mRNA. This is because only the bit of DNA that needs to be expressed will be transcribed. If you wanted to make a cake, but your cookbook couldn't leave your vault (a.k.a. the nucleus), you wouldn't copy the entire cookbook! You would copy only the recipe for the cake.

Since each recipe is a gene, transcription occurs as-needed on a gene-by-gene basis.

The exception to this is prokaryotes because they will transcribe a recipe that can be used to make several proteins. This is called a **polycistronic transcript**. Eukaryotes tend to have one gene that gets transcribed to one mRNA and translated into one protein. Our transcripts are **monocistronic.**

Transcription involves three phases: initiation, elongation, and termination. As in DNA replication, the first initiation step in transcription is to unwind and unzip the DNA strands using helicase. Transcription begins at special sequences of the DNA strand called **promoters**. You can think of a promoter as a docking site or a runway. We will talk about how promoters are involved in regulating transcription later in the chapter.

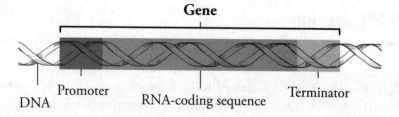

Gene

DNA | Promoter | RNA-coding sequence | Terminator

Because RNA is single-stranded, we have to copy only *one* of the two DNA strands. The strand that serves as the template is known as the **antisense strand,** the **non-coding strand, minus-strand**, or the **template strand**. The other strand that lies dormant is the **sense strand,** or the **coding strand**.

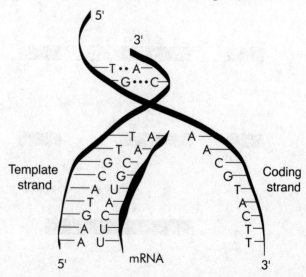

Just as DNA polymerase builds DNA, **RNA polymerase** builds RNA, and just like DNA polymerase, RNA polymerase adds nucleotides only to the 3' side, therefore building a new molecule from 5' to 3'. This means that RNA polymerase must bind to the 3' end of the template strand first (which would be the 5' end of future mRNA).

RNA polymerase doesn't need a primer, so it can just start transcribing the DNA right off the bat. The promoter region is considered to be "upstream" of the actual coding part of the gene. This way the polymerase can get set up before the bases it needs to transcribe, like a staging area before an official parade starting point. The official starting point is called the **start site**.

When transcription begins, RNA polymerase travels along and builds an RNA that is complementary to the template strand of DNA. It is just like DNA replication except when DNA has an adenine, the RNA can't add a thymine since RNA doesn't have thymine. Instead, the enzyme adds a uracil.

Once RNA polymerase finishes adding on nucleotides and reaches the termination sequence, it separates from the DNA template, completing the process of transcription. The new RNA is now a transcript or a copy of the sequence of nucleotides based on the DNA strand. Note that the freshly synthesized RNA is complementary to the template strand, but it is identical to the coding strand (with the substitution of uracil for thymine).

RNA Processing

In prokaryotes, the mRNA is now complete, but in eukaryotes the RNA must be processed before it can leave the nucleus.

In eukaryotes, the freshly transcribed RNA is called an hnRNA (heterogeneous nuclear RNA) and it contains both coding regions and noncoding regions. The regions that express the code that will be turned into protein are **exons**. The noncoding regions in the mRNA are **introns**.

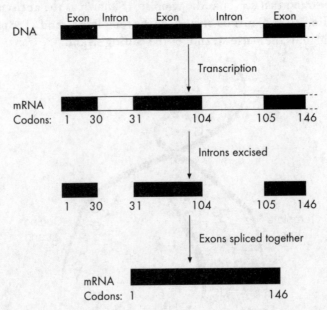

The introns—the intervening sequences—must be removed before the mRNA leaves the nucleus. This process, called **splicing**, is accomplished by an RNA-protein complex called a **spliceosome**. This process produces a final mRNA that is shorter than the transcribed RNA. The way that a transcript is spliced can vary, and alternative splicing variants will result with different exons included.

In addition to splicing, a **poly(A) tail** is added to the 3' end and a **5' GTP cap** is added to the 5' end.

Translation

Translation is the process of turning an mRNA into a protein. Remember, each protein is made of amino acids. The order of the mRNA nucleotides will be read in the ribosome in groups of three. Three nucleotides is called a **codon**. Each codon corresponds to a particular amino acid. The genetic code is redundant, meaning that certain amino acids are specified by more than one codon.

The mRNA attaches to the ribosome to initiate translation and "waits" for the appropriate amino acids to come to the ribosome. That's where tRNA comes in. A tRNA molecule has a unique three-dimensional structure that resembles a four-leaf clover:

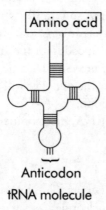

Amino acid

Anticodon

tRNA molecule

One end of the tRNA carries an amino acid. The other end, called an **anticodon**, has three nitrogenous bases that can complementarily base pair with the codon in the mRNA. Usually, the normal rules of base pairing are set in stone, but tRNA anticodons can be a bit flexible when they bind with a codon on an mRNA, especially the third nucleotide in an anticodon. The third position is said to experience **wobble pairing**. Things that don't normally bind will pair up, like guanine and uracil.

Transfer RNAs are the "go-betweens" in protein synthesis. Each tRNA becomes charged and enzymatically attaches to an amino acid in the cell's cytoplasm and "shuttles" it to the ribosome. The charging enzymes involved in forming the bond between each amino acid and tRNA require ATP.

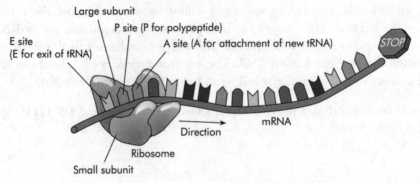

Note: Each site E, P, or A is a groove for the tRNA to fit in.

Translation also involves three phases: **initiation**, **elongation**, and **termination**. Initiation begins when a ribosome attaches to the mRNA.

What does the ribosome do? It helps the process along by holding everything in place while the tRNAs assist in assembling polypeptides.

Initiation

Ribosomes contain three binding sites: an **A site**, a **P site**, and an **E site**. The mRNA will shuffle through from A to P to E. As the mRNA codons are read, the polypeptide will be built. In all organisms, the **start codon** for the initiation of protein synthesis is A–U–G, which codes for the amino acid methionine. Proteins can have AUGs in the rest of the protein that code for methionine as well, but the special first AUG of an mRNA is the one that will kick off translation. The tRNA with the complementary anticodon, U–A–C, is methionine's personal shuttle; when the AUG is read on the mRNA, methionine is delivered to the ribosome.

Elongation

Addition of amino acids is called elongation. Remember that the mRNA contains many codons, or "triplets," of nucleotide bases. As each amino acid is brought to the mRNA, it is linked to its neighboring amino acid by a peptide bond. When many amino acids link up, a polypeptide is formed.

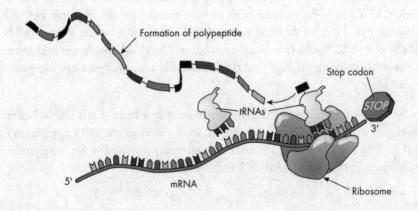

Termination

How does this process know when to stop? The synthesis of a polypeptide is ended by **stop codons**. A codon doesn't always code for an amino acid; there are three that serve as a stop codon. Termination occurs when the ribosome runs into one of these three stop codons.

How about a little review?

- In transcription, mRNA is created from a particular gene segment of DNA.
- In eukaryotes, the mRNA is "processed" by having its introns, or noncoding sequences, removed. A 5' cap and a 3' tail are also added.
- Now, ready to be translated, mRNA proceeds to the ribosome.
- Free-floating amino acids are picked up by tRNA and shuttled over to the ribosome, where mRNA awaits.
- In translation, the anticodon of a tRNA molecule carrying the appropriate amino acid base pairs with the codon on the mRNA.
- As new tRNA molecules match up to new codons, the ribosome holds them in place, allowing peptide bonds to form between the amino acids.
- The newly formed polypeptide grows until a stop codon is reached.
- The polypeptide or protein folds up and is released into the cell.

At the Same Time!

In prokaryotes, transcription and translation can be occurring at the same time. This is because prokaryotes lack a nucleus and their transcription and translation are both occurring in the same place, the cytoplasm.

1st Position (5' End)	U	C	A	G	3rd Position (3' End)
U	Phe	Ser	Tyr	Cys	U
	Phe	Ser	Tyr	Cys	C
	Leu	Ser	Stop	Stop	A
	Leu	Ser	Stop	Trp	G
C	Leu	Pro	His	Arg	U
	Leu	Pro	His	Arg	C
	Leu	Pro	Gln	Arg	A
	Leu	Pro	Gln	Arg	G
A	Ile	Thr	Asn	Ser	U
	Ile	Thr	Asn	Ser	C
	Ile	Thr	Lys	Arg	A
	Met	Thr	Lys	Arg	G
G	Val	Ala	Asp	Gly	U
	Val	Ala	Asp	Gly	C
	Val	Ala	Glu	Gly	A
	Val	Ala	Glu	Gly	G

Gene Regulation

What controls gene transcription, and how does an organism express only certain genes? Regulation of gene expression can occur at different times. The largest point is before transcription, or **pre-transcriptional regulation**. It can also occur post-transcriptionally or post-translationally.

The start of transcription requires DNA to be unwound and RNA polymerase to bind at the promoter. The process is usually a bit more complicated than that because there is a group of molecules called **transcription factors** that can either encourage or inhibit this from happening. This is often accomplished by making it easier or more difficult for RNA polymerase to bind or to move to the start site.

Sometimes changes to the packaging of DNA will alter the ability of the transcription machinery to access a gene. These types of changes are called **epigenetic changes**, and they usually occur through a modification to a histone protein that is involved in winding up the DNA. A tighter wrap around a histone makes the DNA more difficult to access and a looser wrap makes it easier to access.

The following examples are famous models of regulation. You do not need to memorize them, but you should be familiar with the big picture of how things can come together to regulate transcription. Most of what we know about gene regulation comes from our studies of *E. coli*. In bacteria, a cluster of genes can be under the control of a single promoter; these functioning units of DNA are called **operons**. One of the best-understood operons is the *lac* operon, which controls expression of the enzymes that break down (catabolize) lactose.

The operon consists of four major parts: structural genes, promoter genes, the operator, and the regulatory gene:

- **Structural genes** code for enzymes needed in a chemical reaction. These genes will be transcribed at the same time to produce particular enzymes. In the *lac* operon, three enzymes (beta galactosidase, galactose permease, and thiogalactoside transacetylase) involved in digesting lactose are coded for.
- The **promoter gene** is the region where the RNA polymerase binds to begin transcription.
- The **operator** is a region that controls whether transcription will occur; this is where the repressor binds.
- The **regulatory gene** codes for a specific regulatory protein called the repressor. The repressor is capable of attaching to the operator and blocking transcription. If the repressor binds to the operator, transcription will not occur. On the other hand, if the repressor does not bind to the operator, RNA polymerase moves right along the operator and transcription occurs. In the *lac* operon, the **inducer**, lactose, binds to the repressor, causing it to fall off the operator, and "turns on" transcription.

The two diagrams on the next page show the *lac* operon when lactose is absent (A) and when lactose is present (B).

A No lactose present

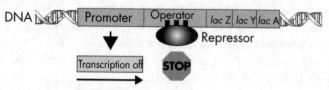

B Lactose present

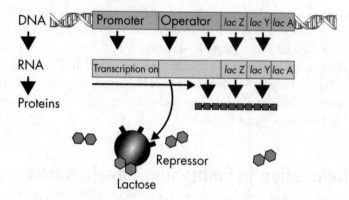

Other operons, such as the *trp* operon, operate in a similar manner except that this mechanism is continually "turned on" and is "turned off" only in the presence of high levels of the amino acid, tryptophan. Tryptophan is a product of the pathway that codes for the *trp* operon. When tryptophan combines with the *trp* repressor protein, it causes the repressor to bind to the operator, which turns the operon "off," thereby blocking transcription. In other words, a high level of tryptophan acts to repress the further synthesis of tryptophan.

The following two diagrams show the *trp* operon when tryptophan is absent (C) and when tryptophan is present (D).

C No tryptophan present

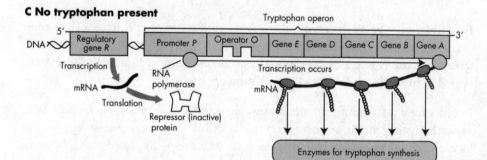

D Tryptophan present

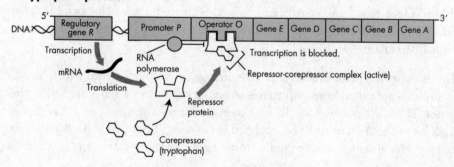

Post-transcriptional regulation occurs when the cell creates an RNA, but then decides that it should not be translated into a protein. This is where RNAi comes into play. RNAi molecules can bind to an RNA via complementary base pairing. This creates a double-stranded RNA (remember RNA is usually single stranded). When a double-stranded RNA is formed, this signals to some special destruction machinery that the RNA should be destroyed. This would prevent it from going on to be translated.

Post-translational regulation can also occur if a cell has already made a protein, but doesn't yet need to use it. This is an especially common regulation for enzymes because when the cell needs them, it needs them ASAP. It is easier to make them ahead of time and then just turn them on or off as needed. This can involve binding with other proteins, phosphorylation, pH changes, cleavage, etc. Remember, even if a protein is made, it might need other things to be made (or not made) in order for the protein to be functional.

Gene Regulation in Embryonic Development

How does a tiny, single-celled egg develop into a complex, multicellular organism? By dividing, of course. The cell changes shape and organization many times by going through a succession of stages. This process is called **morphogenesis**.

When an egg is fertilized by a sperm, it forms a diploid cell called a **zygote.**

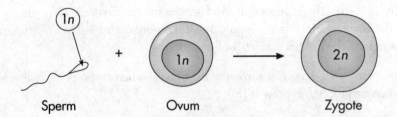

Sperm   Ovum   Zygote

Fertilization triggers the zygote to go through a series of cell divisions. As these occur, the embryo becomes increasingly differentiated, or specialized. An undifferentiated cell is like a blank slate—it can become any type of cell. Once a cell begins to specialize, it is limited in its future options.

In order for a cell to differentiate, it must change. Certain genes will be expressed, and other genes might be turned off. Cells called organizers release signals that let each cell know how they should develop. As an over-simplified example, a cell destined to become a muscle might have an increase in things that make it flexible, while something destined to become bone might have a decrease in those things. Once the change has been made, the future muscle cell can't change its mind and become a bone cell. In other words, cells don't typically go backwards, or dedifferentiate.

The early genes that turn certain cells in the early embryo into future-this or future-that are called **homeotic genes**. A subset of homeotic genes are called **Hox genes**. Timing is essential for the induction (activation) of these genes. Exactly the right bit of the embryo must be modified at exactly the right time. Otherwise, the embryo may develop with the brain in the wrong place, with too many limbs, or with limbs on only one side of the body. Severely damaged embryos will typically stop developing and die.

An interesting additional tool that the developing embryo uses is apoptosis, or programmed cell death. Certain bits of the developing embryo are used as a scaffold for development (like the cells that used to exist in the spaces between your fingers and toes), and when they are no longer necessary, those cells undergo apoptosis. Think of it like an eraser coming along and erasing the webbing tissue between fingers and toes.

MUTATIONS

A **mutation** is an error in the genetic code. Mutations can occur because DNA is damaged and cannot be repaired or because DNA damage is repaired incorrectly. Damage can be caused by chemicals or radiation. It can also occur when a DNA polymerase or an RNA polymerase makes a mistake. DNA polymerases have proofreading abilities, but RNA polymerases do not. This is because RNA is only a temporary molecule and if a mistake is made, it is not usually as problematic. DNA is important now and in the future, since it is passed on from cell to cell in somatic cells and from parent to offspring. If the mistake occurs in a germline cell, that will become a gamete. Mistakes in DNA can last forever.

It is important to know that just having an error in the DNA is not a problem unless that gene is expressed AND the error causes a change in the gene product (RNA or protein). Think of it like an error in a recipe in a cookbook; an error won't harm anyone unless the recipe is actually made AND the error causes a big change in the recipe. As we will see, a mutation can cause different types of effects. Often, the result will be a change to the gene product and the phenotype of the individual. This can be beneficial, detrimental, or neutral to the individual depending on the specific mutation that occurs.

Base Substitution

Base substitution (point) mutations result when a single nucleotide base is substituted for another. There are three different types of point mutations:

There are two types of mutations: **base substitutions** and **gene rearrangements**.

- **Nonsense mutations** cause the original codon to become a stop codon, which results in early termination of protein synthesis.
- **Missense mutations** cause the original codon to be altered and produce a different amino acid.
- **Silent mutations** happen when a codon that codes for the same amino acid is created and therefore does not change the corresponding protein sequence.

Gene Rearrangements

Gene rearrangements involve DNA sequences that have deletions, duplications, inversions, and translocations.

1. **Insertions** and **deletions** result in the gain or loss, respectively, of DNA or a gene. They can involve either the addition (insertion) or removal (deletion) of as little as a single base or much larger sequences of DNA. Insertions and deletions may have devastating consequences during protein translation because the introduction or deletion of bases often results in a change in the sequence of codons used by the ribosome (called a **frameshift mutation**) to synthesize a polyprotein.

Example:

Original DNA Sequence:	ATG	TAT	AAA	CAT	TGA
Original Polyprotein:	Met	Tyr	Lys	His	Stop
DNA Sequence after Insertion:	ATG	**CTA**	TAA	ACA	TTG A
Polyprotein Sequence after Insertion:		Met	Leu	Stop	-	-

In this example, the insertion of an additional cytosine nucleotide resulted in a frameshift and a premature stop codon.

2. **Duplications** can result in an extra copy of genes and are usually caused by unequal crossing-over during meiosis or chromosome rearrangements. This may result in new traits because one copy of the gene can maintain the original function and the other copy may evolve a new function.

3. **Inversions** can result when changes occur in the orientation of chromosomal regions. This may cause harmful effects if the inversion involves a gene or an important regulatory sequence.

4. **Translocations** occur when two different chromosomes (or a single chromosome in two different places) break and rejoin in a way that causes the DNA sequence or gene to be lost, repeated, or interrupted.

5. **Transposons** are gene segments that can cut/paste themselves throughout the genome. The presence of a transposon can interrupt a gene and cause errors in gene expression.

Errors in Meiosis

Nondisjunction occurs when the chromosomes are not separated correctly during meiosis. This can cause entire chromosomes to be duplicated in a gamete, which can cause three copies of a chromosome (triploidy) in offspring. Incorrect numbers of chromosomes often lead to sterility and in plants can cause increased vigor due to polyploidy.

Bacteria and viruses are common pathogens. Bacteriophages are viruses that infect bacteria. Viruses require a host to replicate and sometimes lyse the host cell during infection.

Bacteria

Bacteria, common pathogens, are prokaryotes that come in many shapes and sizes. They can infect many things, and sometimes they cause harm and sometimes they do not. You may have heard of "gut bacteria" before; this is a special colony of bacteria that lives inside each one of us. It helps us with some digestion and makes some things we need. We have a mutualistic relationship with our gut bacteria.

Bacteria divide by fission; however, this does not increase their genetic diversity. Instead, they can perform **conjugation** with other bacterial cells and swap some of their DNA. Genetic variety among bacteria is leading to increased antibiotic resistance.

Viruses

Viruses are nonliving agents capable of infecting cells. Why are viruses considered nonliving? They require a host cell's machinery in order to replicate. A virus consists of two main components: a protein capsid and genetic material made of DNA or RNA, depending on the virus. Viruses are all very specific in which type of cells they infect, and the thing infected by a virus is called a **host.**

Viruses have one goal: replicate and spread. In order to do this, a virus needs to make more genome and make more capsid. They then assemble together into new viral particles.

The viral genome carries genes for building the capsid and anything else the virus needs that the host cannot provide. Sometimes, if two viruses infect the same cell, there will be mixing of the genomes, especially if the viruses have genomes split between several chromosome-like segments. A new virus particle might emerge that is a blend of the two viruses.

A commonly studied virus is a **bacteriophage** (a virus that infects bacteria). Bacteriophages undergo two different types of replication cycles, the **lytic cycle** and the **lysogenic cycle**. In the lytic cycle, the virus immediately starts using the host cell's machinery to replicate the genetic material and create more capsid proteins. These spontaneously assemble into mature viruses and cause the cell to lyse, or break open, releasing new viruses into the environment. In the lysogenic cycle, the virus incorporates itself into the host genome and remains dormant until it is triggered to switch into the lytic cycle. A virus can hide in the genome of a host cell for a very long time. During this time, the cell may divide and replicate the virus as well. By the time the lytic cycle is triggered, the virus may have been replicated many many times as the cell hosting it divides.

The human microbiome is the collection of all the microbes in and on your body. We depend on a huge number of microbes to stay alive. Characterizing and understanding this population is a growing area of research.

Variations on Genetic Variation
Check out Chapter 10 for a more positive look at how an increase in genetic variation makes it more likely that a population will survive a catastrophic event.

When a virus excises from a host genome (becomes unintegrated), it sometimes accidentally takes some of the bacterial cell's DNA with it. Then, the host DNA accidentally gets packaged into new viral particles with the viral genome. The next cell that gets infected is not only getting infected with the viral genome, but also with that stolen chunk of bacterial DNA. If that chunk held a gene for something like antibiotic resistance, the next cell that gets infected will gain that trait. The transfer of DNA between bacterial cells using a lysogenic virus is called **transduction**.

Viruses that infect animals do not have to break their way out of the cell the same way that a bacteriophage does. Since animals cells don't have a cell wall, viruses often just "bud" out of the membrane in a process similar to exocytosis. When a virus does this, it becomes enveloped by a chunk of host cell membrane that it takes with it. Viruses with a lipid envelope are called **enveloped viruses**.

Retroviruses like HIV are RNA viruses that use an enzyme called **reverse transcriptase** to convert their RNA genomes into DNA so that they can be inserted into a host genome. RNA viruses have extremely high rates of mutation because they lack proofreading mechanisms when they replicate their genomes. This high rate of mutation will create lots of variety, which makes these viruses difficult to treat. They evolve quickly, as drug-resistant mutations become naturally selected. New drugs must constantly be identified to treat the resistance.

Viruses can sometimes change quickly if two viruses infect the same cell. The gene segments of each virus can recombine and mix with each other and package into mixed viruses that are a chimera of the original two viruses. This can increase genetic variation among viruses.

BIOTECHNOLOGY

Recombinant DNA

Scientists have learned how to harness the Central Dogma in order to research and cure diseases. **Recombinant DNA** is generated by combining DNA from multiple sources to create a unique DNA molecule that is not found in nature. A common application of recombinant DNA technology is the introduction of a eukaryotic gene of interest (such as insulin) into a bacterium for production. In other words, bacteria can be hijacked and put to work as little protein factories. This branch of technology that produces new organisms or products by transferring genes between cells is called **genetic engineering**.

Polymerase Chain Reaction (PCR)

A few decades ago, it would have taken weeks of tedious experiments to identify and study specific genes. Today, thanks to **polymerase chain reaction (PCR)**, we are able to make billions of identical copies of genes within a few hours. For a PCR, the process of DNA replication is slightly modified. In a small PCR tube,

DNA, specifically designed primers, a powerful and heat-resistant DNA polymerase, and lots of DNA nucleotides (As, Cs, Gs, and Ts) are mixed together.

In a PCR machine, or **thermocycler**, the tube is heated, cooled, and warmed many times. Each time the machine is heated, the hydrogen bonds break, separating the double-stranded DNA. As it is cooled, the primers bind to the sequence flanking the region of the DNA we want to copy. When it is warmed, the polymerase binds to the primers on each strand and adds nucleotides on each template strand. After this first cycle is finished, there are two identical double-stranded DNA molecules. When the second cycle is completed, these two double-stranded DNA segments will have been copied into four. The process repeats itself over and over, creating as much DNA as needed.

Today, a thermocycler is commonplace in science labs. It is regularly used to study small amounts of DNA from crime scenes, determine the origin of our foods, detect diseases in animals and humans, and better understand the inner workings of our cells.

Transformation

Insulin, the protein hormone that lowers blood sugar levels, can now be made for medical purposes by bacteria. Yes, bacteria can be induced to use the universal DNA code to transcribe and translate a human gene! The process of giving bacteria foreign DNA is called **transformation**.

Genes of interest are first placed into a small circular DNA molecule called a **plasmid**. Often, a plasmid is predesigned to have some special helpful genes in it, and together, these are called a **vector**. Plasmid vectors often contain genes for antibiotic resistance and restriction sites. Plasmids and the gene of interest are cut with the same **restriction enzyme** (restriction enzymes cut DNA at very specific predetermined sequences), creating compatible **sticky ends**. When placed together, the gene is inserted into the plasmid creating recombinant DNA.

The bacteria are then transformed (given the recombinant DNA to take up). In most AP Biology courses, this is done by the heat shock method.

Not all bacteria will be transformed, but the ones that did not take up the plasmid are not needed. This is where the antibiotic resistance gene on the vector comes in. By growing all bacteria in the presence of an antibiotic, only those with the resistance gene (a.k.a. those that have been transformed) will survive.

Not only has this laboratory technique been used to safely mass-produce important proteins used for medicine, such as insulin, but it also plays an important role in the study of gene expression. A technique similar to transformation is **transfection**, which is putting a plasmid into a eukaryotic cell, rather than a bacteria cell. This is a bit trickier than transformation. Because eukaryotic cells do not grow like bacteria do, transfection is not as useful for making large quantities of something, but sometimes it is important to use eukaryotic cells.

Gel Electrophoresis

An important part of DNA technology is the ability to observe differences in different preparations of DNA.

DNA fragments can be separated according to their molecular weight and charge with **gel electrophoresis**. Because DNA and RNA are negatively charged, they migrate through a gel toward the positive pole of the electrical field. The smaller the fragments, the faster they move through the gel. Restriction enzymes are also used to create a molecular fingerprint. The patterns created after cutting with restriction enzymes are unique for each person because each person's DNA is slightly different. Some people might have sequences that are cut many times, resulting in many tiny fragments, and others might have sequences that are cut infrequently, producing a few large fragments. When restriction fragments between individuals of the same species are compared, the fragments differ in length because of polymorphisms, which are slight differences in DNA sequences. These fragments are called **restriction fragment length polymorphisms**, or RFLPs. In **DNA fingerprinting**, RFLPs produced from DNA left at a crime scene are compared to RFLPs from the DNA of suspects.

DNA Sequencing

An important tool of modern scientists is the process of **DNA sequencing**. This allows scientists to determine the order of nucleotides in a DNA molecule. By knowing this, scientists could design their own DNA plasmid and use it to study a gene of interest.

KEY TERMS

nucleotide
five-carbon sugar
phosphate
nitrogenous base
deoxyribose
adenine
guanine
cytosine
thymine
phosphodiester bonds
double helix
Watson, Crick, and Franklin
base pairing
complementary
antiparallel
hydrogen bonds
gene
genome
chromosome
histone
nucleosome
euchromatin
heterochromatin
DNA replication
helicase
replication fork
origins of replication
topoisomerase
DNA polymerase
RNA primase
RNA primer
leading strand
lagging strand
Okazaki fragments
DNA ligase
semiconservative replication
telomeres
Avery, MacLeod, and McCarty
Hershey-Chase
transcription
translation
Central Dogma of Biology
ribose

uracil
messenger RNA (mRNA)
ribosomal RNA (rRNA)
transfer RNA (tRNA)
RNA interference
    (RNAi)/silencing RNA (siRNA)
polycistronic transcript
monocistronic transcript
promoter
antisense/non-coding/template
    strand
sense/coding strand
RNA polymerase
start site
exons
introns
splicing
spliceosome
poly(A) tail
5' GTP cap
codon
anticodon
wobble pairing
initiation
elongation
termination
A site, P site, E site
start codon
stop codons
pre-transcriptional regulation
transcription factors
epigenetic changes
operon
structural genes
promoter genes
operator
regulatory gene
inducer
post-transcriptional regulation
post-translational regulation
morphogenesis
zygote
fertilization

homeotic genes
Hox genes
mutation
base substitution
nonsense mutation
missense mutation
silent mutation
gene rearrangements
insertions
deletions
frameshift mutation
duplications
inversions
translocations
transposons
bacteria
conjugation
viruses
host
bacteriophage
lytic cycle
lysogenic cycle
transduction
enveloped virus
retrovirus
reverse transcriptase
recombinant DNA
genetic engineering
polymerase chain reaction (PCR)
thermocycler
transformation
plasmid
vector
restriction enzyme
sticky end
transfection
gel electrophoresis
restriction fragment length poly-
    morphism (RFLP)
DNA fingerprinting
DNA sequencing

Summary

o   DNA is the genetic material of the cell.

o   DNA structure is composed of two antiparallel complementary strands, including:
   • Deoxyribose sugar and phosphate backbone connected by phosphodiester bonds with nitrogenous bases hydrogen bonded together pointing toward the middle. The two strands are oriented antiparallel with 3' and 5' ends and twist into a double helix.
   • Adenine always pairs with thymine, and guanine always pairs with cytosine.

o   All the DNA in a cell is the genome. It is divided up into chromosomes, which are divided up into genes that each encode a particular genetic recipe.

o   DNA replication provides two copies of the DNA for cell division, which involves the following steps:
   • Helicase begins replication by unwinding DNA and separating the strands at the origin of replication.
   • Topoisomerases reduce supercoiling ahead of the replication fork.
   • RNA primase places an RNA primer down on the template strands.
   • DNA polymerase reads DNA base pairs of a strand as a template and lays down complementary nucleotides to generate a new complementary partner strand.
   • Replication proceeds continuously on the leading strand.
   • The lagging strand forms Okazaki fragments, which must be joined later by DNA ligase.

o   Transcription involves building RNA from DNA.
   • RNA polymerase reads one of the DNA strands (the antisense/non-coding/template strand) and adds RNA nucleotides to generate a single strand of mRNA that will be identical to the other strand of DNA (the coding/sense strand) except it will have uracil instead of thymine.
   • mRNA, rRNA, tRNA, and RNAi are all types of RNA that can be created.
   • mRNA gets processed in eukaryotic cells: introns are removed (spliced out), and a 5' GTP cap and a 3' poly(A) tail are added.

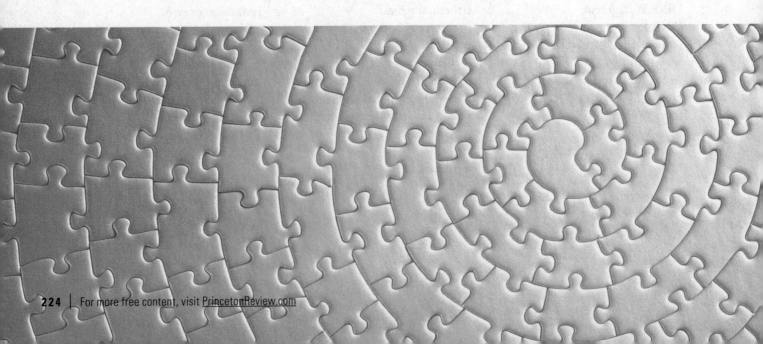

- Translation occurs in the cytoplasm as mRNA carries the message to the ribosome.
  - mRNA passes through three sites on the ribosome, the A-site, the P-site, and the E-site.
  - The mRNA is read in triplets of three nucleotides, called codons. Each tRNA has a region called the anticodon that is complementary to a codon. Sometimes the pairing is not the normal base-pairing. This is called wobble pairing.
  - When a tRNA binds, it brings the corresponding amino acid to add to the growing polypeptide.

- Gene expression is regulated primarily by transcription factors influencing transcription (pre-transcriptional regulation), RNAi after transcription (post-transcriptional regulation), and by various reactions and regulators after translation (post-translational regulation). This regulation is dynamic and can either increase or decrease gene expression, RNA levels, and protein levels according to the needs of the cell.

- Mutations can result from changes in the DNA message or the mRNA message.

- Mutations can be small (single nucleotide swaps, additions, or deletions) or large (big chunks or entire chromosomes are swapped, duplicated, or deleted).

- Some examples of biotechnology are:
  - recombinant DNA
  - polymerase chain reaction (PCR)
  - transformation of bacteria
  - gel electrophoresis

- Bacteria and viruses are common pathogens. Bacteriophages are viruses that infect bacteria. Viruses require a host to replicate and sometimes lyse the host cell during infection.

Chapter 9 Drill

Answers and explanations can be found in Chapter 15.

1. A geneticist has discovered a yeast cell, which encodes a DNA polymerase that may add nucleotides in both the 5′ to 3′ and 3′ to 5′ directions. Which of the following structures would this cell NOT likely generate during DNA replication?

   (A) RNA primers
   (B) Okazaki fragments
   (C) Replication fork
   (D) Nicked DNA by topoisomerases

2. A eukaryotic gene, which does not normally undergo splicing, was exposed to benzopyrene, a known carcinogen and mutagen. Following exposure, the protein encoded by the gene was shorter than before exposure. Which of the following types of genetic rearrangements or mutations was likely introduced by the mutagen?

   (A) Silent mutation
   (B) Missense mutation
   (C) Nonsense mutation
   (D) Duplication

3. DNA replication occurs through a complex series of steps involving several enzymes. Which of the following represents the correct order beginning with the earliest activity of enzymes involved in DNA replication?

   (A) Helicase, ligase, RNA primase, DNA polymerase
   (B) DNA polymerase, RNA primase, helicase, ligase
   (C) RNA primase, DNA polymerase, ligase, helicase
   (D) Helicase, RNA primase, DNA polymerase, ligase

4. If a messenger RNA codon is UAC, which of the following would be the complementary anticodon triplet in the transfer RNA?

   (A) ATG
   (B) AUC
   (C) AUG
   (D) ATT

5. During post-translational modification, the polypeptide from a eukaryotic cell typically undergoes substantial alteration that results in

   (A) excision of introns
   (B) addition of a poly(A) tail
   (C) formation of peptide bonds
   (D) a change in the overall conformation of a polypeptide

6. Which of the following represents the maximum number of amino acids that could be incorporated into a polypeptide encoded by 21 nucleotides of messenger RNA?

   (A) 3
   (B) 7
   (C) 21
   (D) 42

7. A researcher uses molecular biology techniques to insert a human lysosomal membrane protein into bacterial cells to produce large quantities of this protein for later study. However, only small quantities of this protein result in these cells. What is a possible explanation for this result?

   (A) The membrane protein requires processing in the ER and Golgi, which are missing in the bacterial cells.
   (B) Bacteria do not make membrane proteins.
   (C) Bacteria do not use different transcription factors than humans, so the gene was not expressed.
   (D) Bacteria do not have enough tRNAs to make this protein sequence.

8. *Bam*HI is a restriction enzyme derived from *Bacillus amyloliquefaciens* that recognizes short palindromic sequences in DNA. When the enzyme recognizes these sequences, it cleaves the DNA. What purpose would restriction enzymes have in a bacterium like *Bacillus*?

(A) They are enzymes that no longer have a purpose because evolution has produced better enzymes.

(B) They destroy extra DNA that results from errors in binary fission.

(C) They protect *Bacillus* from invading DNA due to viruses.

(D) They prevent, or restrict, DNA replication when the cell isn't ready to copy its DNA.

9. Viruses and bacteria have which of the following in common?

(A) Ribosomes
(B) Nucleic acids
(C) Flagella
(D) Metabolism

10. Griffith was a researcher who coined the term *transformation* when he noticed that incubating nonpathogenic bacteria with heat-killed pathogenic bacteria produced bacteria that ultimately became pathogenic, or deadly, in mice. What caused the transformation in his experiment?

(A) DNA from the nonpathogenic bacteria revitalized the heat-killed pathogenic bacteria.

(B) Protein from the pathogenic bacteria was taken up by the nonpathogenic bacteria.

(C) DNA from the pathogenic bacteria was taken up by the nonpathogenic bacteria.

(D) DNA in the nonpathogenic bacteria turned into pathogenic genes in the absence of pathogenic bacteria.

11. A biologist systematically removes each of the proteins involved in DNA replication to determine the effect each has on the process. In one experiment, after separating the strands of DNA, she sees many short DNA/RNA fragments as well as some long DNA pieces. Which of the following is most likely missing?

(A) Helicase
(B) DNA polymerase
(C) DNA ligase
(D) RNA primase

REFLECT

Respond to the following questions:

- Which topics from this chapter do you feel you have mastered?

- Which content topics from this chapter do you feel you need to study more before you can answer multiple-choice questions correctly?

- Which content topics from this chapter do you feel you need to study more before you can effectively compose a free response?

- Was there any content that you need to ask your teacher or another person about?

Chapter 10
Natural Selection

Natural selection acts on individuals, but evolution acts on populations.

All of the organisms we see today arose from earlier organisms. This process, known as **evolution**, can be described as a change in a population over time. Interestingly, however, the driving force of evolution, called **natural selection**, operates on the level of the individual. In other words, evolution is defined in terms of populations but occurs in terms of individuals.

NATURAL SELECTION

What is the basis of our knowledge of evolution? Much of what we now know about evolution is based on the work of **Charles Darwin**. Darwin was a 19th-century British naturalist who sailed the world in a ship named the HMS *Beagle*. Darwin developed his theory of evolution based on natural selection after studying animals in the Galápagos Islands and other places.

He observed that there were similar animals on the various islands, but the beaks varied in length on finches, and necks varied in length on tortoises. There must be a reason why animals in different areas had different traits. Darwin concluded that it was impossible for the finches and tortoises of the Galápagos simply to "grow" longer beaks or necks as needed. Rather, those traits were inherited and passed on from generation to generation. So, he decided there must be a way for populations to evolve and change their traits (i.e., a population of finches developing longer beaks).

He decided that there must have been a variety of beak lengths originally, but only the longest one was particularly helpful. Since those finches could eat better, survive better, and reproduce better, they were more likely to contribute offspring to the next generation. Over many years, long-beak finches became more and more plentiful with each generation until long beaks were the "norm." In another example, on the first island Darwin studied, there must once have been short-necked tortoises. Unable to reach higher vegetation, these tortoises eventually died off, leaving only those tortoises with longer necks. Consequently, evolution has come to be thought of as "the survival of the fittest": only those organisms most fit to survive will survive and reproduce.

Darwin elaborated his theory in a book entitled *On the Origin of Species*. In a nutshell, here's what Darwin observed:

- Each species produces more offspring than can survive.
- These offspring compete with one another for the limited resources available to them.
- Organisms in every population vary.
- The fittest offspring, or those with the most favorable traits, are the most likely to survive and therefore produce a second generation.

Lamarck and the Long Necks

Darwin was not the first to propose a theory explaining the variety of life on Earth. One of the most widely accepted theories of evolution in Darwin's day was that proposed by **Jean-Baptiste de Lamarck**.

In the 18th century, Lamarck had proposed that *acquired* traits were inherited and passed on to offspring. For example, in the case of Lamarck's giraffes, Lamarck's theory said that the giraffes had long necks because they were constantly reaching for higher leaves while feeding. This theory is referred to as the "law of use and disuse," or, as we might say now, "use it or lose it." According to Lamarck, giraffes have long necks because they constantly *use* them.

We know now that Lamarck's theory was wrong: acquired changes—that is, changes at a "macro" level in an organism's regular (somatic) body cells—will not appear in gamete cells. For example, if you were to lose one of your fingers, your children would not inherit this trait. The gametes that your body makes include copies of your regular old genome, not a new version of a genome that is determined by how careful you are with a table saw.

Don't be Lamarck!
Remember, Lamarck was wrong. Do not take a Lamarckian position on the AP test.

Evidence for Evolution

In essence, nature "selects" which living things survive and reproduce. Today, we find support for the theory of evolution in several areas:

- **Paleontology**, or the study of fossils: paleontology has revealed to us both the great variety of organisms (most of which, including trilobites, dinosaurs, and the woolly mammoth, have died off) and the major lines of evolution. Fossils can be dated by:

    i.   The age of the rocks where a fossil is found
    ii. The rate of decay of isotopes including carbon-14
    iii. Geographical data

- **Biogeography**, or the study of the distribution of **flora** (plants) and **fauna** (animals) in the environment: scientists have found related species in widely separated regions of the world. For example, Darwin observed that animals in the Galápagos have traits similar to those of animals on the mainland of South America. One possible explanation for these similarities is a common ancestor. As we'll see below, there are other explanations for similar traits. However, when organisms share multiple traits, it's pretty safe to say that they likely also shared a common ancestor.

- **Embryology**, or the study of development of an organism: if you look at the early stages in vertebrate development, all embryos look alike! For example, all vertebrates—including fish, amphibians, reptiles, birds, and even mammals such as humans—show fishlike features called gill slits.

- **Comparative anatomy**, or the study of the anatomy of various animals: scientists have discovered that some animals have similar structures that serve different functions. For example, a human's arm, a dog's leg, a bird's wing, and a whale's fin are all the same appendages, though they have evolved to serve different purposes. These structures, called **homologous structures**, also point to a common ancestor.

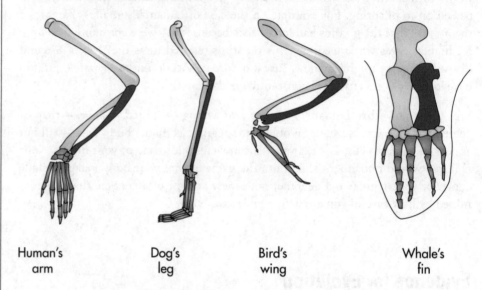

| Human's arm | Dog's leg | Bird's wing | Whale's fin |

In contrast, sometimes animals have features with the same function but that are structurally different. A bat's wing and an insect's wing, for example, are both used to fly. They therefore have the same function, but they have evolved totally independently of one another. These are called **analogous structures**. Another classic example of an analogous structure is the eye. Though scallops, insects, and humans all have eyes, these three different types of eyes are thought to have evolved entirely independently of one another. They are therefore analogous structures.

- **Molecular Biology:** Perhaps the most compelling proof of all is the similarity at the molecular level. Today, scientists can examine nucleotide and amino acid sequences of different organisms. From these analyses, we've discovered that organisms that are closely related have a greater proportion of sequences in common than distantly related species. For example, humans don't look much like chimpanzees. However, by some estimates, as much as 99 percent of our genetic code is identical to that of a chimp. All eukaryotes have membrane-bound organelles, linear chromosomes, and genes that contain introns.

- **Continuing Evolution:** There is evidence that evolution is constantly occurring. We can see consistent small changes in DNA and changes in the fossil record. We can see evolution of resistance to antibiotics, pesticides, herbicides, or chemotherapy drugs since these are particularly strong selective pressures. We can also see how fast replicating pathogens evolve and cause emergent diseases never before seen.

Across every domain, there are certain universal molecules and structures. Many metabolic pathways are also conserved and vary only a little between species.

COMMON ANCESTRY

By using the evidence mentioned above, scientists can get a good handle on how evolution of certain species occurred. It all comes down to who has common ancestors. Life started somewhere. Some original life-form is the **common ancestor** to all life, but where things went from there can become quite convoluted.

Scientists use charts called **phylogenetic trees,** or **cladograms**, to study relationships between organisms. Phylogenetic trees are built using data from the fossil record or molecular record, but the molecular record provides more specific details. Trees show the amount of change over time, but cladograms just show the change. In other words, cladograms are often drawn with even spacing between species, but phylogenetic trees are often drawn with different distances between species and as a result they look more like a tree with uneven branches. Both types are hypotheses and thus they constantly change as more data is uncovered to establish the relationships between species.

They always begin with the common ancestor and then branch out. Anytime there is a fork in the road, it is called a common ancestor node. These common ancestors likely do not exist anymore, but they are the point at which evolution went in two directions. One direction eventually led to one species, and the other eventually led to another species.

For example, we have a common ancestor with chimps. This doesn't mean that our great-great...great-grandfather was a chimp. It means that chimps and we have the same great-great...great-grandfather, who was neither a human nor a chimp. Instead, he was some other species completely, and a speciation event must have occurred where his species was split and evolved in different directions. One lineage became chimps, and the other linage became humans.

This is an example of a phylogenetic tree.

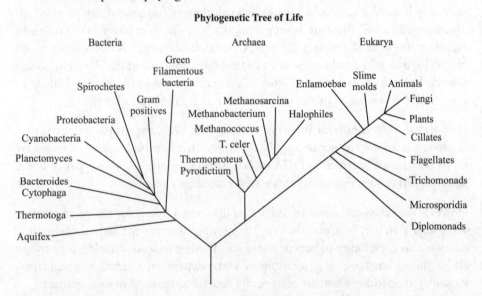

Phylogenetic Tree of Life

In this tree, the three main domains of life (bacteria, archaea, and eukarya) are seen. You can see that plants, animals, and fungi are much more closely related to each other than they are to bacteria. Archaea and eukarya have a common ancestor with bacteria, and then they also have another common ancestor that bacteria doesn't have.

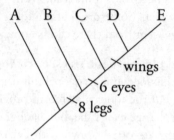

In the phylogenetic tree above, species A is an **out-group** because it is the least related to all of the other species. Species B, C, D, and E all have the shared character of 8 legs. Wings are a shared character for species D and E, but they could also be considered a derived character since they separate species D and E from the rest of the species that do no have wings.

GENETIC VARIABILITY

As you know, no two individuals are identical. The differences in each person are known as **genetic variability**. All this means is that no two individuals in a population have identical sets of alleles (except, of course, identical twins). In fact, survival of a species is dependent on this genetic variation; it allows a species to survive in a changing environment. If a population were all the same, then they would have the same weaknesses and the same strengths. Natural selection occurs only if some individuals have more evolutionary fitness and can be selected. The more variations there are among a population, the more likely that a trait will exist that might be the perfect lifesaver. One famous example is Rudolph the Red-Nosed Reindeer. His red nose was a random mutation, but under the right conditions, it became the best nose to have. A more diverse population is more likely to survive and evolve when things are constantly changing around them.

How did all this wonderful variation come about? Well, the simple answer is that random mutations are occurring all the time. This can be because of errors by DNA polymerase, changes to DNA caused by transposons, or other types of DNA damage. Either way, mutations create new variations and alleles.

However, just the occurrence of mutations does not account for the great amount of genetic variation in a population. The mixing of genes through sexual reproduction also contributes to genetic variation. During meiosis, crossing-over mixes alleles among homologous chromosomes, and independent assortment when chromosomes are packaged further adds to the genetic uniqueness of each gamete.

It might be hard to think of it in this way, but genetic variation is the very foundation of evolution, as we'll soon see. Now that we've reintroduced genes, we can refine our definition of evolution. More specifically:

> Evolution is the change in the gene pool of a population over time.

The Peppered Moths

Let's look at an example. During the 1850s in England, there was a large population of **peppered moths**. In most areas, exactly half of them were dark and carried alleles for dark coloring. The other half were light and carried alleles for light coloring. This 1:1 ratio of phenotypes was observed until air pollution, due primarily to the burning of coal, changed the environment. What happened?

Imagine two different cities: City 1 (in the south) and City 2 (in the north). Prior to the Industrial Revolution, both of these cities had unpolluted environments. In both of these environments, dark moths and light moths lived comfortably side by side. For simplicity's sake, let's say our proportions were a perfect fifty-fifty, half dark and half light. However, at the height of the Industrial Revolution, City 2, our northern city, was heavily polluted, whereas City 1, our southern city, was unchanged. In the north, where all the trees and buildings were covered with soot, the light moths didn't stand a chance. They were impossible for a predator to miss! As a result, the predators gobbled up light-colored moths just as fast as they could reproduce, sometimes even before they reached an age in which they *could* reproduce. However, the dark moths with their dark alleles were just fine. With all the soot around, the predators couldn't even see them; the dark moths continued doing their thing—above all, *reproducing*. And when they reproduced, they had more and more offspring carrying the dark allele.

After a few generations, the peppered moth gene pool in City 2 changed. Although our original moth gene pool was 50 percent light and 50 percent dark, excessive predation changed the population's genetic makeup. By about 1950, the gene pool reached 90 percent dark alleles and only 10 percent light alleles. This occurred because the light moths didn't stand a chance in an environment where they were so easy to spot. The dark moths, on the other hand, multiplied just as fast as they could.

In the southern city, you'll remember, there was very little pollution. What happened there? Things remained pretty much the same. The gene pool was unchanged, and the population continued to have roughly equal proportions of light moths and dark moths.

Humans Are Great Selectors
Humans hold great influence over many species in the world. What we select for (for various reasons) is an example of artificial selection. Colors of plants and sizes of dogs and many other things are greatly influenced by humans who are orchestrating breeding to create traits that we desire. This makes them selected for even if they are not increasing the reproductive fitness of their species.

Fittest of the Fit
Don't confuse evolutionary fitness with physical fitness. Often, things like strength and speed will increase evolutionary fitness, but anything that increases survival and reproducing will add to evolutionary fitness, even if sometimes those individuals are slow or weak. Keep in mind that fitness depends on the environment. A trait may be beneficial in one circumstance and detrimental in another. Selection takes place one moment at a time in one environment at a time.

Phenotypes Not Genotypes
Remember, it is a phenotype that can be selected for/against. It is not the genotype. Genotypes are hidden away inside the DNA; only the phenotype will be exposed to the environmental pressures.

CAUSES OF EVOLUTION

Natural selection, the evolutionary mechanism that "selects" which members of a population are best suited to survive and which are not, works both "internally" and "externally": *internally* through random mutations and *externally* through environmental pressures. Biotic and abiotic factors can fluctuate and this can affect the direction of evolution. With each generation, different traits might be selected for depending on environmental fluctuations.

> Natural selection requires genetic variation and an **environmental pressure** that gives some individuals an advantage.

To see how this process unfolds in nature, let's return to the moth case. Why did the dark moths in the north survive? Because they were dark-colored. But how did they become dark-colored? The answer is, through **random mutation**. One day, long before the coal burning, a moth was born with dark-colored wings. As long as a mutation does not kill an organism before it reproduces (most mutations, in fact, do), it may be passed on to the next generation. Over time, this one moth had offspring. These, too, were dark. The dark- and the light-colored moths lived happily side by side until something from the outside—in our example, the environment—changed all that.

The initial variation came about by chance. This variation gave the dark moths an edge. However, that advantage did not become apparent until something made it apparent. In our case, that something was the intensive pollution from coal burning. The abundance of soot made it easier for predators to spot the light-colored moths, thus effectively removing them from the population. Therefore, dark color is an **adaptation**, a variation favored by natural selection.

Survival of the fittest is the name of the game, and any trait that causes an individual to reproduce better gives that individual **evolutionary fitness**.

These traits are often things that simply help something to survive. After all, survival is essential for reproduction to occur. Strength, speed, height, camouflage, and many other things can be helpful. Sometimes odd things can be helpful as well. For example, a peacock's tail has been selected for because females choose to mate with males that have a large and beautiful tail. The tail doesn't help them to survive because it actually makes it easier for predators to catch them, but it is essential for the females to find them attractive. This is an example of **sexual selection**.

Let's go back to our moth example. We ended up with different gene pools in the two populations (north vs. south). Eventually, over long stretches of time, these two different populations might change so much that they could no longer reproduce together. At that point, we would have two different species, and we could say, definitively, that the moths had evolved. Evolution occured as a consequence of random mutation and the pressure put on the population by an environmental change. Catastrophic events can speed up natural selection and adaptation. When the rules change, sometimes odd variations are selected for. Earth has undergone several mass extinction events. The tree of life would look much different if these events had not occurred.

Genetic drift is something that causes a change in the genetics of a population, but it is not natural selection. Instead, it is caused by random events that drastically reduce the number of individuals in a population. Often called the **bottleneck effect** or the **founder effect**, this occurs when only a few individuals are left to mate and regrow a population. If those individuals were the only survivors due to random luck, that means that the alleles that are present to regrow the population are just random alleles. They are not necessarily the fittest alleles. For example, if a hurricane swept through and only 4 birds were left on an island, the traits that they have will be the traits of the population going forward, even if those traits were not previously being selected for.

Gene flow can occur between different populations of the same species if individuals migrate to/from the populations. In other words, new alleles are entering/exiting and the genetic diversity will change. Small populations are more susceptible than large populations to this type of change.

Types of Selection

The situation with our moths is an example of **directional selection**. One of the phenotypes was favored at one of the extremes of the normal distribution.

In other words, directional selection "weeds out" one of the phenotypes. In our case, dark moths were favored, and light moths were practically eliminated. Here's one more thing to remember: directional selection can happen only if the appropriate allele—the one that is favored under the new circumstances—is already present in the population. Two other types of selection are **stabilizing selection** and **disruptive selection**.

Stabilizing selection means that organisms in a population with extreme traits are eliminated. This type of selection favors organisms with average or medium traits. It "weeds out" the phenotypes that are less adaptive to the environment. A good example is birth weight in human babies. An abnormally small baby has a higher chance of having birth defects; conversely, an abnormally large baby will have a challenge in terms of a safe birth delivery.

Disruptive selection, on the other hand, does the reverse. It favors both the extremes and selects *against* common traits. For example, females are "selected" to be small and males are "selected" to be large in elephant seals. You'll rarely find a female or male of intermediate size.

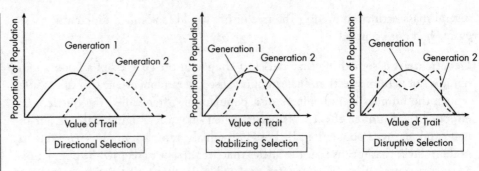

Types of Selection

SPECIES

What Is a Species?
To be considered the same species, two individuals must be able to mate and produce viable offspring that would then be capable of mating and producing offspring.

A dog and a bumblebee obviously cannot produce offspring. Therefore, they are different **species**. However, a chihuahua and a Great Dane can reproduce (at least in theory). We would not say that they are different species; they are merely different breeds. In order for these to become different species of dogs, they would have to become **reproductively isolated** from each other. This would then allow the two groups to undergo natural selection and evolve differently. With different variation and different environmental pressures, they could each change in different ways and no longer be able to mate. This is called **divergent evolution**. Divergent evolution that occurs quickly after a period of stasis (little evolution) is called **punctuated equilibrium**. This is often due to a major event. A change that comes about after many smaller changes is called **gradualism** and it takes hundreds or millions of years. You should also be aware of **adaptive radiation**, which occurs when a species rapidly diversifies due to an abundance of available ecological niches suddenly opening up.

Divergent evolution results in closely related species with different behaviors and traits. As with our example, these species often originate from a common ancestor. More often than not, the "engine" of evolution is cataclysmic environmental change, such as pollution in the case of the moths. Geographic barriers, new stresses, disease, and dwindling resources are all factors in the process of evolution. Pre- and post-zygotic barriers also prevent organisms of two different species from mating to produce viable offspring. **Pre-zygotic barriers** prevent fertilization. Examples of this kind of barrier include temporal isolation, which occurs when two species reproduce at different times of the year. A **post-zygotic barrier** is related to the inability of a hybrid organism to produce offspring. For example, a horse and a donkey can mate to produce a mule, but mules are sterile and therefore cannot produce a second generation.

Convergent evolution is the process by which two unrelated and dissimilar species come to have similar (analogous) traits, often because they have been exposed to similar selective pressures. Examples of convergent evolution include aardvarks, anteaters, and pangolins. They all have strong, sharp claws and long snouts with sticky tongues to catch insects, yet they evolved from three completely different mammals.

There are two types of **speciation**: **allopatric speciation** and **sympatric speciation**. Allopatric speciation simply means that a population becomes separated from the rest of the species by a geographic barrier so that the two populations can't interbreed. An example would be a mountain that separates two populations of ants. In time, the two populations might evolve into different species. If, however, new species form without any geographic barrier, it is called sympatric speciation. This type of speciation is common in plants. Two species of plants may evolve in the same area without any geographic barrier. Often plants undergo speciation due to **polyploidy**, which occurs when they get doubles of their chromosomes.

POPULATION GENETICS

Mendel's laws can extend to the population level. Suppose you caught a bunch of fruit flies—about 1,000. Let's say that 910 of them were red-eyed and 90 were green-eyed. If you allowed the fruit flies to mate and counted the next generation, we'd see that the ratio of red-eyed to green-eyed fruit flies would remain approximately the same: 91 percent red-eyed and 9 percent green-eyed. That is, the allele frequency would remain constant. At first glance you may ask, how could that happen?

Hardy-Weinberg Equilibrium

The **Hardy-Weinberg law** states that even with all the shuffling of genes that goes on, the relative frequencies of genotypes in a population are constant over time. Alleles don't get lost in the shuffle. The dominant gene doesn't become more prevalent, and the recessive gene doesn't disappear.

Hardy-Weinberg law says that a population will be in genetic equilibrium only if it meets these five conditions: (1) a large population, (2) no mutations, (3) no immigration or emigration, (4) random mating, and (5) no natural selection.

When these five conditions are met, the gene pool in a population is pretty stable. Any departure from them results in changes in allele frequencies in a population. For example, if a small group of your fruit flies moved to a new location, the allele frequency may be altered and result in evolutionary changes.

Hardy-Weinberg Equations

Let's say that the allele for red eyes, R, is dominant over the allele for green eyes, r. Red-eyed fruit flies include homozygous dominants, RR, and heterozygous, Rr. The green-eyed fruit flies are recessive, rr. The frequency of each allele is described in the equation below. The allele must be either R or r. Let "p" represent the frequency of the R allele and "q" represent the frequency of the r allele in the population.

$$p + q = 1$$

This sum of the frequencies must add up to one. If you know the value of one of the alleles, then you'll also know the value of the other allele. This makes sense because Hardy-Weinberg works ONLY for populations with 2 alleles that have normal dominant-recessive behavior. So, if you know there are only 2 alleles possible and you know that 70% are the dominant allele, then the remaining 30% must be the recessive allele.

We can also determine the frequency of the *genotypes* in a population using the following equation.

$$p^2 + 2pq + q^2 = 1$$

Don't Forget the Heterozygotes!
Remember that the number of individuals with the dominant phenotype includes those with the homozygous dominant genotype and those with the heterozygous genotype.

In this equation, p^2 represents the homozygous dominants, $2pq$ represents the heterozygotes, and q^2 represents the homozygous recessives.

Both of these equations are listed on the AP Biology Equations and Formulas sheet. But how do you use them? Use proportions in the population to figure out both the allele and genotype frequencies. Let's calculate the frequency of the genotype for green-eyed fruit flies. If 9% of the fruit flies are green-eyed, then the *genotype* frequency, q^2, is 0.09. You can now use this value to figure out the frequency of the recessive allele in the population. The allele frequency for green eyes is equal to the square root of 0.09—that's 0.3. If the recessive allele is 0.3, the dominant allele must be 0.7. That's because 0.3 + 0.7 equals 1.

Using the second equation, you can calculate the genotypes of the homozygous dominants and the heterozygotes. The frequency for the homozygous dominants, p^2, is 0.7 × 0.7, which equals 0.49. The frequency for the heterozygotes, $2pq$, is 2 × 0.3 × 0.7, which equals 0.42. If you include the frequency of the recessive genotype—0.09—the numbers once again add up to 1.

Extinction

There have been many extinctions throughout Earth's history. The rate of extinction can vary and is usually accelerated during ecological stress. Humans have been a large ecological stressor and have likely contributed to extinction events. A population with high genetic diversity is more protected from extinction events because there is a higher chance that an individual will have the characteristics required to survive the ecological stress leading to the extinction event. When a species becomes extinct, the niche it occupied can become available to other species.

ORIGINS OF THE EARTH

The Earth formed approximately 4.6 billions years ago. Prior to 3.9 billion years ago, the Earth was likely an extremely hostile place. The earliest evidence of life is approximately 3.5 billions years ago, but how did life arise? This is still a hotly debated topic among scientists. Most scientists believe that the earliest precursors of life arose from nonliving matter (basically, gases) in the primitive oceans of the earth. But this theory didn't take shape until the 1920s. Two scientists, **Alexander Oparin** and **J. B. S. Haldane**, proposed that the primitive atmosphere contained mostly inorganic molecules and was rich in the following gases: methane (CH_4), ammonia (NH_3), hydrogen (H_2), and water (H_2O). Interestingly enough, there was almost no free oxygen (O_2) in this early atmosphere. They believed that these gases collided, producing chemical reactions that eventually led to the organic molecules we know today.

This theory didn't receive any substantial support until 1953. In that year, **Stanley Miller** and **Harold Urey** simulated the conditions of primitive Earth in a laboratory. They put the gases theorized to be abundant in the early atmosphere into a flask, struck them with electrical charges in order to mimic lightning, and organic compounds similar to amino acids appeared!

But how do we make the leap from simple organic molecules to more complex compounds and life as we know it? Since no one was around to witness the process, no one knows for sure how (or when) it occurred. It is likely that the original life-forms were simply molecules of RNA. This is called the **RNA-world hypothesis**. RNA can take many shapes because it is not restricted to be a double helix. It is possible that RNA molecules capable of replicating and thus passing along themselves (their genome) were the first life-forms. Complex organic compounds (such as proteins) must have formed via dehydration synthesis. Simple cells then used organic molecules as their source of food. Over time, simple cells evolved into complex cells.

Let's finish by defining a few more key terms. All organisms capture and store free energy for use in biological processes. This can be done two different ways. Living organisms that rely on organic molecules for food are called **heterotrophs** (consumers). For example, we're heterotrophs, but the earliest heterotrophs were simple unicellular life-forms. Heterotrophs metabolize carbohydrates, lipids, and proteins, and then hydrolyze them as sources of free energy. Eventually, some life-forms found a way to make their own food—most commonly through photosynthesis. These organisms capture free energy present in sunlight and are called **autotrophs** (producers). Early autotrophs (most likely cyanobacteria) are responsible for Earth's oxygenated atmosphere.

The heterotroph hypothesis suggests that the first cells were likely heterotrophic and would have fed on organic molecules that had been made without cells. These cells likely survived by performing processes similar to glycolysis and fermentation. Once autotrophs were producing and releasing oxygen, aerobic respiration followed.

Alternative Hypothesis
An alternative hypothesis is that organic molecules were transported to Earth on a meteorite or due to some other celestial event.

For a refresher on biochemical processes, refer back to Chapter 6.

KEY TERMS

evolution
natural selection
Charles Darwin
Jean-Baptiste de Lamarck
paleontology
biogeography
flora
fauna
embryology
comparative anatomy
homologous structures
analogous structures
molecular biology
continuing evolution
phylogenetic tree (cladogram)
out-group
genetic variability
peppered moths
environmental pressure
random mutation
adaptation
evolutionary fitness
sexual selection
genetic drift

bottleneck (founder effect)
gene flow
directional selection
stabilizing selection
disruptive selection
species
reproductively isolated
divergent evolution
punctuated equilibrium
gradualism
adaptive radiation
pre-zygotic barriers
post-zygotic barriers
convergent evolution
speciation
allopatric speciation
sympatric speciation
polyploidy
Hardy-Weinberg law
Alexander Oparin and J. B. S. Haldane
Stanley Miller and Harold Urey
RNA-world hypothesis
heterotrophs (consumers)
autotrophs (producers)

Summary

- o Charles Darwin made the following key observations that led to his theory of natural selection:
  - Each species produces more offspring than can survive.
  - Offspring compete with each other for limited resources.
  - Organisms in every population vary.
  - The offspring with the most favorable traits are most likely to survive and reproduce.

- o Evidence for evolution includes:
  - fossils
  - biogeography
  - comparison of developmental embryology
  - comparative anatomy, including homologous and analogous structures
  - molecular biology (sequences of genes are conserved across many types of species)

- o Members of a species are defined by the ability to reproduce fertile offspring.

- o Evolution and speciation occur when environmental pressures favor traits that permit survival and reproduction of selected individuals in a varied population. The favored traits may evolve to cause divergent or convergent evolution.

- o Adaptations are random variations in traits that end up being selected for until they become prominent in the population.

- o Genetic drift occurs when a population's traits change due to random events rather than natural selection.

- o Selection can be stabilizing if an average phenotype is preferred, disruptive if the extremes are preferred, and directional if one extreme is preferred.

- o Hardy-Weinberg equations can be used to determine genetic variation within a population. These equations are as follows:
  - $p + q = 1$ Frequency of the dominant (p) and recessive (q) alleles
  - $p^2 + 2pq + q^2 = 1$ Frequency of the homozygous dominants (p^2), heterozygotes ($2pq$), and homozygous recessives (q^2)

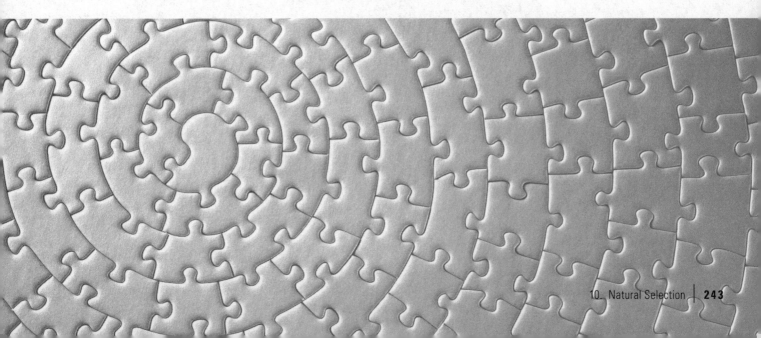

○ Hardy-Weinberg equations describe a population that is not evolving but is instead said to be in genetic equilibrium. Such a population will be so only if it is not violating any of the following five conditions (almost all populations are evolving due to a violation of one or more of these factors):

- large population
- no mutations
- no immigration or emigration
- random mating
- no natural selection

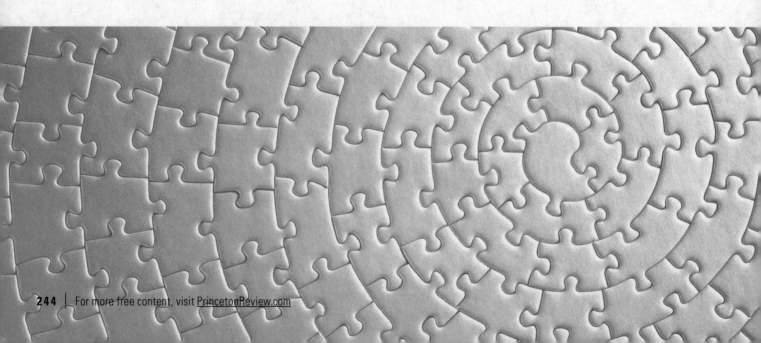

Chapter 10 Drill

Answers and explanations can be found in Chapter 15.

1. The eye structures of mammals and cephalopods such as squid evolved independently to perform very similar functions and have similar structures. This evolution is an example of which of the following?

   (A) Allopatric speciation
   (B) Sympatric association
   (C) Divergent evolution
   (D) Convergent evolution

Questions 2–5 refer to the following graph and paragraph.

During the Industrial Revolution, a major change was observed in many insect species due to the mass production and deposition of ash and soot around cities and factories. One of the most famous instances was within the spotted moth population. An ecological survey was performed in which the number of spotted moths and longtail moths were counted in 8 different urban settings over a square kilometer in 1802. A repeat experiment was performed 100 years later in 1902. The results of the experiment are shown below.

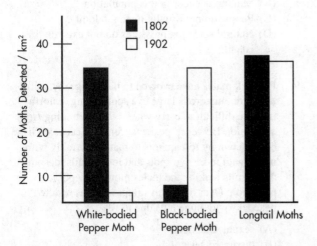

2. What type of selection is represented by the results of this study?

   (A) Stabilizing selection
   (B) Directional selection
   (C) Disruptive selection
   (D) Divergent selection

3. Which of the following statements best explains the data?

   (A) As time passed from 1802 to 1902, the frequency of white-bodied pepper moths increased and black-bodied pepper moths decreased.
   (B) As time passed from 1802 to 1902, the frequency of white-bodied pepper moths decreased and black-bodied pepper moths increased.
   (C) As time passed from 1802 to 1902, the frequency of white-bodied pepper moths and black-bodied pepper moths both increased.
   (D) As time passed from 1802 to 1902, the frequency of white-bodied pepper moths and black-bodied pepper moths both decreased.

4. Why was the population of longtail moths also surveyed in this study?

   (A) Variations in the environment were expected to alter the population of longtail moths.
   (B) Longtail moths were included as a control because they were not expected to change appreciably due to changes associated with the Industrial Revolution.
   (C) The peppered moth did not have a long enough tail to visualize.
   (D) Longtail moths were poisoned by the coal dust and suffered drastic population loss.

5. How would the results of this study have been different if factories produced white or light gray ash and soot rather than black?

   (A) There would be no change to the results of the experiment.
   (B) There would have been added selection pressure for more white-bodied spotted moths and against black-bodied spotted moths.
   (C) There would have been added selection pressure for more black-bodied spotted moths and against white-bodied spotted moths.
   (D) There would have been an increase in the frequency of both black-bodied and white-bodied spotted moths.

6. The Middle East blind mole rat (*Nannosplalax ehrenbergi*) lives in the Upper Galilee Mountains of Israel. Two groups of these mole rats live in the same region, but scientists discovered that there is a 40% difference in mitochondrial DNA between these two groups. These rats do not seem to interbreed in the wild. This is an example of

(A) allopatric speciation
(B) sympatric speciation
(C) pre-zygotic barrier
(D) hybrid zone

Questions 7 and 8 refer to the following scenario.

A cruise ship with 500 people on board crashes on a deserted island. They like the island so much that they decide to stay. There is a small number of people on board with red hair, an autosomal recessive trait. Assume that red hair provides no increase in fitness.

7. What is most likely to occur with respect to the number of red-haired individuals on the island?

(A) There will be no change in the overall number of red-haired people in the population over time.
(B) Convergent evolution will occur, with all people eventually becoming red-haired.
(C) It is likely that over many generations, the proportion of red-haired people in the population will increase.
(D) Red-haired people will probably decrease in the population over many generations.

8. Evolution does not occur when a population is said to be in equilibrium by the Hardy-Weinberg definition. In the situation above, which of the following does NOT support the claim that evolution will occur in this population?

(A) The situation describes a small population.
(B) Mutations occur in this population.
(C) People do not usually mate randomly.
(D) Natural selection pressures do not exist on this island.

9. Peacock males have evolved to have huge, beautiful tails with numerous large eye spots. Long, colorful tails are difficult to carry when you are running from a predator. However, peahens (female peacocks) have been shown by researchers to mate preferably with males with more eye spots and longer tails, despite these traits making the males more susceptible to predation. Over time, this preference has resulted in peacocks with huge tails due to

(A) sexual selection
(B) disrupted selection
(C) divergent evolution
(D) sympatric speciation

REFLECT

Respond to the following questions:

- Which topics from this chapter do you feel you have mastered?

- Which content topics from this chapter do you feel you need to study more before you can answer multiple-choice questions correctly?

- Which content topics from this chapter do you feel you need to study more before you can effectively compose a free response?

- Was there any content that you need to ask your teacher or another person about?

Chapter 11
Ecology

In the preceding chapters, we looked at individual organisms and the ways they solve life's many problems: acquiring nutrition, reproducing, and so on. Now let's turn to how organisms deal with their environments.

INTERACTING WITH THE ENVIRONMENT

Organisms are constantly interacting with their environments. As the environmental stimuli change so must the behavior of the organism. This is important for gathering the most energy while using the least energy and reproducing as efficiently as possible.

A general rule is that larger animals have a slower metabolic rate per unit mass than smaller animals. In other words, if you took the same amount of tissue from a small animal and a large animal, the tissue from the small animal would have a higher rate of metabolism.

When animals have excess energy, they grow and store energy. When they are deficient, then they use up their stores and lose mass and eventually die.

Body temperature is one way that animals must respond to their environment. **Endotherms** are animals that generate their own body heat through metabolism. **Ectotherms** lack an internal mechanism to control body temperature. Instead they regulate their temperature behaviorally by seeking out cool or warm places. One example is a lizard on a warm rock.

A big part of behavior is communication. This is important between members of the same species to coordinate group behavior. Animals use visual, auditory, chemical, tactile, and electrical signals to indicate dominance, find food, establish territory, and reproduce successfully. Communication is also important between members of different species that are engaged in cooperative behaviors.

BEHAVIOR

Some animals behave in a programmed way to specific stimuli, while others behave according to some type of learning. We humans can do both. For the AP Biology Exam, you'll have to know a little about these different types of **behavior**—that is, how organisms cope with their environments.

Instinct

Instinct is an inborn, unlearned behavior. Sometimes the instinctive behavior is triggered by environmental signals called releasers. The releaser is usually a small part of the environment that is perceived. For example, when a male European robin sees another male robin, the sight of a tuft of red feathers on the male is a releaser that triggers fighting behavior. In fact, because instinct underlies all other behavior, it can be thought of as the circuitry that guides behavior.

For example, hive insects, such as bees and termites, never learn their roles; they are born knowing them. On the basis of this inborn knowledge, or instinct, they carry out all their other behaviors: a worker carries out "worker tasks," a drone "drone tasks," and a queen "queen tasks." Another example of instinct is the "dance" of the honey bee, which is used to communicate the location of food to other members of the beehive.

There are other types of instinct that last for only a part of an animal's life and are gradually replaced by "learned" behavior. For example, human infants are born with an ability to suck from a nipple. If it were not for this instinctual behavior, the infant would starve. Ultimately, however, the infant will move beyond this instinct and learn to feed itself. What exactly, then, is instinct?

For our purposes:

> Instinct is the inherited "circuitry" that directs and guides behavior.

Learning

Another form of behavior is **learning**. Learning refers to a change in a behavior brought about by an experience (which is what you're doing this very moment). Animals learn in a number of ways.

Imprinting

Have you ever seen a group of goslings waddling along after their mother? How is it that they recognize her? Well, the mother arrives and gives out a call that the goslings recognize. The goslings, hearing the call, know that this is their mother, and follow her around until they are big enough to head out on their own.

Now imagine the same goslings, newly hatched. If the mother is absent, they will accept the first moving object they see as their mother. This process is known as **imprinting**.

Animals undergo imprinting within a few days after birth in order to recognize members of their own species. While there are different types of imprinting, including parent, sexual, and song imprinting, they all occur during a **critical period**—a window of time when the animal is sensitive to certain aspects of the environment.

Innate Selection
Natural selection favors certain behaviors that increase the survival of offspring. Some of these behaviors are instincts. Many learned behaviors are also essential for survival indicating that the capability to learn might in fact be a trait that is naturally selected for.

Remember:

> Imprinting is a form of learning that occurs during a brief period of time, usually early in an organism's life.

Habituation

Habituation is another form of learning. It occurs when an animal learns not to respond to a stimulus. For example, if an animal encounters a stimulus over and over again without any consequences, the response to it will gradually lessen and may altogether disappear. For example, if a certain hallway always smells weird, you will notice at first but eventually learn not to notice it. The basic response (smell) has not been lost, merely modified by learning. If you walk down a new hallway and it has the same weird smell, you will notice it.

Internal Clocks: The Circadian Rhythm

There are other instinctual behaviors that occur in both animals and plants. One such behavior deals with time. Have you ever wondered how roosters always know when to start crowing? The first thought that comes to mind is that they've caught a glimpse of the sun. Yet many crow even before the sun has risen.

Roosters do have internal alarm clocks. Plants have them as well. These internal clocks, or cycles, are known as **circadian rhythms**. Circadian sounds like the word circle, just like these rhythms have a circular nature.

Watch out though: seasonal changes, like the loss of leaves by deciduous trees or the hibernation of mammals, are not examples of circadian rhythms. *Circadian* refers only to daily rhythms. Need a mnemonic? Just think how bad your jet lag would be after a trip around the world. In other words: circling the globe screws up your circadian rhythm.

The Science Behind Your Jet Lag

If you've ever flown overseas, you know all about circadian rhythms. They're the cause of jet lag. Our bodies tell us it's one time, while our watches tell us it's another. The sun may be up, but our body's internal clock is crying, "Sleep!" This sense of time is purely instinctual: you don't need to know how to tell time in order to feel jet lag.

HOW ANIMALS COMMUNICATE

Some animals use signals as a way of communicating with members of their species. These signals, which can be chemical, visual, electrical, or tactile, are often used to influence mating and social behavior.

Chemical signals are one of the most common forms of communication among animals. **Pheromones**, for instance, are chemical signals between members of the same species that stimulate olfactory receptors and ultimately affect behavior. For example, when female insects give off their pheromones, they attract males from great distances.

Visual signals also play an important role in the behavior observed among members of a species. For example, fireflies produce pulsed flashes that can be seen by other fireflies far away. The flashes are sexual displays that help male and female fireflies identify and locate each other in the dark.

Other animals use electrical channels to communicate. Some fish generate and receive weak electrical fields. Others use tactile signals and have mechanoreceptors in their skin to detect prey. For instance, cave-dwelling fishes use mechanoreceptors in their skin for communicating with other members and detecting prey.

Social Behavior

Many animals are highly social species, and they interact with each other in complex ways. Social behaviors can help members of the species survive and reproduce more successfully. Several behavioral patterns for animal societies are summarized below:

- **Agonistic behavior** is aggressive behavior that occurs as a result of competition for food or other resources. Animals will show aggression toward other members that tend to use the same resources. An example of agonistic behavior is fighting among competitors.

- **Dominance hierarchies** (pecking orders) occur when members in a group have established which members are the most dominant. The more dominant male will often become the leader of the group and will usually have the best pickings of the food and females in the group. Once the dominance hierarchy is established, competition and tension within the group is reduced.

- **Territoriality** is a common behavior when food and nesting sites are in short supply. Usually, the male of the species will establish and defend his territory (called a home range) within a group in order to protect important resources. This behavior is typically found among birds.

- **Altruistic behavior** is defined as unselfish behavior that benefits another organism in the group at the individual's expense because it advances the genes of the group. For example, when ground squirrels give warning calls to alert other squirrels of the presence of a predator, the calling squirrel puts itself at risk of being found by the predator.

Cyanobacteria are bacteria that can perform photosynthesis. They contain chlorophyll pigments but do not contain chloroplast organelles. Algae are eukaryotic protists that can perform photosynthesis.

Symbiotic Relationships

Many organisms that coexist exhibit some type of **symbiotic relationship.** These include remoras, or "sucker fish," which attach themselves to the backs of sharks, and lichen—the fuzzy, mold-like stuff that grows on rocks. Lichen appears to be one organism, when in fact it is two organisms—a fungus and a photosynthetic organism such as an alga or cyanobacterium—living in a complex symbiotic relationship.

Overall, there are three basic types of symbiotic relationships:

- **Mutualism**—in which both organisms win (for example, the lichen components)
- **Commensalism**—in which one organism lives off another with no harm to the host organism (for example, the remora)
- **Parasitism**—in which the organism actually harms its host

There are many special types of relationships between multicellular organisms and unicellular organisms. One important example is bacteria living inside the guts of many mammals. We have a mutualistic relationship. Humans and other mammals provide the bacteria a nice home with nutrients, and they help us in many ways by breaking things down and helping us make things. It is also good that we are filled with "good" bacteria because otherwise we might be susceptible to dangerous pathogenic bacteria.

Most interactions are performed to gain energy. This can be by directly gaining resources or by doing something else more efficiently so that more energy is available to the organism.

PLANT BEHAVIOR

Plants have also evolved specific ways to respond to their environment. The plant behaviors covered on the AP Biology Exam are photoperiodism and tropisms. Plants flower in response to changes in the amount of daylight and darkness they receive. This is called **photoperiodism**. Although you'd think that plants bloom based on the amount of sunlight they receive, they actually flower according to the amount of uninterrupted darkness.

Tropisms

Plants need light. Notice that all the plants in your house tip toward the windows. This movement of plants toward the light is known as phototropism. As you know, plants generally grow up and down: the branches grow upward, while the roots grow downward into the soil, seeking water. This tendency to grow toward or away from the earth is called gravitropism. All of these tropisms are examples of behavior in plants.

A **tropism** is a turning in response to a stimulus.

There are three basic tropisms in plants. You can remember them because their prefixes represent the stimuli to which plants react:

- **Phototropism** refers to the way plants respond to sunlight—for example, bending toward light.

- **Gravitropism** refers to the way plants respond to gravity. Stems exhibit negative gravitropism (they grow up, away from the pull of gravity), whereas roots exhibit positive gravitropism (they grow downward into the earth).
- **Thigmotropism** refers to the way plants respond to touch. For example, ivy grows around a post or trellis.

These responses are initiated by hormones. The major plant hormones you need to know belong to a class called **auxins**. Auxins serve many functions in plants. They can promote growth on one side of the plant. For example, in phototropism, the side of the plant that faces away from the sunlight grows faster, thanks to the plant's auxins, making the plant bend toward the light.

Generally speaking, auxins are in the tip of the plant, because this is where most growth occurs. Auxins are also involved in cell elongation and fruit development.

Other plant hormones that regulate the growth and development of plants are **gibberellins, cytokinins, ethylenes,** and **abscisic acid**. Here's a summary of the functions of plant hormones:

FUNCTIONS OF PLANT HORMONES	
Hormone	**Function**
Gibberellins	Promote stem elongation, especially in dwarf plants
Cytokinins	Promote cell division and differention
Ethylene	Induces leaf abscission and promotes fruit ripening
Abscisic acid	Inhibits leaf abscission and promotes bud and seed dormancy
Auxins	Promote plant growth and phototropism

We've already discussed how important cell communication is to organisms, and plants are no exception. These hormones play an important role in allowing plants to grow, develop, survive, and adapt to environmental cues. They act as signaling molecules, bind to receptors, and function using signal transduction. These cell communication pathways are similar to what happens in animal cells.

ECOLOGY

The study of interactions between living things and their environments is known as **ecology**. We've spent most of our time discussing individual organisms. However, in the real world, organisms are in constant interaction with other organisms and the environment. The best way for us to understand the various levels of ecology is to progress from the big picture, the biosphere, down to the smallest ecological unit, the population.

There is a hierarchy within the world of ecology. Each of the following terms represents a different level of ecological interaction:

- **Biosphere**: The entire part of the earth where living things exist. This includes soil, water, light, and air. In comparison to the overall mass of the earth, the biosphere is relatively small. If you think of the earth as a basketball, the biosphere is equivalent to a coat of paint over its surface
- **Ecosystem**: The interaction of living and nonliving things
- **Community**: A group of populations interacting in the same area
- **Population**: A group of individuals that belong to the same species and that are interbreeding

Biosphere

There are several major biomes: tundra, taiga, temperate deciduous forest, grasslands, deserts, and tropical rainforests.

The biosphere can be divided into large regions called **biomes**. Biomes are massive areas that are classified mostly on the basis of their climates and plant life. Because of different climates and terrains on the earth, the distribution of living organisms varies.

Remember that biomes tend to be arranged along particular latitudes. For instance, if you hiked from Alaska to Kansas, you would pass through the following biomes: tundra, taiga, temperate deciduous forests, and grasslands.

Ecosystem

Ecosystems are self-contained regions that include both living and nonliving factors. For example, a lake, its surrounding forest, the atmosphere above it, and all the organisms that live in or feed off the lake would be considered an ecosystem. Living factors are called **biotic factors** and nonliving factors are called **abiotic factors.** Abiotic factors include water, humidity, temperature, soil/atmosphere composition, light, and radiation. As you probably know, there is an exchange of materials between the components of an ecosystem.

On the next page, take a look at the flow of carbon through a typical ecosystem.

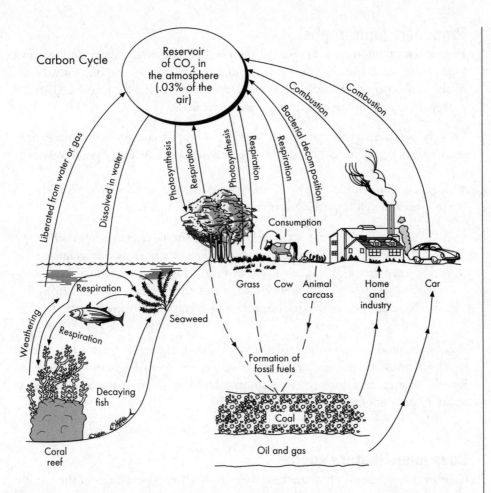

Carbon Cycle

You'll notice how carbon is recycled throughout the ecosystem—this is called the **carbon cycle**. In other words, carbon flows through ecosystems. There are also cycles for nitrogen and water. The interactions between the biotic and the abiotic factors are crucial for balance of the ecosystem.

Community

The next smaller level is the community. A community refers to a group of interacting plants and animals that show some degree of interdependence. For instance, you, your dog, and the fleas on your dog are all members of the same community.

Each organism has its own **niche**—its position or function in a community. When two organisms occupy the same niche, they will compete for the resources within that niche. If a species can occupy an unoccupied niche, it will usually thrive without competition. These connections are shown in the **food chain**. A food chain describes the way different organisms depend on one another for food. There are basically four levels to the food chain: producers, primary consumers, secondary consumers, and tertiary consumers.

Producers (autotrophs)

Producers, or autotrophs, have all of the raw building blocks to make their own food. From water and the gases that abound in the atmosphere and with the aid of the sun's energy, photosynthetic autotrophs convert light energy to chemical energy. They accomplish this through photosynthesis.

Primary productivity is the rate at which autotrophs convert light energy into chemical energy in an ecosystem. There are two types of **primary productivity:**

Chemosynthetic

There is another group of autotrophs that are not photosynthetic and instead are chemosynthetic and can take up nutrients from their surroundings and make their own energy in the absence of oxygen.

1. The gross productivity just from photosynthesis cannot be measured because cell respiration is occurring at the same time.

2. Net productivity measures only organic materials that are left over after photosynthetic organisms have taken care of their own cellular energy needs. This is calculated by measuring oxygen production in the light, when both photosynthesis and cell respiration are occurring. The equation for this relationship is listed on the Biology Equations and Formulas sheet.

Autotrophs produce all of the available food. They make up the first trophic (feeding) level. They also possess the highest **biomass** (the total weight of all the organisms in an area) and the greatest numbers. Did you know that plants make up about 99 percent of the earth's total biomass?

Consumers (heterotrophs)

Consumers, or heterotrophs, are forced to find their energy sources in the outside world. Basically, heterotrophs digest the carbohydrates, lipids, proteins, and nucleic acids of their prey into carbon, hydrogen, oxygen, etc. and use these molecules to make organic substances.

The bottom line is that heterotrophs, or consumers, get their energy from the things they consume.

Primary consumers are organisms that directly feed on producers. A good example is a cow. These organisms are also known as **herbivores**. They make up the second trophic level.

The next level consists of organisms that feed on primary consumers. They are the **secondary consumers**, and they make up the third trophic level. Above these are **tertiary consumers**.

Decomposers

A food chain shows a linear network of links, from producer to apex predator species. A food web consists of many food chains and shows many paths that connect plants and animals.

All organisms at some point must finally yield to decomposers. **Decomposers** are the organisms that break down organic matter into simple products. Generally, fungi and bacteria are the decomposers. They serve as the "garbage collectors" in our environment.

- Producers make their own food.

- Primary consumers (herbivores) eat producers.

- Secondary consumers (carnivores and omnivores) eat producers and primary consumers.

- Tertiary consumers eat all of the above.

- Decomposers break things down.

Keystone Species

Sometimes one organism is particularly important to an ecosystem. Maybe it is the only producer or maybe it is the only thing that keeps a particularly deadly predator in check. Important species like this are called **keystone species**. If a keystone species is removed from the ecosystem, the whole balance can be undone very quickly.

The 10% Rule

In a food chain, only about 10 percent of energy is transferred from one level to the next—this is called the **10% rule**. The other 90 percent is used for things like respiration, digestion, reproduction, running away from predators—in other words, it's used to power the organism doing the eating! This energy is eventually changed into heat energy. As a level, producers have the most energy in an ecosystem; the primary consumers have less energy than producers; secondary consumers have less energy than the primary consumers; and tertiary consumers have the least energy of all.

Not For Individuals
This idea about energy does not apply to individuals. It means the level as a whole has less energy if you add up all the members at that level.

The energy flow, biomass, and numbers of members within an ecosystem can be represented in an **ecological pyramid**. Organisms that are higher up on the pyramid have less biomass and energy, and they are fewer in number.

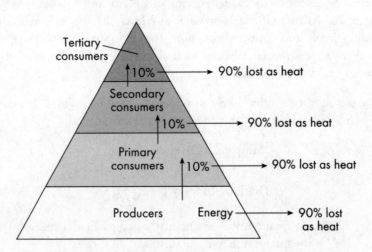

Buildup of Toxins

The downside of food pyramids is that when consumers at the top eat something beneath them, it is like they are eating that thing AND the thing that thing ate AND and the thing that thing ate, and so on. So, if there is a toxin in the environment, the consumers at the top are getting the most of it because it becomes more concentrated at each level.

> Toxins in an ecosystem are more concentrated and thus more dangerous for animals further up the pyramid.

This simply means that if a toxin is introduced into an ecosystem, the animals most likely to be affected are those at the top of the pyramid. This occurs because of the increasing concentration of such toxins. The classic example of this phenomenon is DDT, an insecticide initially used to kill mosquitoes. Although the large-scale spraying of DDT resulted in a decrease in the mosquito population, it also wound up killing off ospreys.

Ospreys are aquatic birds of prey whose diet consists primarily of fish. The fish that ospreys consumed had, in turn, been feeding on contaminated insects (**bioaccumulation**). Because fish eat thousands of insects, and ospreys hundreds of fish, the toxins grew increasingly concentrated (**biomagnification**). Though the insecticide seemed harmless enough, it resulted in the near extinction of certain osprey populations. What no one knew was that in sufficient concentrations, DDT weakened the eggshells of ospreys. Consequently, eggs broke before they could hatch, killing the unborn ospreys.

Such environmental tragedies still occur. All ecosystems, small or great, are intricately woven, and any change in one level invariably results in major changes at all the other levels.

Disrupting the Balance

Ecosystems are never exactly stable, existing in a precarious balance. Many things can disrupt the balance. If the biomass at any level changes, this will affect the surrounding levels and spread throughout the ecosystem. Geological changes, weather changes, new species, diseases, lack of resources, new resources, and many other things can disrupt the balance.

A diverse community is more likely to withstand extreme events. The Simpson's Diversity Index is a measure of the diversity of a community.

Simpson's Diversity Index

$$\text{Diversity Index} = 1 - \Sigma \left(\frac{n}{N} \right)^2$$

n = the total number of organisms of a particular species
N = the total number of organisms of all species

DDT stands for dichloro-diphenyltrichloroethane. You can see why an abbreviation is better here!

Ecosystem Interconnectedness
Because all levels of an ecosystem are closely connected, a disruption in one level due to environmental changes, for example, affects all other aspects of that ecosystem.

Invasive species are particularly dangerous since they can severely disrupt the balance of the ecosystem. This is often because they have no natural predator or competition and soon expand and take resources from other species.

POPULATION ECOLOGY

Population ecology is the study of how populations change. Whether these changes are long-term or short-term, predictable or unpredictable, we're talking about the growth and distribution patterns of a population.

When studying a population, you need to examine four things: the size (the total number of individuals), the density (the number of individuals per area), the distribution patterns (how individuals in a population are spread out), and the age structure.

For example, the graph below shows there is a high death rate among the young of oysters, but those that survive do well. On the other hand, there is a low death rate among the young of humans, but, after age 60, the death rate is high.

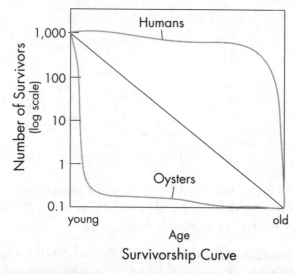

Survivorship Curve

Population growth can be represented as the number of births minus the number of deaths divided by the size of the population.

$$r = (\text{births} - \text{deaths})/N$$

(r is the reproductive rate, and N is the population size.)

Population growth can also be calculated in the following way:

$$\frac{Change\ in\ Population\ Size}{Change\ in\ Time} = Birth\ Rate - Death\ Rate$$

Each population has a **carrying capacity**—the maximum number of individuals of a species that a habitat can support. Most populations, however, don't reach their carrying capacity because they're exposed to limiting factors.

One important factor is **population density**. The factors that limit a population are either density-independent or density-dependent. **Density-independent factors** affect the population, regardless of the density of the population. Some examples are severe storms and extreme climates. On the other hand, **density-dependent**

factors are those with effects that depend on population density. Resource depletion, competition, and predation are all examples of density-dependent factors. In fact, these effects become even more intense as the population density increases.

Exponential Growth

The growth rates of populations also vary greatly. There are two types of growth: exponential growth and logistic growth. Equations for both are included on the AP Biology Equations and Formulas sheet. **Exponential growth** occurs when a population is in an ideal environment. Growth is unrestricted because there are lots of resources, space, and no disease or predation. Here's an example of exponential growth. Notice that the curve arches sharply upward—the exponential increase.

Exponential Growth Equation

Change in Population Size/Change in Time = maximum growth rate × Population Size

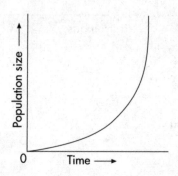

Exponential growth occurs very quickly, resulting in a J-shaped curve. A good example of exponential growth is the initial growth of bacteria in a culture. There's plenty of room and food, so they multiply rapidly.

However, as the bacterial population increases, the individual bacteria begin to compete with each other for resources. When the resources become limited, the population reaches its carrying capacity, and the curve levels off.

Logistic Growth Equation

Change in Population Size/Change in Time = maximum growth rate × Population Size × ((Carrying Capacity − Population Size)/Carrying Capacity)

Essentially it's the same as the exponential equation except when the population size approaches the carrying capacity, the growth rate approaches zero.

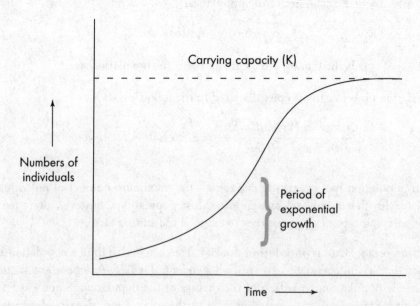

The population becomes restricted in size because of limited resources. This is referred to as **logistic growth**. Notice from the graph on the previous page that the growth forms an S-shaped curve.

These growth patterns are associated with two kinds of life-history strategies: r-selected species and k-selected species.

R-strategists tend to thrive in areas that are barren or uninhabited. Once they colonize an area, they reproduce as quickly as possible. Why? They know they've got to multiply before competitors arrive on the scene! The best way to ensure their survival is to produce lots of offspring. Typical examples are common weeds, dandelions, and bacteria.

At the other end of the spectrum are the **k-strategists**. These organisms are best suited for survival in stable environments. They tend to be large animals, such as elephants, with long lifespans. Unlike r-strategists, they produce few offspring. Given their size, k-strategists usually don't have to contend with competition from other organisms.

ECOLOGICAL SUCCESSION

Communities of organisms don't just spring up on their own; they develop gradually over time. **Ecological succession** refers to the predictable procession of plant communities over a relatively short period of time (decades or centuries).

Centuries may not seem like a short time to us, but if you consider the enormous stretches of time over which evolution occurs, hundreds of thousands or even millions of years, you'll see that it is pretty short.

The process of ecological succession in which no previous organisms have existed is called **primary succession**.

How does a new habitat full of bare rocks eventually turn into a forest? The first stage of the job usually falls to a community of lichens. Lichens are hardy organisms. They can invade an area, land on bare rocks, and erode the rock surface, over time turning it into soil. Lichens are considered **pioneer organisms**.

Once lichens have made an area more habitable, they've set the stage for other organisms to settle in. Communities establish themselves in an orderly fashion. Lichens are replaced by mosses and ferns, which in turn are replaced by tough grasses, then low shrubs, then evergreen trees, and, finally, deciduous trees. Why are lichens replaced? Because they can't compete with the new plants for sunlight and minerals.

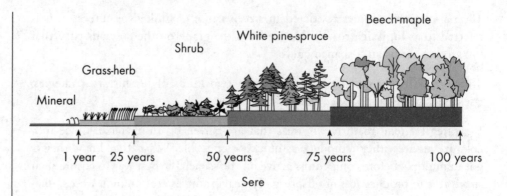

The entire sequence is called a **sere**. The final community is called the **climax community**. The climax community is the most stable. In our example, deciduous trees are part of the climax community.

Now what happens when a forest is devastated by fire? The same principles apply, but the events occur much more rapidly. The only exception is that the first invaders are usually not lichens, but grasses, shrubs, saplings, and weeds. When a new community develops where another community has been destroyed or disrupted, this event is called **secondary succession**. The existing community can be cleared by events such as fire, tornado, or human impact. These impacts leave soil intact, explaining how development occurs more rapidly.

HUMAN IMPACT ON THE ENVIRONMENT

Unfortunately, humans have disturbed the existing ecological balance, and the results are far-reaching. Soils have been eroded and various forms of pollution have increased. Potential consequences on the environment are summarized below.

- **Greenhouse effect:** Atmospheric concentrations of carbon dioxide have increased by burning fossil fuels, and forests have contributed to warming of the earth. Higher temperatures may cause polar ice caps to melt and flooding to occur. Other potential effects of global warming include changes in precipitation patterns, changes in plant and animal populations, and detrimental changes in agriculture.

- **Ozone depletion:** Pollution has also led to depletion of the atmospheric ozone layer by such chemicals as chlorofluorocarbons (CFCs), which are used in aerosol cans. Ozone (O_3) forms when UV radiation reacts with O_2. Ozone protects the earth's surface from excessive ultraviolet radiation. Its loss could have major genetic effects and could increase incidence of cancer.

- **Acid rain:** Burning fossil fuels produces pollutants such as sulfur dioxide and nitrogen dioxide. When these compounds react with droplets of atmospheric water in clouds, they form sulfuric and nitric acids, respectively. The rain that falls from these clouds is weakly acidic and is called acid rain. Acid rain lowers the pH of aquatic ecosystems and soil, which damages water systems, plants, and soil. For example, change in soil pH causes calcium and other nutrients to leach out, which damages plant roots and stunts their growth. Furthermore, useful microorganisms that release nutrients from decaying organic matter into the soil are also killed, resulting in fewer nutrients being available for the plants. Low pH also kills fish, especially those that have just hatched.

- **Desertification:** When land is overgrazed by animals, it turns grasslands into deserts and reduces available habitats for organisms.

- **Deforestation:** When forests are cleared (especially by the slash and burn method), erosion, floods, and changes in weather patterns can occur.

- **Pollution:** Another environmental concern is toxic chemicals in our environment. One example is DDT, a pesticide used to control insects. DDT was overused at one time and later found to damage plants and animals worldwide. DDT is particularly harmful because it resists chemical breakdown, and it can still be found in tissues of nearly every living organism. The danger with toxins such as DDT is that as each trophic level consumes DDT, the substance becomes more concentrated by the process of biomagnification.

- **Reduction in biodiversity:** As different habitats have been destroyed, many plants and animals have become extinct. Some of these plants could have provided us with medicines and products that may have been beneficial.

- **Introduction and Spread of Disease:** Humans travel and disrupt habitats and bring diseases with them. These diseases can devastate all types of living things and act as immediate selective pressures, leaving the landscape of an ecosystem forever changed.

Acid rain, deforestation, pollution, soil erosion, and depletion of the ozone layer are just a few of the problems plaguing the environment today.

KEY TERMS

endotherms
ectotherms
behavior
instinct
learning
imprinting
critical period
habituation
circadian rhythms
pheromones
agonistic behavior
dominance hierarchy
territoriality
altruistic behavior
symbiotic relationship
mutualism
commensalism
parasitism
photoperiodism
tropism
phototropism
gravitropism
thigmotropism
auxins
gibberellins
cytokinins
ethylenes
abscisic acid
ecology
biosphere
ecosystem
community
population
biomes
biotic factors
abiotic factors
carbon cycle
niche

food chain
producers (autotrophs)
primary productivity
biomass
consumers (heterotrophs)
primary consumers
herbivores
secondary consumers
tertiary consumers
decomposer
keystone species
10% rule
ecological pyramid
bioaccumulation
biomagnification
Simpson's Diversity Index
population growth
carrying capacity
population density
density-independent factors
density-dependent factors
exponential growth
logistic growth
r-strategists
k-strategists
ecological succession
primary succession
pioneer organisms
sere
climax community
secondary succession
greenhouse effect
ozone depletion
acid rain
desertification
deforestation
pollution
reduction in biodiversity

Summary

Behavior

o Behavior is an organism's response to the environment. Behavior can be instinctual (inborn), learned, or social:
  - Types of social behaviors include agonistic behavior, dominance hierarchies, territoriality, and altruistic behavior.

o There are also plant-specific behaviors known as tropisms. The three basic tropisms are phototropism, gravitropism, and thigmotropism.

Ecology

o There are several major biomes (tundra, taiga, temperate deciduous forest, grasslands, deserts, tropical rainforests) that make up the biosphere. Each biosphere contains ecosystems.

o Within an ecosystem are communities, which consist of organisms fulfilling one of three main roles:
  - Producers, or autotrophs, convert light energy to chemical energy via photosynthesis.
  - Consumers, or heterotrophs, acquire energy from the things they consume. Their digestion of carbohydrates produces carbon, hydrogen, and oxygen, which are then used to make organic substances.
  - Decomposers form fossil fuels from the detritus of other organisms in the ecosystem.
  - The 10% rule says that only 10% of the energy consumed from one level will be retained by the higher level that consumed it. The other energy will be spent to perform normal daily activities.

o The smallest unit of ecology is the population. The growth of a population can be found with the equation $(r) = (\text{births} - \text{deaths}) / N$.

o The carrying capacity is the maximum number of individuals that can be supported by a habitat. Most populations do not reach carrying capacity due to factors such as population density (density-independent factors and density-dependent factors).

o Exponential growth (J-shaped curve) occurs when a population is in an ideal environment. Logistic growth (S-shaped curve) of a population occurs when there are limited resources in an environment.

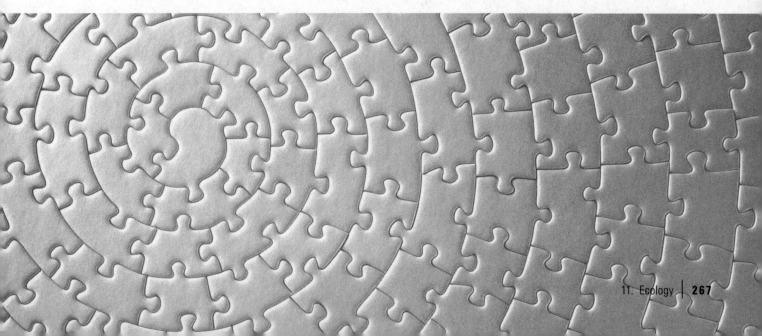

o   Organisms are generally either r-strategists or k-strategists. R-strategists ensure their survival by producing lots of offspring. K-strategists, which are usually large animals, produce few offspring but have longer life spans and less competition from other organisms for resources.

o   Succession describes the way in which ecosystems recover after a disturbance in terms of pioneers, sere, climax community, and secondary succession.

o   Human impact on the planet includes the following issues:
  • greenhouse effect
  • ozone depletion
  • acid rain
  • desertification
  • deforestation
  • pollution
  • reduction in biodiversity
  • introduction and spread of disease

Chapter 11 Drill

Answers and explanations can be found in Chapter 15.

1. A chimpanzee stacks a series of boxes on top of one another to reach a bunch of bananas suspended from the ceiling. This is an example of which of the following behaviors?

   (A) Operant learning
   (B) Imprinting
   (C) Instinct
   (D) Insight

2. Viruses are obligate intracellular parasites, requiring their host cells for replication. Consequently, viruses generally attempt to reproduce as efficiently and quickly as possible in a host. Below is a graph depicting the initial growth pattern of a bacteriophage within a population of *E. coli*. This reproductive strategy is most similar to which of the following?

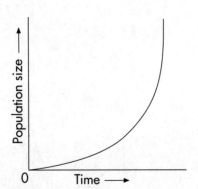

   (A) An r-strategist, because it aims to produce a large abundance of offspring to ensure survival
   (B) A k-strategist, because it aims to produce a large abundance of offspring to ensure survival
   (C) An r-strategist, because it is best suited to thrive in stable environments and over a long life span
   (D) A k-strategist, because it is best suited to thrive in stable environments and over a long life span

3. In a pond ecosystem, spring rains trigger an expansion of species at levels of the food chain. Runoff from nearby hills brings nutrients which, when combined with warming temperatures, trigger an algae bloom. The populations of small protozoans such as plankton expand by ingesting the algae. These plankton, subsequently, are consumed by small crustaceans such as crayfish, which ultimately become prey for fish such as catfish or carp. In this ecosystem, which of the following accurately describes the crayfish?

   (A) They are producers.
   (B) They are primary consumers.
   (C) They are secondary consumers.
   (D) They are tertiary consumers.

Questions 4–6 refer to the following table, chart, and paragraph.

Table 1: Minimum pH Tolerance of Common Aquatic Organisms

Animal	pH Minimum
Brook Trout	4.9
American Bullfrog	3.8
Yellow Perch	4.6
Tiger Salamander	4.9
Crayfish	5.4
Snails	6.1
Clams	6.0

Chart 1: Change in Populations of Aquatic Organisms in Richard Creek (1980–2000)

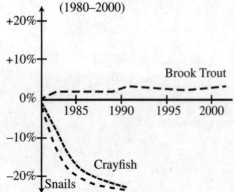

Over the period of 1980 to 2000, the average pH in Richard Creek changed drastically. An ecological survey was performed to evaluate the effect of detectable decreases in pH on the aquatic life of the creek. Four times a year, ecological surveys were performed to identify the number of snails (a primary consumer), crayfish (a secondary consumer), and brook trout (a tertiary consumer) present at five different locations. The percent change relative to 1980 is shown in **Chart 1** above. Many aquatic organisms cannot live in low pH conditions. The minimum pH necessary for common aquatic organisms to sustain life is shown in **Table 1**.

4. According to the data in Chart 1 and Table 1, the average pH in the creek is most nearly which of the following?

   (A) 5.9
   (B) 5.5
   (C) 5.1
   (D) 4.7

5. Which of the following is the most likely explanation for the abrupt decrease in pH in Richard Creek?

   (A) Greenhouse effect
   (B) Deforestation
   (C) Acid rain
   (D) Pollution

6. In order for snails to return to Richard Creek, the pH of the creek must exceed

   (A) 5.4
   (B) 5.7
   (C) 6.0
   (D) 6.1

7. *Mimosa pudica* is a plant often called the "sensitive plant" because when you touch the leaves, they immediately close up. One theory about the purpose of this type of movement is that herbivores avoid the plant due to this movement. The movement of *Mimosa pudica* is an example of

    (A) phototropism
    (B) gravitropism
    (C) aquatropism
    (D) thigmotropism

Question 8 refers to the following graph.

Foodprints by Diet Type: $t\ CO_2^e$/person

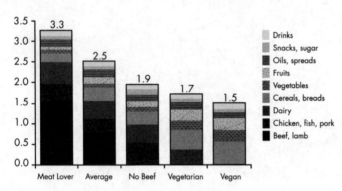

8. A proponent of the vegan diet writes a blog post that claims that the vegan diet is the only way to "reduce your carbon footprint, with respect to food, because it cuts your carbon footprint in half" and shows the graph above. Which of the following directly supports this claim?

    (A) According to the 10% rule, only 10% of energy is passed from producers to consumers, so when humans consume animals, the energy harvested is greatly reduced.
    (B) Livestock produce methane emissions and consume large amounts of plants every year.
    (C) Deforestation for grazing livestock contributes to the loss of plants for carbon sequestration.
    (D) Beef and lamb livestock—not chicken, fish, or pork—has the most dramatic impact on carbon footprint.

9. Costa Rica is plentiful with many beautiful tropical rainforests. For a long time, cattle farmers cleared much of the forests for their livestock. However, in recent decades, the government has placed more emphasis on reforestation. Several fields are now being reclaimed by the rainforest. One such field has many long grasses, a few tall palms, and some dense shrubbery along the periphery of the field. This field can be said to be

    (A) invaded by pioneer species
    (B) a product of secondary succession
    (C) a climax community
    (D) filled with invasive species

10. Probiotics are beneficial bacteria or yeast applied to maintain intestinal health in humans or livestock. In theory, these organisms inhabit the intestines and aid in digestion of food. Also, these organisms have bioprotective effects. How can addition of bacteria to the intestine protect the host from disease?

    (A) Probiotic bacteria crowd the intestine, preventing pathogenic bacteria from growing due to density-dependent limitation on their population.
    (B) Probiotic bacteria crowd the intestine, preparing a niche for pathogenic bacteria to grow.
    (C) Bacteria populations grow exponentially until the carrying capacity is reached.
    (D) Pathogenic bacteria do not recognize the intestinal cells because they are coated with probiotic bacteria.

REFLECT

Respond to the following questions:

- Which topics from this chapter do you feel you have mastered?

- Which content topics from this chapter do you feel you need to study more before you can answer multiple-choice questions correctly?

- Which content topics from this chapter do you feel you need to study more before you can effectively compose a free response?

- Was there any content that you need to ask your teacher or another person about?

Chapter 12
Quantitative Skills and Biostatistics

Questions related to quantitative skills and biostatistics pop up all over the place. Even if you think you are a whiz, check out the formula pages near the back of this book. Expect at least a couple questions using formulas from these sheets.

It is often easy to see patterns in data, but it is not always easy to determine if a pattern is valid or significant. Quantitative data analysis is the first step in figuring this out. In this chapter, we will review how data can be summarized, presented, and tested for validity. This depends on what kind of question was being asked at the beginning of the experiment. When you're reading this chapter, pay attention to which techniques are used when. On the AP Biology Exam, you should be able to justify which techniques are best.

Don't forget to check out the formula sheet!

This list is a good reference for the types of quantitative questions you should be able to tackle. We will not walk though all of these in detail, but you will find some examples in the end-of-chapter drill and in the practice tests. Use the formula sheets to familiarize yourself with these concepts.

> ## Must-Know Types of Quantitative Problems
>
> ☐ Hardy-Weinberg Equilibrium
>
> ☐ Water Potential/Osmosis
>
> ☐ Energy Pyramids/Biomass
>
> ☐ Chi-squared Analysis
>
> ☐ Gene Linkage Analysis
>
> ☐ Inheritance I Probability
>
> ☐ Reading from Graphs
>
> ☐ Predicting from Graphs
>
> ☐ Dilutions of Solutions
>
> ☐ Population Growth

SUMMARIZING AND PRESENTING DATA

Instead of presenting raw data, scientists use **descriptive statistics** and graphs to summarize large datasets, present patterns in data, and communicate results. Descriptive statistics summarize and show variation in the data. There are many types of descriptive statistics, including measures of central tendency (such as mean, median, and mode) and measures of variability (such as standard deviation, standard error, and range). The AP Biology Equations and Formulas sheet contains definitions for mean, median, mode, and range. It also includes equations for mean, standard deviation, and standard error. That said, you will not need to calculate standard deviation or standard error. Instead, focus on understanding how these values are used.

Descriptive statistics are used to summarize the data collected from samples, but may also describe the entire population you're trying to study. Experiments are designed to include a **sample**, which is a subset of the **population** being studied. The best experiments use **random sampling**, which makes sure there is no bias in picking which individuals from the population will be included in the sample. It is important that the sample is big enough that the data from the experiment is a good representation of what would happen if the whole population were measured.

Graphs are visual representations of data and are often used to reveal trends that might not be obvious by looking at a table of numbers. There are six types of graphs you should be familiar with:

1. Bar graph
2. Pie graph
3. Histogram
4. Line graph
5. Box-and-whisker plot
6. Scatterplot

Each of these is described more in the sections that follow. Pay attention to which graph is used when. On the AP Biology Exam, you may need to decide which graph is best and then create it. You may also have to analyze graphs given to you on the exam.

A good graph must include the following things:

- a good title that describes what the experiment was and what was measured
- axes labeled with numbers, labels and units, and index marks
- a frame or perimeter
- data points that are clearly marked (e.g., easily identifiable lines or bars)

TYPES OF DATA

Data can be **quantitative** (based on numbers or amounts that can be measured or counted) or **qualitative** (data that is descriptive, subjective, or difficult to measure). For example, behavioral observations are qualitative. Most data on the AP Biology Exam will be quantitative, with either counts or measurements.

Count Data

Count data are generated by counting the number of items that fit into a category. For example, you could count the number of organisms with a particular phenotype or the number of animals in one habitat versus another. Count data also include data that is collected as percentages or the results of a genetic cross. This type of data is usually summarized in a **bar graph** or a **pie graph**. For example, suppose a mixed population of *E. coli* were plated on growth plates. Some cells contained a β-galactosidase marker and grew in blue colonies, while others did not contain the marker and grew in white colonies. The colonies were counted after 24 hours of growth.

E.coli Colony Growth on Plates after 24 hours

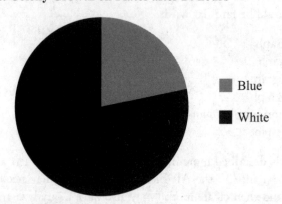

In another example, suppose fly populations were monitored in northern Maine to determine if population size varies over the warm months of the year.

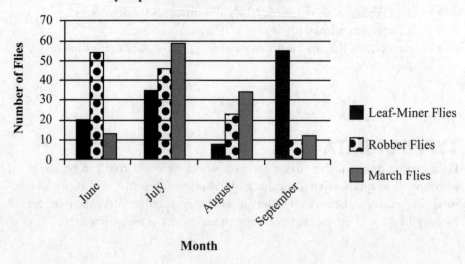

Count data can be analyzed with hypothesis testing, which will be described later in this chapter.

Measurements

Measurements are continuous, meaning there is an infinite number of potential measurements over a given range. Size, height, temperature, weight, and response rate are all measurements. There are two types of measurement data: parametric and nonparametric.

Normal or Parametric Data

Normal, or **parametric**, **data** is measurement data that fits a **normal curve**, or **distribution**, usually when a large sample size is used. For example, if you took a large sample of 17-year-olds in America and graphed the frequency of heights, the results will be normally distributed.

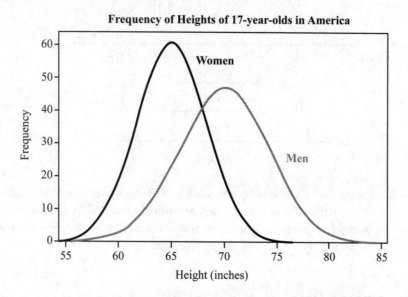

Several descriptive statistics can be used to summarize normal data.

The **sample size** (n) refers to the number of members of the population that are included in the study. Sample size is an important consideration when you're trying to determine how well the data in the study represents a population. Large sample sizes are always better, but for technical reasons, experiments can't be infinitely large.

The **mean** ($\bar{x}$) is the average of the sample, calculated by adding all of the individual values and dividing by the number of values you have. The mean is not necessarily a number provided in the sample. You should be able to recognize what the mean of a given dataset is and be able to calculate the mean using the equation on the AP Biology Equations and Formulas sheet. One limitation of mean is that it is influenced by **outliers**, or numerical observations that are far removed from the rest of the observations.

The **standard deviation** (*s*) can determine if numbers are packed together or dispersed, because it is a measure of how much each individual number differs from the mean. A low standard deviation means the data points are all similar and close to the mean, while a high standard deviation means the data are more spread out. Here is the relationship between a normal distribution and standard deviation (*SD*).

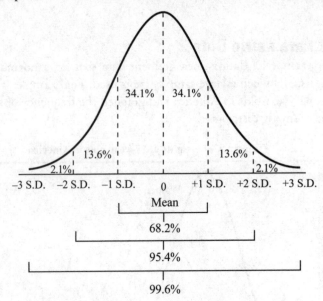

This means that about 70 percent of the data is within one standard deviation of the mean in a normally distributed dataset. **Standard error** (*SE*) can also be used to report how much a given dataset varies and is calculated by dividing the standard deviation by the square root of the sample size.

Nonparametric Data

Nonparametric data often includes large outliers and do not fit a normal distribution. In order to determine if data is parametric or not, you can construct a **histogram**, or **frequency diagram**. These graphs give information on the spread of the data and the central tendencies. Making a histogram is like setting up bins, or intervals with the same range, that cover the entire dataset (the *x*-axis). The measurements in each bin are graphed on the *y*-axis. For example, suppose you counted the number of pine needles on each branch of a *Pinus strobus* tree. A histogram of this data may look like this.

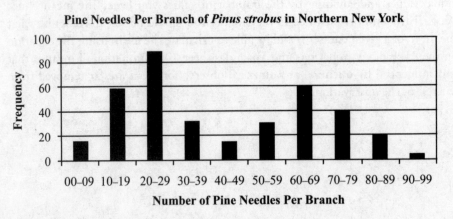

Pine Needles Per Branch of *Pinus strobus* in Northern New York

This dataset does not match a normal distribution, meaning it is a nonparametric dataset. Keep in mind that even if a certain dataset is not normal, the population itself could be. There could have been sampling bias or errors in the data collection, which can lead to sample data that doesn't match population data. The best way to fix this is usually to increase the sample size. Histograms differ from bar graphs in that they show ranges instead of just categories.

Nonparametric data requires a different set of descriptive statistics than parametric data.

The **median** is the middle number in a dataset and is determined by putting the numbers in consecutive order and finding the middle number. If there is an odd number of numbers, there will be a single number that is the median. If there is an even number of numbers, the median is determined by averaging the two middle numbers. Therefore, the median is not necessarily one of the numbers in the dataset. The median is useful in gauging the midpoint of the data, but will not necessarily tell you much about outliers.

The **mode** is the most frequently recurring number in the dataset. If there are no numbers that occur more than once, there is no mode. If there are multiple numbers that occur most frequently, each of those numbers is a mode. The mode must be one of the numbers in the sample, and modes are never averaged.

The **range** is the difference between the smallest and largest number in a sample. This value is less useful than standard deviation, because it does not give any information on individual values or the majority of values.

Example 1: Eleven male laboratory mice were weighed at 32 weeks of age. The data collected was: 32 g, 28 g, 29 g, 34 g, 30 g, 28 g, 32 g, 31 g, 30 g, 32 g, 33 g. What is the mean, median and mode of this dataset?

Solution: Let's start by putting the 11 numbers in the dataset in order, from smallest value to largest value: 28 g, 28 g, 29 g, 30 g, 30 g, 31 g, 32 g, 32 g, 32 g, 33 g, 34 g. The mean is

$$Mean = \frac{28 + 28 + 29 + 30 + 30 + 31 + 32 + 32 + 32 + 33 + 34}{11}$$

$$= \frac{339}{11}$$

$$= 30.8\,g$$

The median is the middle value, or 31 g. The mode is the most frequent number, or 32 g.

Example 2: In the following set of values, what is the range?

Values: –5, 8, 11, –1, 0, 4, 14

Solution: The smallest value in the set above is –5, and the largest is 14. The difference between these two is the range, which is 19.

TYPES OF EXPERIMENTS OR QUESTIONS

You will be tested on scientific experiments and the scientific reasoning behind different aspects of the experiments a LOT. You should be clear with the steps of the scientific method and know how to properly design an experiment and recognize a poorly designed experiment. The following list should remind you:

- **Hypothesis:** A well-thought-out prediction of what the outcome of the experiment will be.
- **Independent Variable:** The factor that you, as the experimenter, will change between the different groups in the experiment. If one plant is grown in pH 5 and the other in pH 7, then pH is the independent variable.
- **Dependent Variable:** The data, or the thing that you measure during the experiment. The height of the plant you are growing might be the dependent variable.
- **Constants (Controlled Variables):** The things that are the same between all your groups. Hint: Everything should be constant except the independent variable.
- **Control Groups:** Any group that is needed simply so you can compare your interesting experimental groups to it. These are often the "no treatment" group.
- **Statistical Significance:** The trustworthiness of the results and the certainty you have in your conclusions. This can always be increased by including more individuals in the groups or including more trials.

Most biological experiments do one of three things:

1. look at how something changes over time
2. compare groups of some sort
3. test for an association

How you present and summarize data depends on what kind of experiment you perform.

Time-Course Experiments

Time-course experiments look at how something changes over time. A **line graph** is usually used to present this type of data. Several lines can be plotted on the same graph, but these must be clearly labeled. Also, each dot on a line graph could represent one data point or a mean of values. If mean values are plotted, standard deviation or standard error can also be shown. More generally, line graphs can also be used to compare the way in which a dependent variable (*y*-axis) changes in relation to an independent variable (*x*-axis).

When making line graphs, the intervals on the axes must be consistent. Also make sure it is clear whether the data starts at the origin (0, 0) or not. The same guidelines apply to scatterplots, which will be discussed below.

Data points should be connected with a solid line. You can extrapolate (extend) the line past the data points, but then you must use a broken line. Finally, the slope on a line graph tells you the rate of change.

Suppose we want to determine how stress hormone levels change after the end of an important exam. If we measure cortisol and epinephrine in the serum of several people every hour for five hours after the end of an exam, we would have several values for each time point. We could plot the mean of each measurement and also show data variance using standard error bars. In this example, the distance above the data point is the standard error, and the distance below the data point is the standard error. We call this "±SE."

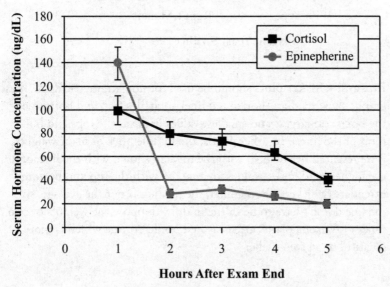

Serum Hormone Concentrations After Exam End

Comparative Experiments

Comparative experiments compare populations, groups, or events. There are several options for graphing data from these experiments:

- **Bar graphs** are helpful to compare categories of data if the data is parametric. It is also a good idea to show variance of the data by including the standard error. This gives information on how different the two means are from each other. Suppose you compare two strains of fission yeast before and after treatment with a drug that inhibits cell wall synthesis. As a measure of viability, you measure how much Myc mRNA was made in each strain with and without treatment. In this case, each bar is a mean of values and the error bars are ±SE, as described above.

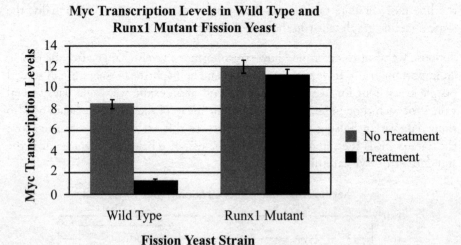

- **Box-and-whisker plots** should be used for nonparametric data. These graphs show median (horizontal line), quartiles (top and bottom of the box), and largest and smallest values (ticks at the tops and bottoms of the lines). In other words, the middle 50% of plant weights is represented by the box, with the median shown with the line, and the highest and lowest values are marked with the top and bottom extenders (whiskers). Remember, the median is not the average, so the line doesn't have to be in the middle. Suppose the weights of two types of Arizona plants are measured, to determine which is more plentiful in a given niche:

Grass and Cactus Weights in the Arizona Desert

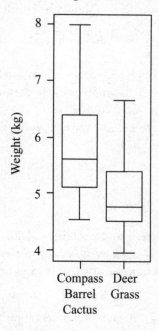

In order to conclude that the samples or groups being compared are different or not, hypothesis testing should be performed. This will be discussed later in this chapter.

Association Experiments

Association experiments look for associations between variables. They attempt to determine if two variables are correlated, and additional tests can demonstrate causation. **Scatterplots** are used to present data from association experiments. Each data point is plotted as a dot. Suppose different substrate concentrations are used in an enzymatic reaction (using the enzyme Kozmase III), and the enzyme efficiency is measured (in percent of maximum). The data could be presented like this.

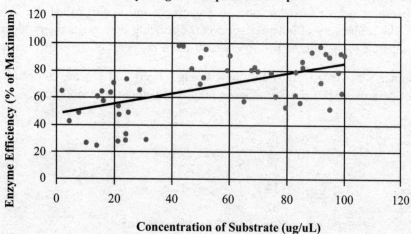

Enzyme Efficiency and Substrate Concentration for Kozmase III at Physiological Temperature and pH

If the relationship looks linear, a linear regression line can be added. There are a few common shapes of scatterplots you should be familiar with:

- A bell-shaped curve is associated with random samples and normal distributions.
- An upward curve (concave in shape) is associated with exponentially increasing functions. A common example of this is the lag phase of bacterial growth in a liquid culture.
- A curve that looks like the sine wave is common when biological rhythms are being studied.

PROBABILITY

You already saw an introduction to probability calculations in Chapter 8 of this book. The **probability** (P) that an event will occur is the number of favorable cases (a) divided by the total number of possible cases (n).

$$P = a \, / \, n$$

This can be determined experimentally by observation (such as when population data is being collected) or by the nature of the event. For example, the probability of getting a two when rolling a die is 1/6, since there are six sides on a die.

Example: What is the probability of drawing the nine of hearts from a deck of cards?

Solution: Since there are 52 cards in a deck and this question is asking about one particular card, $P = 1/52$.

Combining Probabilities

Many genetics problems involve several probabilities, all being considered together to answer a larger biological question. There are three rules that are often used.

Product Rule

The **product rule** is used for independent events and is also called the "AND rule." The probability of independent events occurring together is the product of their separate probabilities. This rule is used when the order of the events matters.

Example: What is the probability of drawing a heart and a spade consecutively from a deck of cards if the first draw was replaced before the second draw?

Solution:

$$P(\text{heart}) = \frac{13}{52} = \frac{1}{4}$$

$$P(\text{spade}) = \frac{13}{52} = \frac{1}{4}$$

$$P(\text{heart and spade}) = P(\text{heart}) \times P(\text{spade})$$

$$P(\text{heart and spade}) = \frac{1}{4} \times \frac{1}{4} = \frac{1}{16}$$

Sum Rule

The **sum rule** is used for studying two mutually exclusive events, and can be thought of as the "EITHER" rule. The probability of either of two events occurring is the sum of their individual probabilities.

Example: What is the probability of drawing a diamond or a heart from a deck of cards?

Solution:

$$P(\text{diamond}) = \frac{13}{52} = \frac{1}{4}$$

$$P(\text{heart}) = \frac{13}{52} = \frac{1}{4}$$

$$P(\text{diamond or heart}) = P(\text{diamond}) + P(\text{heart})$$

$$P(\text{diamond or heart}) = \frac{1}{4} + \frac{1}{4} = \frac{2}{4} = \frac{1}{2}$$

If the events you are studying are not mutually exclusive, such as having brown hair and brown eyes then you could have both occur at the same time. If you are trying to determine the likelihood of EITHER occurring but NOT BOTH, then you must calculate using the sum rule (odds of either occurring) and the product rule (odds of both occurring). Then, take the result using the product rule and subtract it from the result when using the sum rule since you want the odds of either occurring but not both.

HYPOTHESIS TESTING

t-test and *p*-values

Many experiments involve comparing two datasets or two groups, and a **t-test** can be used to calculate whether the means of two groups are significantly different from each other. This test is most often applied to datasets that are normally distributed. A **p-value** equal to or below 0.05 is considered significant in most biology-related fields. *T*-test calculations involve comparing each data point to the group's mean and also take standard deviation and sample size into consideration.

Understanding mathematically how *p*-values are calculated and *t*-tests are performed is not important here. Instead, it's important that you understand how these values are used to interpret data.

Let's work through an example to demonstrate how common statistics are used in biological labs. A researcher has sections of two different types of skin cancers from human patients. She stains them for the protein CD31, which is a marker of endothelial cells. She then takes digital pictures of the immunofluorescent sections and counts the number of blood vessels present in each picture. She takes five pictures of each slide.

Slide A	Slide B
10	3
8	2
7	4
8	4
11	4

(a)   What is the mean, median, and mode for each dataset?

(b)   The standard error of group A is 0.735, and the standard error for group B is 0.400. What does this tell you about the data?

(c)   What assumptions are made about the data in performing these tests?

(d)   What could the researcher do to increase her confidence in the results?

Solutions:

(a)   The means are

$$Mean_{Group\ A} = \frac{10 + 8 + 7 + 8 + 11}{5} = 8.8$$

$$Mean_{Group\ B} = \frac{3 + 2 + 4 + 4 + 4}{5} = 3.4$$

Remember, the median is the middle number, so it's best to put the data in order first.

Slide A	Slide B
7	2
8	3
8	4
10	4
11	4

The median of group A is 8, and the median of group B is 4.

The mode is the most frequent value. For group A, this is 8. For group B, the mode is 4.

(b)   Standard error is the standard deviation divided by the square root of the sample size. The sample size is five for both group A and group B because five pictures were taken for each slide. Because the standard error of group A is larger than that of group B and because the two groups have the same sample size, the standard deviation of group A must also be larger than that of group B. Note that this may or may not be the case if the sample sizes were different. A larger standard deviation means the data is more variable, so you can conclude that the data from group A has a larger spread than the data from group B.

(c)   As with most datasets, the researcher is assuming the data fits a normal distribution and that her sample size is large enough to be meaningful.

(d)   More data allows for more confident conclusions. The researcher could therefore take more pictures from each slide to increase the sample size.

Chi-Square (χ^2) Tests

A **chi-square test** is a statistical tool used to measure the difference between observed and expected data. For example, suppose you roll a die 120 times. You would expect a 1:1:1:1:1:1 ratio between all the numbers that come up. In other words, you would expect to see each side 20 times. However, because this is an experiment and the sample size is only 120, you will probably see some variation in the data. If you rolled 24 sixes for example, the disagreement between observed (24) and expected (20) is small, and could have occurred by chance. If you rolled 40 sixes, however, there is a large disagreement from the expected. Maybe it is a weighted die. The region in between these two values is trickier. Where is the line drawn between expected/normal variations and results that are so different that they must tell a different story altogether?

Chi-square tests are one way to make this decision. They start with a **null hypothesis** (H_o), which represents a set numerical outcome that you want to compare your data with. The chi-square test will determine if your actual data is close to the null hypothesis outcome and could have occurred due to a fluke (i.e., rolling a six 24 times) or if it is so different that it is probably not a fluke (i.e., 40 rolled sixes) and the null hypothesis outcome must not correctly describe what is occurring. In the dice case, rolling a six 40 times would cause the null hypothesis of expecting a 1:1:1:1:1:1 to be rejected and the die can be considered to be weighted. In the calculations, expected results if the null hypothesis were true are compared to the actual values, and an x^2 value is calculated. This value is compared to a **critical value**, and a decision is made to reject or accept the null hypothesis.

For example, suppose data was collected from 100 families with three children each. Fourteen families had three female children, 36 families had two females and a male, 30 families had one female and two males, and 20 families had three male children. We would expect offspring sex to segregate independently. In other words, the sex of a given child shouldn't be affected by the sex of previous children. Let's test whether this is true for this dataset.

Since we expect sex to segregate independently, this is the null hypothesis (H_o). Each family has three children, so there are four possible outcomes for each family:

1. three daughters
2. two daughters and a son
3. one daughter and two sons
4. three sons

Using some advanced math (that you don't need to worry about), the probabilities for each of these is:

1. P(three females) = 12.5%
2. P(two females, one male) = 37.5%
3. P(one female, two males) = 37.5%
4. P(three males) = $(1)(0.50)^3$ = 12.5%

Because there are 100 families, we expect 12.5 to have three females, 37.5 to have two females and a male, 37.5 to have one female and two males, and 12.5 to have three males. This can be calculated by multiplying the probability of each event by the total (100 in this case).

Next, we compare this expected data (E) to what actually happened (the observed data, O). Note that the total numbers for observed and expected columns must be the same. To calculate an χ^2 value, the formula $(O - E)^2/E$ is calculated for each row of the table, and summed.

Family Type	Observed (O)	Expected (E)	$(O - E)^2/E$
3 females	14	12.5	0.18
2 females 1 male	36	37.5	0.06
1 female 2 males	30	37.5	1.5
3 males	20	12.5	4.5
Total	**100**	**100**	**6.24**

The calculated χ^2 value is 6.24. Next, this value needs to be compared to a critical value (CV). This is obtained from a table like this one.

	Chi-Square Table							
	Degrees of Freedom							
p-value	1	2	3	4	5	6	7	8
0.05	3.84	5.99	7.82	9.49	11.07	12.59	14.07	15.51
0.01	6.64	9.21	11.34	13.28	15.09	16.81	18.48	20.09

This chart is included on the AP Biology Equations and Formulas sheet. In order to use this table, you must know the **degrees of freedom** (DF), which represent the number of independent variables in the data. In most cases, this is the number of possibilities being compared minus one. In this example, DF = 4 − 1, because we are comparing four different family types.

You also need a reference p-value. This is the probability of observing a deviation from the expected results due to chance. For most biological tests, a p-value of 0.05 is used.

Since we are using DF = 3 and p = 0.05, the critical value is 7.82 (according to the chart above). Next, the critical value is compared to the χ^2 value; if $\chi^2 <$ CV, you accept H_o. If $\chi^2 >$ CV, you reject H_o. We calculated an χ^2 value of 6.24, which is smaller than 7.82. Since $\chi^2 <$ CV, we accept H_o and conclude that offspring sex could, indeed, be segregating independently in this dataset.

KEY TERMS

descriptive statistics
sample
population
random sampling
quantitative data
qualitative data
count data
bar graph
pie graph
normal, or parametric, data
normal curve, or distribution
sample size
mean
outliers
standard deviation
standard error
nonparametric data
histogram, or frequency diagram
median
mode
range
hypothesis

independent variable
dependent variable
contants (contolled variables)
control groups
statistical significance
time-course experiment
line graph
comparative experiments
bar graph
box-and-whisker plot
association experiments
scatterplot
probability
product rule
sum rule
t-test
p-value
chi-square test
null hypothesis
critical value
degrees of freedom

Summary

- Descriptive statistics (such as mean, standard deviation, standard error, median, mode, range) and graphs are used to summarize data and show patterns and conclusions:
  - Normal (parametric) measurement data are summarized by using mean and standard deviation or standard error.
  - Parametric data are summarized by using median, mode, and range.
  - Histograms can be used to see if a dataset has a standard deviation or not.

- Count data is summarized in a bar graph or a pie graph.

- Time course experiments look at how something changes over time; summarize these using a line graph.

- Experiments that compare groups are summarized in a bar graph (if the dataset has a standard deviation) or a box-and-whisker plot (if it does not).

- Scatterplots summarize association experiments, and regression lines can be used to determine if the relationship is linear.

- Probability can be calculated with the sum rule or the product rule.

- Hypothesis testing is used to determine if two groups are significantly different from each other. They start with a null hypothesis, which is rejected or accepted, depending on how a calculated p-value or chi-square value compares to a standard value.

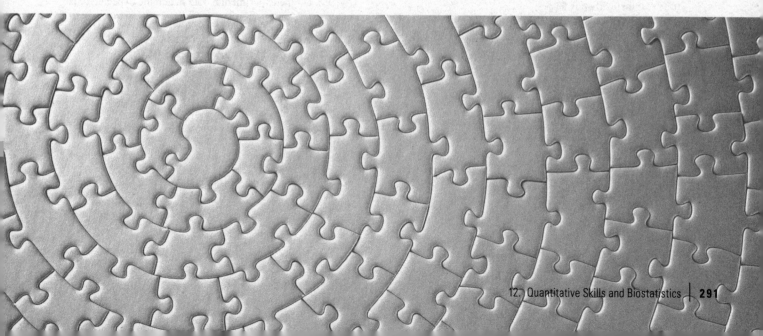

Chapter 12 Drill

Answers and explanations can be found in Chapter 15.

1. Five subjects were weighed before and after an 8-week exercise program. What is the average amount of weight lost in pounds for all five subjects, rounded to the nearest pound?

Subject	Starting Weight (pounds)	Final Weight (pounds)
1	184	176
2	200	190
3	221	225
4	235	208
5	244	225

(A) 12 pounds
(B) 13 pounds
(C) 14 pounds
(D) 15 pounds

2. The height of six trees is measured. Is plant 6 taller than the median for all six trees?

Plant	Height (inches)
1	67
2	61
3	72
4	71
5	66
6	68

(A) Yes, the median is 67.3.
(B) No, the median is 67.3.
(C) Yes, the median is 67.5.
(D) No, the median is 67.5.

3. In the following set of test scores, what is the mode and what is the range?

Test Scores: 71, 67, 75, 65, 66, 32, 69, 70, 72, 82, 73, 68, 75, 68, 75, 78

(A) Mode: 68; Range: 75
(B) Mode: 69; Range: 50
(C) Mode: 75; Range: 70.5
(D) Mode: 75; Range: 50

4. Given the cross $AaBbCc \times AaBbCc$, what is the probability of having an $AABbCC$ offspring?

(A) $\dfrac{1}{4}$

(B) $\dfrac{1}{8}$

(C) $\dfrac{1}{16}$

(D) $\dfrac{1}{32}$

5. Given the cross $AaBb \times aabb$, what is the probability of having an $Aabb$ or $aaBb$ offspring?

(A) $\dfrac{1}{2}$

(B) $\dfrac{1}{4}$

(C) $\dfrac{1}{16}$

(D) 0

6. Two pea plants are crossed, and a ratio of 3 yellow plants to 1 green plant is expected in the offspring. It is found that out of 100 plants phenotyped, 84 are yellow and 16 are green. Do the experimental data match the expected data?

(A) Yes, the χ^2 value is greater than 3.84.
(B) Yes, the χ^2 value is smaller than 3.84.
(C) No, the χ^2 value is greater than 3.84.
(D) No, the χ^2 value is smaller than 3.84.

7. A mating is set up between two pure breeding strains of plants. One parent has long leaves and long shoots. The other parent has short leaves and stubby shoots. F_1 plants are collected, and all have long leaves and long shoots. F_1 plants are self-crossed, and 1,000 F_2 plants are phenotyped. The data is as follows:

Phenotype	# of F_2
Long leaves, long shoots	382
Long leaves, stubby shoots	109
Short leaves, long shoots	112
Short leaves, stubby shoots	397
Total	1,000

Are the genes for leaf and shoot length segregating independently?

(A) Yes; the degrees of freedom are 3, and the calculated χ^2 value is small.

(B) No; the degrees of freedom are 3, and the calculated χ^2 value is large.

(C) Yes; the degree of freedom is 1, and the calculated χ^2 value is small.

(D) No; the degree of freedom is 1, and the calculated χ^2 value is large.

REFLECT

Respond to the following questions:

- Which topics from this chapter do you feel you have mastered?

- Which content topics from this chapter do you feel you need to study more before you can answer multiple-choice questions correctly?

- Which content topics from this chapter do you feel you need to study more before you can effectively compose a free response?

- Was there any content that you need to ask your teacher or another person about?

Chapter 13
Sample
Free-Response
Questions

SECTION II: FREE RESPONSE

Section II of the AP Biology Exam contains six questions, consisting of two long free-response questions and four short free-response questions. You will have a total of 90 minutes to complete all questions. The directions will look something like this (read carefully!).

> **Directions:** Questions 1 and 2 are long free-response questions that should require about 25 minutes each to answer and are worth 8–10 points each. Questions 3 through 6 are short free-response questions that should require about 10 minutes each to answer. Questions 3 through 6 are worth 4 points each.
>
> Read each question carefully and completely. Write your response in the space provided following each question. Only material written in the space provided will be scored. Answers must be written out in paragraph form. Outlines, bulleted lists, or diagrams alone are not acceptable unless specifically requested.

Remember the Questions

For a full list of the six types of questions and how they're scored, refer to page 52.

Free-Response Tips and Strategies

AP Biology Readers (the people who grade your free-response essays) expect students to interpret data and apply knowledge to new situations in the free-response questions. Some important things to keep in mind when writing your answers to the free-response questions are provided in the following list:

- **Answer the question.** Do not become distracted or waste time writing about tangential topics that you know more about than what is being asked. Focus your response.

- **Think quantitatively.** Use data whenever possible to explain your answer. This is especially necessary when the question features a graph, table, or other type of diagram (which many free-response questions now do).

- **Be specific.** Offer specific detailed examples where appropriate.

- **Use vocabulary correctly and often.** Hone your understanding of the words provided in the Key Terms lists in this book and the various biological ideas and situations to which they can be applied. This will help you use the correct terminology when writing your responses.

- **Remember the Big Ideas.** Know the four big ideas covered in the AP Biology course (page 49), as they are reflected in many free-response questions. Understanding these ideas and being able to articulate them will help you craft your responses.

Okay, now it's time for some practice. Take about 25 minutes and try to write a long free-response to the following question. This is an example of Long Question #2.

The regulation of the expression of gene products is crucial to the development of an embryo and the maintenance of homeostasis. The table below illustrates the mRNA levels, total protein levels, and total enzyme activity levels for three different enzymes. These enzymes are encoded by genes A, B, and C after being expressed in both the absence and presence of "Protein X" or "Protein Y."

	Untreated			+ Protein X			+ Protein Y		
	Gene A	Gene B	Gene C	Gene A	Gene B	Gene C	Gene A	Gene B	Gene C
mRNA Levels	1,200	2,000	1,400	2,100	250	1,400	1,200	2,000	1,400
Protein Levels	400	1,100	300	750	20	300	400	1,100	300
Total Enzyme Activity Levels	20,000	32,000	18,000	35,000	5	18,000	20,000	50,000	200
Std Error mRNA	100	132	154	172	23	120	132	186	144
Std Error Protein	43	101	22	37	2	38	20	89	21
Std Error Total Enzyme Activity	1200	1500	1000	2600	1	900	875	3200	28
Total Enzyme Activity Standard Deviation	3,000	5,200	3,200	4,000	3	2,000	2,500	800	30

a. **Describe** differences in maintaining homeostasis in unicellular and multicellular organisms.

b. **Construct** a graph of the Total Enzyme Levels for each of the 3 genes. Be sure to show the results in the presence and absence of Protein X and Protein Y.

c. **Analyze** the data and **describe** the influence of Protein X and Protein Y on the gene expression of Enzyme A and Enzyme B. Include at which point in gene expression the regulation likely occurs.

d. Another scientist repeats the experiment with Protein Y and Enzyme C, and adds a known inhibitor to Protein Y to the tube. **Predict** the effect on mRNA, protein, and total enzyme levels as a result.

SCORING GUIDELINES

To help you grade this sample free response, we've put together a checklist that you can use to calculate the number of points that should be assigned to each part of this question. We'll first explain the important points on the checklist and then give you sample student responses to show you how test reviewers would evaluate them.

Free-Response Checklist

a.   **Describe** differences in maintaining homeostasis in unicellular and multicellular organisms. **(2pts total)**
   i.   Describe homeostasis in unicellular organisms
   ii. Describe homeostasis in multicellular organisms

b.   **Construct** a graph of the Total Enzyme Levels for each of the 3 genes. Be sure to show the results in the presence and absence of Protein X and Protein Y. **(4pts total)**
   i.   Proper data selection/data placement
   ii.   Proper axes
   iii. Proper error bars
   iv.   Proper labels

c.   **Analyze** the data and **describe** the influence of Protein X and Protein Y on the gene expression of Enzyme A and Enzyme B. Include at which point in gene expression the regulation likely occurs. **(2pts total)**
   i.   Expression of Enzyme A appears to be stimulated by Protein X and is unaffected by Protein Y. The mRNA, protein, and enzyme activity increased with Protein X, so the regulation must be pre-transcriptional.
   ii. The expression of Enzyme B appears to be inhibited by Protein X (mRNA, protein, and enzyme activity reduced) so this seems like pre-transcriptional inhibition. The expression of Enzyme B appears to be stimulated by Protein Y. The mRNA and Protein Levels are the same, but the Enzyme Level is increased. This implies post-translational activation.

d.   Another scientist repeats the experiment with Protein Y and Enzyme C, and adds a known inhibitor to Protein Y to the tube. **Predict** the effect on mRNA, protein, and total enzyme levels as a result. **(2pts total)**
   i.   The mRNA and protein levels would be unchanged. The Total Enzyme Activity Levels would likely increase to near 18,000.

SAMPLE ESSAYS
The following response would earn full points.

a) Homeostasis is important in all living things. In unicellular things, temperature must be maintained and osmotic pressure must be controlled and pH must be regulated at a cellular level. These same things are important in multicellular things. In addition, a multicellular thing like a plant or an animal has to coordinate their cells to work together. Individual cells must have homeostasis, but the overall living thing must have it too. With more complex systems comes more things to regulate like blood pressure and heart rate.

b)

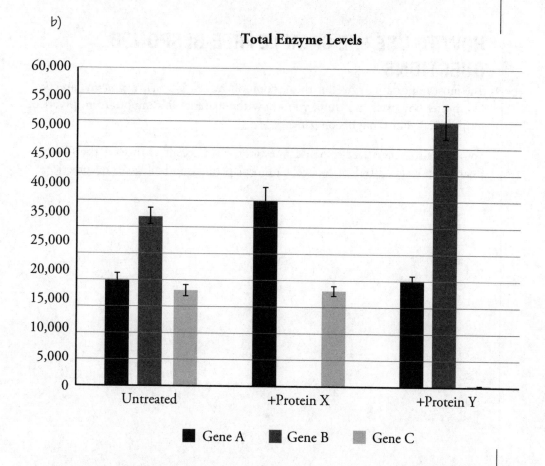

Total Enzyme Levels

c) Gene A expression has no change with the addition of Protein Y, but it has increased transcription, protein, and enzyme activity with the addition of Protein X. It is likely that Protein X is an enhancer of transcription of gene A. Perhaps it binds directly to the promoter region of gene A and assists RNA polymerase to bind. It might also remove a repressor of transcription or work together with other transcription factors. The increase in mRNA would lead to increased protein and increased enzyme activity down the road.

Gene B has a decrease in mRNA, protein, and enzyme activity with Protein X. It has increased total activity with Protein Y. It is likely that Protein X is a repressor of transcription for gene B. It might bind to the promoter of gene B or to another transcription factor to prevent RNA polymerase binding or prevent transcription from occurring in another way. Since the enzyme activity increases with Protein Y, but the mRNA and the protein levels did not change, this means that the enzyme activity was somehow enhanced on the same amount of enzyme that was present without Protein Y. Protein Y might be an enzyme activator of Enzyme B. It might bind to Enzyme B directly or maybe the protein phosphorylates it or cleaves it to make the enzyme active.

d) It looks like Enzyme C is unaffected by Protein Y at the mRNA and protein levels, but it looks like the enzyme activity is turned off. With an inhibitor of Protein Y the enzyme level would likely return to the untreated level of arond 18,000. The mRNA and the protein would not change with the inhibitor of protein Y.

HOW TO USE THE SAMPLE FREE-RESPONSE QUESTIONS

The most important advice for this section of the test is *practice, practice, practice!* No matter how well you think you know the material, it's important to practice formulating your thoughts on paper.

You can also access free-response questions from recent exams on the College Board website at apstudent.collegeboard.org/apcourse/ap-biology/exam-practice.

Summary

- The free-response section features six questions, two of which require long responses. You will have 90 minutes to complete the entire section.

- Each question is worth a certain number of points. The long-response questions are worth 8–10 points, and the short-response questions are worth 4 points each.

- Your responses must be written in paragraph form; outlines are not acceptable. Diagrams can be used to supplement your writing but will not receive credit on their own (unless otherwise specified).

- Remember to keep your response focused and use specific data and examples (where applicable) to justify your answers.

- Familiarize yourself with the Key Terms in this book as well as the Big Ideas for the AP Biology course. Understanding these will help you write clear, focused, and accurate responses.

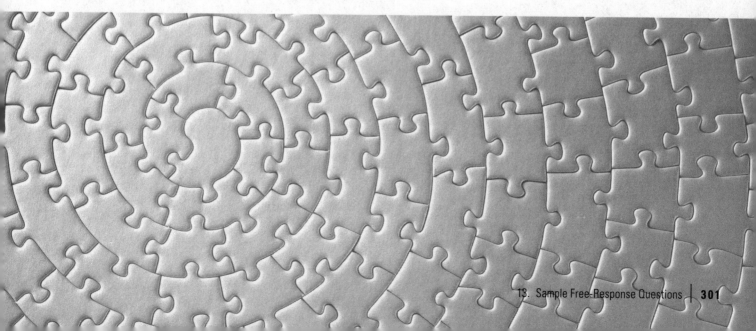

REFLECT

Respond to the following questions:

- Which topics from this chapter do you feel you have mastered?

- Which content topics from this chapter do you feel you need to study more before you can effectively compose a free response?

- Was there any content that you need to ask your teacher or another person about?

Chapter 14
Laboratory

All AP Biology courses have a laboratory component that gives students hands-on experience regarding some of the biology topics covered in class. Through these laboratory exercises, you can learn the scientific method, lab techniques, and problem-solving skills. There is not an official set of labs that must be completed, and AP teachers get to choose the ones for their classroom. This means that you will not be tested on memorizing specific protocols or data from the labs. However, all labs teach the same basic skills of critical thinking, developing hypotheses, judging results, making conclusions, adapting experiments, and understanding variables. So, those are the skills that you are expected to fully understand. The exam will likely present you with example experiments and you will need to make decisions and answer questions about them. There are 13 labs that are very popular to do in AP Biology classes, and these are the same labs that often come up on the AP test. We have given you short summaries of those 13 labs. This way, if you encounter a question about one of these labs, you will be one step ahead of students that have never even heard of it before.

LAB 1: ARTIFICIAL SELECTION

This lab explores artificial selection—the process by which humans decide traits to enhance or diminish in other species by crossing individuals with the desired phenotype. You need to understand the basic principles of natural selection and how natural selection drives evolution, listed below:

- Variation is present in any population.

- Natural selection, or differential reproduction in a population, is a major mechanism in evolution.

- Natural selection acts on phenotypic variations in populations. Some organisms will have traits that are more favorable to the environment and will survive to reproduce more than other individuals, causing the genetic makeup of the population to change over time.

This lab specifically deals with Wisconsin Fast Plants. In order to artificially select for certain traits, plants with the desired traits can be crossed, changing the genetic makeup of the population. For example, if you wanted to select for height, you could cross only the tallest plants with one another. The new population will have a higher mean height than the previous generation.

LAB 2: MATHEMATICAL MODELING: HARDY-WEINBERG

In order to understand everything you need for the AP Biology Exam, you need to be able to use the Hardy-Weinberg principles and equations to determine allele frequencies in a population. One way to study evolution is to study how the frequencies of alleles change from generation to generation:

- Know how to calculate the allele and genotype frequencies using the two Hardy-Weinberg equations: $p + q = 1$ and $p^2 + 2pq + q^2 = 1$. Don't forget: if the population obeys Hardy-Weinberg's rules, these frequencies remain constant over time.

- Review the discussion of the Hardy-Weinberg principle in this book. Know the five conditions of the Hardy-Weinberg equilibrium: (1) large population, (2) no mutations, (3) no immigration or emigration, (4) random mating, and (5) no natural selection.

- Review natural selection and how it can lead to changes in the genetic makeup of a population.

LAB 3: COMPARING DNA SEQUENCES TO UNDERSTAND EVOLUTIONARY RELATIONSHIPS WITH BLAST

This laboratory uses BLAST (Basic Local Alignment Search Tool), a database that allows you to input a DNA sequence for a gene to look for similar or identical sequences present in other species. In order to understand everything you need for the AP Exam you need to be able to do the following:

- Look at a phylogenic tree, which is a way to visually represent evolutionary relatedness. Endpoints of each branch correspond to a specific species, and each junction on the tree represents a common ancestor. Species that are closer on a phylogenic tree are more closely related.

- Understand BLAST scoring. Most phylogenic trees are constructed by examining nucleotide sequences; the more identical two species' sequences are for a specific gene, the more closely related they are. BLAST is able to analyze different sequences to tell you how similar they are to one another based on a score. The higher the score, the closer the two sequences align.

LAB 4: DIFFUSION AND OSMOSIS

This lab investigates the process of diffusion and osmosis in a semipermeable membrane as well as the effects of solute concentration and water potential on these processes.

What are the general concepts you really need to know?

Fortunately, this lab covers the same concepts about diffusion and osmosis that are discussed in this book. Just remember that osmosis is the movement of water across a semipermeable membrane, from a region of high water concentration to one of low water concentration, or from a hypotonic region (low solute concentration) to a hypertonic region (high solute concentration).

- Be familiar with the concept of water potential, which is simply the free energy of water. It is a measure of the tendency of water to diffuse across a membrane. Water moves across a selectively permeable membrane from an area of higher water potential to an area of lower water potential.

- Be familiar with the effects of water gain in animal and plant cells. In animals, the direction of osmosis depends on the concentration of solutes both inside and outside the cell. In plants, osmosis is also influenced by turgor pressure—the pressure that develops as water presses against a cell wall. If a plant cell loses water, the cell will shrink away from the cell wall and plasmolyze.

Another important concept to understand is the importance of surface area and volume in cells. There are several questions about each of these topics, but the necessary formulas are listed on the AP Biology Equations and Formulas sheet. Cells maintain homeostasis by regulating the movement of solutes across the cell membrane. Small cells have a large surface area-to-volume ratio; however, as cells become larger, this ratio becomes smaller, giving the cell relatively less surface area to exchange solutes. A cell is limited in size by the surface area-to-volume ratio. There are many organisms that have evolved strategies for increasing surface area, like root hairs on plants and villi in the small intestines of animals.

LAB 5: PHOTOSYNTHESIS

The chemical equation for photosynthesis is:

$$6CO_2 + 6H_2O \rightarrow C_6H_{12}O_6 + 6O_2$$

Because plants consume some of this energy during photosynthesis, measuring the oxygen produced by a plant can tell us about the net photosynthesis that is occurring. In this laboratory, photosynthesis rates are measured by using leaf discs that begin to float as photosynthesis is carried out, allowing you to see that photosynthesis is occurring.

There are several properties that affect the rates of photosynthesis, including:

- light intensity, color, and direction
- temperature
- leaf color, size, and type

Be able to hypothesize about the effects of these variables: for example, as light intensity increases, so does the rate of photosynthesis. Remember, both plants and animals contain mitochondria and carry out cell respiration!

LAB 6: CELLULAR RESPIRATION

In this lab, the respiratory rate of germinating and nongerminating seeds and small insects is investigated. The equation for cellular respiration is:

$$C_6H_{12}O_6 + O_2 \rightarrow 6CO_2 + 6H_2O$$

Germinating seeds respire and need to consume oxygen in order to continue to grow. Non-germinating seeds do not respire actively. In this lab, the amount of oxygen consumed by these types of seeds is measured with a respirometer. The experiment is also conducted at two temperatures, 25°C and 10°C, because seeds consume more oxygen at higher temperatures. You should know the following:

- Oxygen is consumed in cellular respiration.

- Germinating seeds have a higher respiratory rate than non-germinating seeds, which have a very low respiratory rate.

- Know how to design a study to determine the effect of temperature on cell respiration.

- Know the significance of a control using glass beads. A control is a condition held constant during an experiment. In this case, glass beads are used as a control because they will not consume any oxygen.

LAB 7: MITOSIS AND MEIOSIS

This lab highlights the differences between mitosis and meiosis. In this lab, slides of onion root tips are prepared to study plant mitosis. The important information and skills to review in this lab include the following:

- Mitosis produces two genetically identical cells, while meiosis produces haploid gametes.

- Cell division is highly regulated by checkpoints that depend, in part, on complexes of proteins called cyclins with other proteins called cyclin-dependent kinases. One such example is the mitosis promoting factor (or MPF) that has its highest concentration during cell division and is thought to usher a cell into mitosis.

- Nondisjunction, or the failure of chromosomes to separate correctly, can lead to an incorrect number of chromosomes (too many or too few) in daughter cells.

- Know what each phase of the cell cycle looks like under a microscope. Chapter 8 contains diagrams of each phase.

In one section of this lab, the sexual life cycle of the fungus *Sordaria fimicola* is examined. Sexual reproduction in this fungus involves the fusion of two nuclei—a (+) strain and a (–) strain—to form a diploid zygote. This zygote immediately undergoes meiosis to produce asci, which contain eight haploid spores each.

- During meiosis, crossing-over can occur to increase genetic variation. If crossing-over, or recombination, has occurred, different genetic combinations will be observed in the offspring when compared with the parent strain.

- These offspring with new genetic combinations are called recombinants. By examining the numbers of recombinants with the total number of offspring, an estimate of the linkage map distance between two genes can be calculated with the following equation.

Map distance (in map units) = [(# recombinants) / (# total offspring)] × 100.

LAB 8: BIOTECHNOLOGY: BACTERIAL TRANSFORMATION

In this lab, the principles of genetic engineering are studied. Biotechnologists are able to insert genes into an organism's DNA in order to introduce new traits or phenotypes, like inserting genes into a corn genome that help the crops ward off pests. This process is very complicated in higher plants and animals, but relatively simple in bacteria. You are responsible for knowing the ways in which bacteria can accept fragments of foreign DNA:

- Conjugation: the transfer of genetic material between bacteria via a pilus, a bridge between the two cells

- Transformation: a process in which bacteria take up foreign genetic material from the environment

- Transduction: a process in which a bacteriophage (a virus that infects bacteria) transfers genetic material from one bacteria to another

In addition, DNA can also be inserted into bacteria by using plasmids, which are small, circular DNA fragments that can serve as a vector to incorporate genes into the host's chromosome. Plasmids are key elements in genetic engineering. The concepts you need to know about plasmids include the following:

- One way to incorporate specific genes into a plasmid is to use restriction enzymes, which cut foreign DNA at specific sites, producing DNA fragments. A specific fragment can be mixed together with a plasmid, and this recombinant plasmid can then be taken up by *E. coli*.

- Plasmids can give a transformed cell selective advantage. For example, if a plasmid carries genes that confer resistance to an antibiotic like ampicillin, it can transfer these genes to the bacteria. These bacteria are then said to be transformed. That means if ampicillin is in the culture, only transformed cells will grow. This is a clever way scientists can find out which bacteria have taken up a plasmid.

- In order to make a bacteria cell take up a plasmid, you must (1) add $CaCl_2$, (2) heat shock the cells, and (3) incubate them in order to allow the plasmid to cross the plasma membrane.

LAB 9: BIOTECHNOLOGY: RESTRICTION ENZYME ANALYSIS OF DNA

This laboratory introduces you to the technique of gel electrophoresis. This technique is used in genetic engineering to separate and identify DNA fragments. You need to know the steps of this lab technique for the AP exam:

1. DNA is cut with various restriction enzymes.

2. The DNA fragments are loaded into wells on an agarose gel.

3. As electricity runs through the gel, the fragments move according to their molecular weights. DNA is a negatively charged molecule; therefore, it will migrate toward the positive electrode. The longer the DNA fragment, the slower it moves through the gel.

4. The distance that each fragment has traveled is recorded.

Restriction mapping allows scientists to distinguish between the DNA of different individuals. Since a restriction enzyme will cut only a specific DNA sequence, it will cause each individual to have a unique set of fragments called restriction fragment length polymorphisms, or RFLPs for short. This technology is used at crime scenes to help match DNA samples to suspects.

LAB 10: ENERGY DYNAMICS

This lab examines energy storage and transfer in ecosystems. Almost all organisms receive energy from the sun either directly or indirectly:

- Producers, or autotrophs, are organisms that can make their own food using energy captured from the sun. They convert this into chemical energy that is stored in high-energy molecules like glucose.

- Consumers, or heterotrophs, must obtain their energy from organic molecules in their environment. They can then take this energy to make the organic molecules they need to survive.

- Biomass is a measure of the mass of living matter in an environment. It can be used to estimate the energy present in an environment.

LAB 11: TRANSPIRATION

This lab investigates the mechanisms of transpiration, the movement of water from a plant to the atmosphere through evaporation. What do you need to take away from this lab?

- The special properties of water that allow it to move through a plant from the roots to the leaves include polarity, hydrogen bonding, cohesion, and adhesion.

- Know the vascular tissues that are involved in transport in plants. Xylem transports water from roots to leaves, while phloem transports sugars made by photosynthesis in the leaves down to the stem and roots.

- Stomata are small pores present in leaves that allow CO_2 to enter for photosynthesis and are also a major place where water can exit a plant during transpiration.

LAB 12: FRUIT FLY BEHAVIOR

In this lab, fruit flies are given the choice between two environments by using a choice chamber, which allows fruit flies to move freely between the two environments. Typically, fruit flies prefer an environment that provides either food or a place to reproduce. They also respond to light and gravity:

- Taxis is the innate movement of an organism based on some sort of stimulus. Movement toward a stimulus is called positive taxis, while movement away from a stimulus is negative taxis. In this lab, fruit flies exhibited a negative gravitaxis (or a movement opposite to the force of gravity) and positive phototaxis (a movement toward a light source).

LAB 13: ENZYME ACTIVITY

This lab demonstrates how an enzyme catalyzes a reaction and what can influence rates of catalysis. In this lab, the enzyme peroxidase is used to catalyze the conversion of hydrogen peroxide to water and oxygen.

$$H_2O_2 + peroxidase \rightarrow 2H_2O + O_2 + peroxidase$$

The following are the major concepts you need to understand for the AP exam:

- Enzymes are proteins that increase the rate of biological reactions. They accomplish this by lowering the activation energy of the reaction.

- Enzymes have active sites, which are pockets that the substrates (reactants) can enter that are specific to one substrate or set of substrates.

- Enzymes have optimal temperature and pH ranges at which they catalyze reactions. Enzyme concentration and substrate concentration can also influence the rates of reaction.

- If you're asked to design an experiment to measure the effect of these four variables on enzyme activity, keep all the conditions constant except for the variable of interest. For example, to measure the effects of pH in an experiment, maintain the temperature, enzyme concentration, and substrate concentration constant as you change pH.

REFLECT

Respond to the following questions:

- Which topics from this chapter do you feel you have mastered?

- Which content topics from this chapter do you feel you need to study more before you can answer multiple-choice questions correctly?

- Which content topics from this chapter do you feel you need to study more before you can effectively compose a free response?

- Was there any content that you need to ask your teacher or another person about?

Chapter 15
Chapter Drill
Answers and
Explanations

CHAPTER 4 DRILL: CHEMISTRY OF LIFE

1. **C** Water has many unique properties that favor life, including (A), a high specific heat, (B), high surface tension and cohesive properties, and (D), high intermolecular forces due to hydrogen bonding. However, water is a very polar molecule and is an excellent solvent, making (C) inaccurate and the correct answer.

2. **B** The pH scale is logarithmically based, meaning that each difference of 1 on the log scale is indicative of a tenfold difference in the hydrogen (H^+) ion concentration. According to the question, the pH values of the cytoplasm of cells and gastric juices are approximately 7.4 and 1.5, respectively. Therefore, the pH values vary by 6, and the hydrogen (H^+) ion concentrations must vary by 6 orders of ten (10^6), or 1,000,000-fold. Since lower pH values have more hydrogen ions, they are also more acidic.

3. **C** All amino acids share a carboxylic acid group, COOH, labeled (B), and an amino-group, NH_2, labeled (D). They also share a hydrogen atom bound to the central carbon (A). Differences in amino acids are defined by variations in the fourth position called the R-group, labeled (C).

4. **B** The hypothesis that life may have arisen from formation of complex molecules from the primordial "soup" of Earth is not supported by the absence of nucleic acids. All life is DNA-based, yet no nucleic acid molecules were detected. The presence of carbon molecules, amino acids, and sugars, which are common compounds and compose life, supports the hypothesis, so eliminate (A), (C), and (D). Choice (B) is correct.

5. **C** The amino acids cysteine and methionine contain the element sulfur (as indicated by the *S* in the amino acid structure shown). However, no sulfur-based compounds were included in the Miller-Urey experiment, so it was impossible to form these two amino acids under the conditions of the experiment.

6. **B** Silica is a mineral form of glass, is not a common component of life-forms, and is largely chemically inert. Since oxygen is already present in several compounds included in the experiment, the addition of this compound does not provide any additional elements or chemical substrates, which would permit generation of additional amino acids or synthesis of nucleic acids. The addition of sulfur compounds and phosphorus is necessary to generate some amino acids and all nucleic acids, which eliminates (A), (C), and (D).

7. **D** Hydrolysis adds water to a polymer to break the linkages between monomers. Therefore, (A), free water, will not result because water is being broken down in the reaction. Free water would result from the opposite reaction, for condensation reactions. Choice (B), adenine, results from the hydrolysis of DNA, as it is a monomer of DNA. Choice (C), cholesterol, is a steroid and therefore does not undergo hydrolysis. Dipeptides (D) could result from the hydrolysis of proteins, which are composed of polypeptides.

8. **B** Phospholipids have a long fatty acid tail and a polar head group. The fatty acid tails associate in the membrane, and the phospholipid head group associates with water at the boundary of the

membrane. Nucleotides, (A), are mostly polar and found in DNA and RNA. Water, (C), is polar and not found in plasma membranes. Amino acids, (D), are zwitterionic, that is, they have both a positive and a negative charge. Free amino acids are not found in the membrane.

9.   **C** $(CH_2O)_3 = C_{18}H_{36}O_{18}$, as in (A), but the dehydration reactions to produce two glycosidic bonds between the three monosaccharides would remove two H_2O molecules, or four hydrogens and two oxygens. Therefore, the correct answer is (C).

10. **A**   All isotopes of hydrogen will contain one proton, by definition. All atoms that have only one proton are identified as hydrogen. Tritium is an isotope that has one proton and two neutrons, for a total of three particles in its nucleus. Choice (A) is correct. Choice (B) is not correct because two protons would mean that the element is helium, not hydrogen. Choice (C) is not correct because the atomic number, one, will never change without changing the identity of the atom. Finally, (D) is not correct because radioactive atoms do not give off electrons.

CHAPTER 5 DRILL: CELL STRUCTURE AND FUNCTION

1.   **C**   Active transport is the movement of substances across a membrane against their concentration gradient through the use of energy (ATP). The sodium-potassium pump is a critical structure which uses ATP hydrolysis to move sodium and potassium ions against their respective concentration gradients. Diffusion of oxygen is an example of simple diffusion and can occur without the need of channels or pores. Water uses aquaporins to travel across the membrane, down its concentration gradient; therefore, it also does not require energy. Movement of sodium ions by a voltage-gated ion channel, (B), is an example of facilitated diffusion. Because the sodium ions are still undergoing diffusion from high to low concentrations without the need of ATP, this does not represent a form of active transport.

2.   **C**   Bacterial cells can be visualized using light microscopy. In fact, back in the 17th century, some of the earliest studies using primitive microscopes recorded the shape and organization of bacteria. Choices (A), (B), and (D) are incorrect because virus and cell organelle structures are too small to be observed using light microscopy and require electron microscopy.

3.   **D**   The cell wall is a structure that is present in bacteria but absent in animal cells. Consequently, this structure is targeted by several leading classes of antibiotics and would be an effective target of therapeutics against *V. cholerae*. Cytoplasm, plasma membrane, and ribosomes—(A), (B), and (C), respectively—are all structures that are present in animal cells.

4.   **C**   Based on the microscopy data, the organism has a cell wall and lacks mitochondria. The presence of a cell wall suggests that the organism is not a protozoan, so eliminate (B). The absence of mitochondria is indicative of prokaryotic structure because they lack organelles, eliminating (A) and (D). The most likely conclusion is that this organism is a bacterial species.

5. **B** Although many of the mentioned organelles work closely together, the only choice that can be correct is (B). Ribosomes translate and manufacture proteins, often on the rough endoplasmic reticulum. Those proteins are then packaged in the Golgi bodies into vesicles, which are then fused with the plasma membrane. Choice (A) is incorrect because the nuclear envelope and nucleolus do not interact with vacuoles; they interact with the centrioles only during mitosis. Choice (C) is incorrect because ribosomes don't often interact with mitochondria, and lysosomes interact only with mitochondria and chloroplasts when they degrade those organelles. Choice (D) is incorrect because the nucleolus doesn't interact with the smooth endoplasmic reticulum, lysosomes, or centrioles.

6. **C** Estrogen is said to be lipid-soluble. This means it can slip through the membrane and bind to an intracellular receptor. A noncompetitive inhibitor would bind to an estrogen receptor, reducing the effectiveness of this binding. Choice (A) is incorrect because estrogen binds to intracellular receptors, not receptors at the plasma membrane. Choice (B) is incorrect because testosterone and estrogen have different effects, but testosterone would not change estrogen effectiveness. Choice (D) is also incorrect because wiping out the ovaries would eliminate, rather than reduce, estrogen levels.

7. **B** You would need a cell that makes and secretes a lot of protein or hormones since the ER moves things out of the cell. Bacteria, (C), do not have organelles. Choice (D), blood cells, is not the best answer because leukocytes do not secrete large amounts of hormones or proteins. Choice (A), neurons, is a potential answer, but (B) is the best answer because the pancreas is a gland that secretes copious amounts of protein and hormones, such as insulin.

8. **A** Choice (A) is correct because contractile vacuoles expel water that accumulates in cells in a hypotonic environment. The strategy described in (B) is wrong because it would increase water intake in *Paramecium*. Choice (C) is true for *Paramecium* but does not help with water gain. Choice (D) would help it find even more dangerously hypotonic environments.

9. **B** ATP is consumed by the Na^+K^+ pump, so (B) is correct. The sodium-potassium pump actively pumps both ions. A cotransporter requires one to be passive and one to be active. Choice (A) is not true. ATP is hydrolyzed and not produced. Choice (C) is not true. Choice (D) is incorrect since ATP is hydrolyzed.

CHAPTER 6 DRILL: CELLULAR ENERGETICS

1. **B** The Krebs cycle occurs primarily in the matrix of the mitochondria. The inner membrane and the intermembrane space—(A) and (C), respectively—are used in oxidative phosphorylation.

2. **D** Inhibitor Y is binding at a site outside the active site and is inducing a conformational change in the enzyme structure. By binding outside the active site, it must be an allosteric inhibitor, eliminating (A) and (C). Because the inhibitor is binding outside the active site, it is not competing with the substrate for binding, so it is considered a noncompetitive inhibitor.

3. **B** Enzymes are biological catalysts, which lower the activation energy (the energy threshold that must be met to proceed from reactant to product). The reaction coordinate diagram must reflect a decrease in the activation energy, eliminating (C). Furthermore, the enzyme does not alter the energy of the reactants or products, eliminating (A) and (D).

4. **B** Based on the pathway provided, consumption of one glucose and two ATP results in production of four ATP. In other words, each glucose results in a net gain of two ATP. Therefore, two glucose molecules would result in a net gain of four ATP.

5. **B** In fermentation, pyruvic acid is converted into either ethanol or lactic acid. During this process, NADH is recycled into NAD^+.

6. **B** Glycolysis results in the production of ATP (energy), so it is considered an exergonic process.

7. **B** Thermophilic bacteria live in hot environments. Also, a DNA polymerase replicates DNA. Therefore, in the PCR technique, the stage in which DNA is elongated by a DNA polymerase is part three of the cycle, at 72°C. Therefore, bacteria growing in hydrothermal vents between 70–75°C is the answer. The conditions described in (A) and (D) do not reflect the data, which focuses on temperature. Choice (C) describes hot springs, but the temperature does not reflect the enzyme activity most likely for *Taq* polymerase.

8. **D** The free energy does not change between catalyzed and uncatalyzed reactions; therefore, (A) is incorrect. Choice (B) is a correct statement, but it does not answer the question. Choice (C) is incorrect because enzymes catalyze reactions.

9. **B** ATP production will increase in the treated mitochondria because the low pH provides more H^+ ions in the solution. Also, oxygen provides a terminal electron acceptor for oxidative phosphorylation.

10. **D** Choices (A) and (B) are clearly wrong, as they defy physics. Choice (C) is not correct because the anabolic and catabolic reactions are not necessarily equal nor are they direct opposites. Choice (D) is the correct answer because it explains the increase in entropy. Even though organisms build and develop as ordered systems, heat is lost continuously. Additionally, organisms exhale gases and produce waste products that balance the effect of order.

CHAPTER 7 DRILL: CELL COMMUNICATION AND CELL CYCLE

1. **C** During anaphase, the chromatids are separated by shortening of the spindle fibers. Chemically blocking the shortening of these fibers would arrest the cell in metaphase. The cells are arrested in metaphase as indicated by the alignment of the chromosomes in the center of the cell and their attachment to spindle fibers, eliminating (A) and (D). The chromosomes still seem to be attached to the fibers, so there doesn't appear to be dissociation of the fibers, eliminating (B).

2. **C**   The synthesis, or S phase, of the cell cycle represents the step in which the genetic material is duplicated. The only phase labeled in the experiment that represents an increase is phase B. Based on the time scale on the x-axis, this phase lasts approximately 30 minutes.

3. **D**   Anaphase represents the cell division stage of the cell cycle and would be the phase that occurs right before the amount of genetic material should decrease. Phase D is the phase right before the genetic material would drop, so (D) is the correct answer.

4. **D**   Choices (A) and (B) do not occur, but if they did, they would not impact the gametes. Choice (C) is not likely to occur because the mitotic spindles attach only to kinetochore protein complexes at the centromere. Therefore, (D) is the answer.

5. **B**   Both (A) and (B) have sperm or ovum as the first type of cell. Because these are both gametes, they will both have half of the DNA of other types of cells. Choices (C) and (D) can thus be eliminated. Muscles and neurons are both terminally differentiated in G_0 arrest. However, liver and taste buds potentially differ. Although liver cells can divide, taste buds divide at a higher rate, so they are most likely to be in G_2 phase. Therefore, the answer is (B).

6. **A**   Nondisjunction results in major changes to the genome, so (B) and (C) can be eliminated. Deletion of an enhancer, (D), could affect the gene expression and then phenotype. However, in (A), translocation may not produce any effect, as the same information exists in the genome.

CHAPTER 8 DRILL: HEREDITY

1. **D**   The father and mother are both AB blood type. Since neither parent has a recessive allele, it is impossible for their child to be O blood type.

2. **B**   When the phenotype associated with two traits is mixed, this is considered an example of incomplete dominance. In this case, neither red nor blue is dominant, and the resulting progeny exhibited a mixture of the traits (purple).

3. **B**   Crossing the pea plant that is heterozygous for both traits (*TtGg*) with a plant that is recessive for both traits (*ttgg*) results in the following possible combinations, each of which should occur 25 percent of the time: *TtGg* (tall and green), *Ttgg* (tall and yellow), *ttGg* (short and green), and *ttgg* (short and yellow). Using the rules of probability, there is a 1/2 likelihood of it being tall and also a 1/2 likelihood of it being yellow. Multiply them together to get 1/4.

4. **C**   Because the woman is a carrier, she must have one normal copy of the X chromosome and one diseased copy. Since the boy will receive an X chromosome from his mother, there is a 50 percent chance that he will receive a diseased copy. Because he doesn't have a second X chromosome, he must have the disease if he receives the diseased X chromosome.

5. **D** They both must have a normal copy of the X chromosome. It is possible that the woman may be affected with hemophilia; however, that scenario is extremely unlikely because such a case would require two diseased copies of the X chromosome.

6. **D** Essentially, transmission of hemophilia to a girl born with Turner syndrome would be very similar to the conditions by which a boy would receive the disease. Both boys and girls with Turner syndrome have only one copy of the X chromosome. Therefore, they both would have a 50 percent chance of receiving the diseased copy of the X chromosome.

7. **B** Choice (A) is true, but it doesn't explain why the parents are normal. Choice (B) is a better explanation. Carriers of recessive genetic diseases frequently do not have a recognizable phenotype because one normal allele provides enough functioning protein to avoid the ill effects of the disease allele. Choice (C) is also possible, but (B) is still the better answer. Finally, (D) is not correct because CLN3 is an autosomal gene.

8. **C** Choice (C) describes a testcross that will actually let the breeder know if the male is heterozygous. If any white-spotted pups result from that cross, then the male would have contributed a spotted allele. Choice (B) describes a male that is homozygous for the spot allele, so it's incorrect. Choice (A), stop breeding Speckle, is a way to remove spots from the line, but it will not help you identify males with the spot allele. Choice (D) also will not help you determine which black Labrador males have a spot allele.

9. **A** Choice (B) does not apply to this question. Choices (C) and (D) refer to a gene not following Mendel's Law of Dominance. Choice (A)—genes are linked on the same chromosome—describes a normal reason that certain traits would not follow Mendel's Law of Independent Assortment.

10. **A** Choices (B) and (C) are unlikely to be true and can be eliminated. Choice (D) is true but does not answer the question. Choice (A) is the easiest and most probable explanation.

CHAPTER 9 DRILL: GENE EXPRESSION AND REGULATION

1. **B** Okazaki fragments are generated during DNA replication when the DNA polymerase must create short DNA segments due to its requirement for 5' to 3' polymerization. Since the newly discovered yeast cell has 3' to 5' activity, there would be no lagging strand and likely no Okazaki fragments.

2. **C** Since the gene is much shorter than expected, a stop codon must have been introduced by mutagenesis. This is an example of a nonsense mutation.

3. **D** The order for DNA replication is helicase, RNA primase, DNA polymerase, and ligase.

4. **C** If an mRNA codon is UAC, the complementary segment on a tRNA anticodon is AUG.

5. **D** During post-translational modification, the polypeptide undergoes a conformational change. Choices (A), intron excision, and (B), a poly(A) addition, are examples of post-transcriptional modifications. Formation of peptide bonds (C) occurs during translation, not afterward.

6. **B**   If 21 nucleotides compose a sequence and 3 nucleotides compose each codon, there would be 7 codons and thus a maximum of 7 amino acids.

7. **A**   Choice (B) is incorrect because bacteria make membrane proteins that reside on the plasma membrane. Choice (C) is also incorrect because bacteria use transcription factors. Choice (D) is possible, but (A) is the best answer.

8. **C**   Choice (A) is likely incorrect. Choice (B) would be deleterious for the cell. Choice (D) is also incorrect, as DNA replication occurs only during or preceding binary fission.

9. **B**   Viruses do not have their own ribosomes, flagella, or independent metabolism, so (A), (C), and (D) are incorrect. The only possible answer is (B).

10. **C**   Transformation occurs when bacteria take up DNA from their surroundings. The pathogenic bacteria did not come alive again, as suggested by (A). Protein was not the transforming agent, so (B) is incorrect. Finally, (D) does not make sense, as genes cannot turn into other genes simply by being in a different cell context.

11. **C**   If there are DNA/RNA fragments, then helicase, (A), is present and has opened the strands. RNA primase, (D), is also present because RNA primers have been added. DNA polymerase, (B), must be present because there are small chunks and long chunks. If there was not polymerase, then there wouldn't be small chunks because nothing new would be made. It is likely that DNA ligase (C) is missing and couldn't attach the short Okazaki fragments and the longer leading strand fragments.

CHAPTER 10 DRILL: NATURAL SELECTION

1. **D**   Mammals and cephalopods developed similar eye structures independently due to similar selective pressures. This is an example of convergent evolution.

2. **B**   The data provided show a transition toward one extreme (black) and away from another (white). This is an example of directional selection.

3. **B**   Based on the data, the number of white-bodied pepper moths decreased between 1802 and 1902, and the number of black-bodied pepper moths increased during the same period.

4. **B**   Longtail moths were included in the experiment as a control to compare the effects that are not associated with color.

5. **B**   If the color of ash or soot produced by the Industrial Revolution were white or light gray, this would likely reverse the trend observed, applying additional selection against the black moths.

6. **B**   The mole rats live in the same location, which means there is not a geographic barrier so (A) is incorrect. They do not attempt to breed, so (C) and (D) are incorrect. They have formed two separate species in the same geographic area, so (B) is correct.

7. **D** Choice (A) is incorrect because this is a small population, so genetic drift is likely. Choice (B) is incorrect because this situation does not describe convergent evolution, and the second part of the answer choice does not accurately describe the change of the population. Choice (D) is correct because the number of people that are homozygous recessives will likely become fewer as the red-head allele is mixed with the more dominant hair colors.

8. **D** Choice (D) is the answer because it is the only choice that does not predict evolution occurring on the island, thus supporting a claim for Hardy-Weinberg equilibrium. Choices (A), (B), and (C) all prevent a Hardy-Weinberg equilibrium and predict evolution. The population is small, mutations are inevitable, and humans usually do not mate randomly. This population will definitely evolve or undergo genetic drift.

9. **A** Choice (A) is correct because the trait is being selected based on female mating preferences. Longer, bigger tails indicate reproductive fitness. Peahens choose males who are healthy and strong enough to grow these big tails and will hopefully produce the best offspring. Choices (B), (C), and (D) are incorrect because they do not describe the female mating preferences.

CHAPTER 11 DRILL: ECOLOGY

1. **D** The behavior displayed by the chimpanzee represents insight because the chimpanzee has figured out how to solve the problem without external influence or learning.

2. **A** Viruses would display reproductive strategies most similar to r-strategists because they aim to reproduce as fast as possible and create as many progeny as possible in order to increase their odds of transmission to other hosts.

3. **C** They would be considered secondary consumers because they consume the primary consumers (plankton), which consume algae (the producers).

4. **C** Since brook trout can tolerate pH values as low as 4.9 and do not appear to diminish, the pH of the creek must exceed 4.9, so we can eliminate (D). Since crayfish cannot tolerate pH levels lower than 5.4, the pH of the stream must have dropped lower than this value. Only (C) falls in this range.

5. **C** Only acid rain would directly explain why the pH would drop in the creek over the time period.

6. **D** Based on Table 1, the pH must exceed 6.1 for snails to be able to return to the creek ecosystem.

7. **D** Choices (A) and (B) refer to light and gravity responses of plants. Choice (C) is not an actual tropism—it is a made-up word. Thigmotropism is the term for the way in which plants respond to touch. The correct answer is (D).

8. **D** If (D) is a correct statement, then simply cutting out beef and lamb from your diet can dramatically decrease your carbon footprint—even if you do not decide to eat vegan. Choices (A), (B), and (C) are all true statements with respect to reducing the carbon footprint of food.

9. **B** A community that has been destroyed and then rebuilds is a result of secondary succession. Choice (A) would be the answer only if there were no life present before the new growth. Choice (C) is true only of a mature forest. Choice (D) is incorrect, as the species taking over the field is native to the rainforests on the periphery of the field. Nothing in the question suggests invasive species. Therefore, (B) is correct.

10. **A** If the intestines are occupied with beneficial bacteria, there will not be any room available for pathogenic bacteria to grow by density-dependent limitations on the population. Choice (B) is not accurate because pathogenic bacteria do not require a niche prepared for them by probiotic bacteria. Choice (C) is true, but it does not answer the question. Choice (D) is also incorrect, leaving (A) as your answer.

CHAPTER 12 DRILL: QUANTITATIVE SKILLS AND BIOSTATISTICS

1. **A** In order to answer this question, you must first calculate how much weight each subject lost and then divide by the number of subjects (in this case, five).

Subject	Starting Weight (pounds)	Final Weight (pounds)	Weight Lost (pounds)
1	184	176	8
2	200	190	10
3	221	225	−4
4	235	208	27
5	244	225	19

Note that subject 3 *gained* four pounds. Total weight lost is 60 pounds (remember to subtract 4 pounds for subject 3, not add), divided by 5 subjects is 12 pounds. The average weight lost is 12 pounds ((A) is correct).

2. **C** Plant 6 is 68 inches. All the answer choices list median values smaller than this, so the answer must start with "Yes" (eliminate (B) and (D)). In order to determine the median height for all six plants, their heights must first be organized in ascending order: 61, 66, 67, 68, 70, 72. The middle two numbers are 67 and 68; when averaged, this produces a median of 67.5 inches ((C) is correct). Notice that if all the data points are whole numbers, the median value must end in either .0 or .5, so you can quickly eliminate (A) and (B) in this question. In fact, (A) and (B) are giving the value for mean, not median.

3. **D** The test scores ordered from smallest to largest are:

32, 65, 66, 67, 68, 68, 69, 70, 71, 72, 73, 75, 75, 75, 78, 82

The most frequently recurring number in the set above is 75, so this is the mode (eliminate (A) and (B)). The smallest number is 32 and the largest is 82. The range is the difference between the two, or 50 ((D) is correct).

4. **D** This question is testing the product rule.

$$P(AABbCC) = P(AA) \times P(Bb) \times P(CC)$$

$$P = \frac{1}{4} \times \frac{1}{2} \times \frac{1}{4}$$

$$P = \frac{1}{32}$$

5. **A** This question is testing the sum rule, but you also need to use the product rule.

$$P(Aabb \text{ or } aaBb) = P(Aabb) + P(aaBb)$$

$$P = [P(Aa) \times P(bb)] + [P(aa) \times P(Bb)]$$

$$P = \left(\frac{1}{2}\right)\left(\frac{1}{2}\right) + \left(\frac{1}{2}\right)\left(\frac{1}{2}\right)$$

$$= \frac{1}{4} + \frac{1}{4}$$

$$P = \frac{2}{4} = \frac{1}{2}$$

6. **C** In chi-square tests, a calculated χ^2 value is compared to a critical value (CV) from a chi-square table, like the one on the AP Biology Equations and Formulas sheet. If $\chi^2 < CV$, you accept the null hypothesis (H_o). If $\chi^2 > CV$, you reject H_o. Our H_o is that the observed and expected data match and that the experimental plants have a 3:1 ratio of yellow to green plants. Based on this, the only possible answer choices are (B) and (C); the other choices mix up how χ^2 values and critical values are compared (eliminate (A) and (D)).

One hundred plants were studied, and we expect three-fourths of them to be yellow (75 plants) and one-fourth of them to be green (25 plants). Next, we compare expected (E) and observed (O) data and calculate an χ^2 value.

Plant Phenotype	Observed (O)	Expected (E)	$(O - E)^2/E$
Yellow	84	75	1.08
Green	16	25	3.24
Total	*100*	*100*	*4.32*

Because two possibilities are being compared (green and yellow), the degrees of freedom in this test = 2 − 1 = 1. Using $p = 0.05$, you can determine the critical value to be 3.84 (which you don't need to look up because it is listed in all answer choices). Since $\chi^2 > CV$, you reject H_0. The observed data does not match the expected data, and the correct answer is (C).

7. **B** The first thing you need to do is sort out the alleles for each gene. Since the F_1 generation had long phenotypes, you know these must be dominant to the short phenotypes. Let's use:

L = long leaves
l = short leaves
S = long shoots
s = stubby shoots

The parental cross must have been $LLSS \times llss$, and all F_1 plants were $LlSs$. Our null hypothesis (H_0) is that the two genes are segregating independently. This means the expected ratio of F_2 phenotypes will be $9LS:3Ls:3lS:1ls$. Since 1,000 F_2 plants were generated, we expect:

$$P(\text{long leaves, long shoots}) = P(LS \text{ phenotype}) = \frac{9}{16} \times 1,000 = 562.5$$

$$P(\text{long leaves, stubby shoots}) = P(Ls \text{ phenotype}) = \frac{3}{16} \times 1,000 = 187.5$$

$$P(\text{short leaves, long shoots}) = P(lS \text{ phenotype}) = \frac{3}{16} \times 1,000 = 187.5$$

$$P(\text{short leaves, stubby shoots}) = P(ls \text{ phenotype}) = \frac{1}{16} \times 1,000 = 62.5$$

Next, generate a chart:

F_2 Phenotype	F_2 Genotype	Observed (o)	Expected (e)	$(o − e)^2/e$
Long leaves, long shoots	$L–S–$	382	562.5	57.9
Long leaves, stubby shoots	$L–ss$	109	187.5	32.9
Short leaves, long shoots	$llS–$	112	187.5	30.4
Short leaves, stubby shoots	$llss$	397	62.5	1,790.2
Total		1,000	1,000	1,911.4

$\chi^2 = 1,911.4$ and degrees of freedom = # of possibilities − 1 = 4 − 1 = 3 (eliminate (C) and (D)). Using $p = 0.05$, we get a critical value of 7.82. Since $\chi^2 > CV$, we reject H_0. These two genes are not segregating independently (eliminate (A); (B) is correct).

AP BIOLOGY
EQUATIONS AND FORMULAS

Statistical Analysis and Probability

Mean

$$\bar{x} = \frac{1}{n} \sum_{i=1}^{n} x_i$$

Standard Deviation

$$S = \sqrt{\frac{\sum (x_i - \bar{x})^2}{n-1}}$$

Standard Error of the Mean

$$SE_{\bar{x}} = \frac{S}{\sqrt{n}}$$

Chi-Square

$$\chi^2 = \sum \frac{(o-e)^2}{e}$$

Chi-Square Table

p-value	Degrees of Freedom							
	1	2	3	4	5	6	7	8
0.05	3.84	5.99	7.82	9.49	11.07	12.59	14.07	15.51
0.01	6.64	9.21	11.34	13.28	15.09	16.81	18.48	20.09

$\bar{x}$ = sample mean

n = size of the sample

s = sample standard deviation (i.e., the sample-based estimate of the standard deviation of the population)

o = observed results

e = expected results

$\sum$ = sum of all

Degrees of freedom are equal to the number of distinct possible outcomes minus one.

Laws of Probability

If A and B are mutually exclusive, then:

$$P(A \text{ or } B) = P(A) + P(B)$$

If A and B are independent, then:

$$P(A \text{ and } B) = P(A) \times P(B)$$

Hardy-Weinberg Equations

$p^2 + 2pq + q^2 = 1$ p = frequency of allele 1 in a population

$p + q = 1$ q = frequency of allele 2 in a population

Metric Prefixes

Factor	Prefix	Symbol
10^9	giga	G
10^6	mega	M
10^3	kilo	k
10^{-2}	centi	c
10^{-3}	milli	m
10^{-6}	micro	μ
10^{-9}	nano	n
10^{-12}	pico	p

Mode = value that occurs most frequently in a data set

Median = middle value that separates the greater and lesser halves of a data set

Mean = sum of all data points divided by the number of data points

Range = value obtained by subtracting the smallest observation (sample minimum) from the greatest (sample maximum)

Rate and Growth

Rate

$$\frac{dY}{dt}$$

Population Growth

$$\frac{dN}{dt} = B - D$$

Exponential Growth

$$\frac{dN}{dt} = r_{max}N$$

Logistic Growth

$$\frac{dN}{dt} = {}^{r}max^{N}\left(\frac{K - N}{K}\right)$$

dY = amount of change

dt = change in time

B = birth rate

D = death rate

N = population size

K = carrying capacity

r_{max} = maximum per capita growth rate of population

Simpson's Diversity Index

$$\text{Diversity Index} = 1 - \Sigma\left(\frac{n}{N}\right)^2$$

n = the total number of organisms of a particular species
N = the total number of organisms of all species

Water Potential (ψ)

$$\psi = \psi_P + \psi_S$$

ψ_P = pressure potential

ψ_S = solute potential

The water potential will be equal to the solute potential of a solution in an open container because the pressure potential of the solution in an open container is zero.

The Solute Potential of a Solution

$$\psi_S = -iCRT$$

i = ionization constant (1.0 for sucrose because sucrose does not ionize in water.)

C = molar concentration

R = pressure constant (R = 0.0831 liter bars/mole K)

T = temperature in Kelvin (°C + 273)

$$\mathbf{pH} = -\log_{10}[H^+]$$

Surface Area and Volume

Surface Area of a Sphere
$SA = 4\pi r^2$

Surface Area of a Rectangular Solid
$SA = 2lh + 2lw + 2wh$

Surface Area of a Cylinder
$SA = 2\pi rh + 2\pi r^2$

Surface Area of a Cube
$SA = 6s^2$

Volume of a Sphere
$$V = \frac{4}{3}\pi r^3$$

Volume of a Rectangular Solid
$V = lwh$

Volume of a Cylinder
$V = \pi r^2 h$

Volume of a Cube
$V = s^3$

r = radius

l = length

h = height

w = width

s = length of one side of a cube

SA = surface area

V = volume

Part VI
Additional
Practice Tests

Practice Test 2

AP® Biology Exam

SECTION I: Multiple-Choice Questions

DO NOT OPEN THIS BOOKLET UNTIL YOU ARE TOLD TO DO SO.

At a Glance

Total Time
1 hour and 30 minutes
Number of Questions
60
Percent of Total Score
50%
Writing Instrument
Pencil required

Instructions

Section I of this examination contains 60 multiple-choice questions.

Indicate all of your answers to the multiple-choice questions on the answer sheet. No credit will be given for anything written in this exam booklet, but you may use the booklet for notes or scratch work. After you have decided which of the suggested answers is best, completely fill in the corresponding oval on the answer sheet. Give only one answer to each question. If you change an answer, be sure that the previous mark is erased completely. Here is a sample question and answer.

Sample Question	Sample Answer
Chicago is a	

(A) state
(B) city
(C) country
(D) continent

Use your time effectively, working as quickly as you can without losing accuracy. Do not spend too much time on any one question. Go on to other questions and come back to the ones you have not answered if you have time. It is not expected that everyone will know the answers to all the multiple-choice questions.

About Guessing

Many candidates wonder whether or not to guess the answers to questions about which they are not certain. Multiple-choice scores are based on the number of questions answered correctly. Points are not deducted for incorrect answers, and no points are awarded for unanswered questions. Because points are not deducted for incorrect answers, you are encouraged to answer all multiple-choice questions. On any questions you do not know the answer to, you should eliminate as many choices as you can, and then select the best answer among the remaining choices.

BIOLOGY
SECTION I
60 Questions
Time—90 minutes

Directions: Each of the questions or incomplete statements below is followed by four suggested answers or completions. Select the one that is best in each case and then fill in the corresponding oval on the answer sheet.

1. In general, animal cells differ from plant cells in that only animal cells

   (A) perform cellular respiration
   (B) contain transcription factors
   (C) do not contain vacuoles
   (D) lyse when placed in a hypotonic solution

2. A cell from the leaf of the aquatic plant Elodea was soaked in a 15 percent sugar solution, and its contents soon separated from the cell wall and formed a mass in the center of the cell. All of the following statements are true about this event EXCEPT

   (A) the vacuole lost water and became smaller
   (B) the space between the cell wall and the cell membrane expanded
   (C) the large vacuole contained a solution with much lower water potential than that of the sugar solution
   (D) the concentration of solutes in the extracellular environment is hypertonic with respect to the cell's interior

3. A chemical agent is found to denature acetylcholinesterase in the synaptic cleft, a space between nerve cells that neurotransmitters cross to cause signalling. What effect will this agent have on the neurotransmitter, acetylcholine?

   (A) Acetylcholine will not be released from the presynaptic membrane into the synaptic cleft.
   (B) Acetylcholine will not bind to receptor proteins on the nerve cell across the synaptic cleft.
   (C) Acetylcholine will not diffuse across the cleft to the second cell.
   (D) Acetylcholine will not be degraded in the synaptic cleft.

4. The base composition of DNA varies from one species to another. Which of the following ratios would you expect to remain constant in the DNA?

   (A) Cytosine : Adenine
   (B) Pyrimidine : Purine
   (C) Adenine : Guanine
   (D) Guanine : Deoxyribose

Questions 5–7 refer to the following passage.

Consider the following pathway of reactions catalyzed by enzymes (shown in numbers):

$$A \xrightarrow{1} B \xrightarrow{2} C \xrightarrow{3} D \xrightarrow{4} E \xrightarrow{5} F$$
$$\xrightarrow{6} X \xrightarrow{7} Y$$

5. Which of the following situations represents feedback inhibition?

   (A) Protein D activating enzyme 4
   (B) Protein B stimulating enzyme 1
   (C) Protein 7 inhibiting enzyme C
   (D) Protein X inhibiting enzyme 2

6. An increase in substance F leads to the inhibition of enzyme 3. All of the following are direct or indirect results of the process EXCEPT

   (A) an increase in substance X
   (B) increased activity of enzyme 6
   (C) decreased activity of enzyme 4
   (D) increased activity of enzyme 5

7. If a competitive inhibitor is bound to enzyme 1, which of the following would be decreased?

   I. Protein A
   II. Protein C
   III. Protein X

   (A) I only
   (B) II only
   (C) II and III
   (D) I, II, and III

GO ON TO THE NEXT PAGE.

Questions 8–11 refer to the following passage.

The affinity of hemoglobin for oxygen is reduced by many factors, including low pH and high CO_2. The graph below shows the different dissociation curves that maternal (normal) hemoglobin and fetal hemoglobin have.

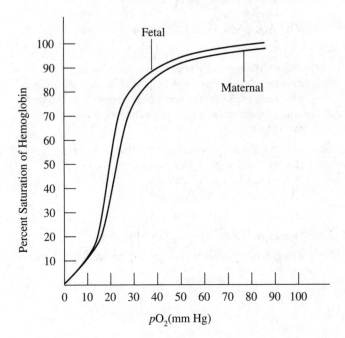

8. Based on the graph, it can be concluded that

(A) fetal hemoglobin surrenders O_2 more readily than maternal hemoglobin

(B) the dissociation curve of fetal hemoglobin is to the right of maternal hemoglobin

(C) fetal hemoglobin has a higher affinity for O_2 than does maternal hemoglobin

(D) fetal and maternal hemoglobin differ in structure

9. Which of the following processes would likely shift the normal dissociation curve to the right?

(A) Photosynthesis

(B) Respiration

(C) Fermentation

(D) Mitosis

10. Hemoglobin's affinity for O_2

(A) decreases as blood pH decreases

(B) increases as H^+ concentration increases

(C) increases as blood pH decreases

(D) decreases as OH^- concentration increases

11. How much pO_2 would it take in an extremely CO_2-rich environment to saturate hemoglobin 90 percent?

(A) 15

(B) 30

(C) 45

(D) 60

12. All of the following are differences between prokaryotes and eukaryotes EXCEPT

(A) eukaryotes have linear chromosomes, while prokaryotes have circular chromosomes

(B) eukaryotes possess double-stranded DNA, while prokaryotes possess single-stranded DNA

(C) eukaryotes process their mRNA, while in prokaryotes, transcription and translation occur simultaneously

(D) eukaryotes contain membrane-bound organelles, while prokaryotes do not

13. In minks, the gene for brown fur (*B*) is dominant over the gene for silver fur (*b*). Which set of genotypes represents a cross that could produce offspring with silver fur from parents that both have brown fur?

(A) *BB* × *BB*

(B) *BB* × *Bb*

(C) *Bb* × *Bb*

(D) *Bb* × *bb*

14. Retroviruses violate the Central Dogma because

(A) they store their genetic material as RNA which is reverse-transcribed to DNA

(B) they do not contain hereditary material but use the host DNA instead

(C) retroviruses do not use the same codon sequences that living organisms do.

(D) retroviruses do not require transcription to make proteins

15. All of the following are examples of hydrolysis EXCEPT

(A) conversion of fats to fatty acids and glycerol

(B) conversion of proteins to amino acids

(C) conversion of starch to simple sugars

(D) conversion of pyruvic acid to glucose

16. In cells, which of the following can catalyze reactions involving hydrogen peroxide, provide cellular energy, and make proteins, in that order?

(A) Peroxisomes, mitochondria, and ribosomes

(B) Peroxisomes, mitochondria, and lysosomes

(C) Peroxisomes, mitochondria, and Golgi apparatus

(D) Lysosomes, chloroplasts, and ribosomes

GO ON TO THE NEXT PAGE.

Questions 17 and 18 refer to the following passage.

The Loop of Henle is a structure within each of the million nephrons within a kidney. As shown in the figure, the two sides have different permeabilities, and there is differential movement across each membrane. The Loop acts as a counter-current multiplier that makes the medulla of the kidney very osmotic. The longer the loop, the higher and more powerful the osmolarity gradient that is created. The gradient is required for the reclamation of water from the urine collecting duct. On the right side of the figure is the urine collecting duct. If the body needs to retain water, anti-diuretic hormone makes this region permeable to water via the introduction of aquaporins, and the osmotic pull of the medulla reclaims the water, out of the collecting duct, which makes the urine more concentrated.

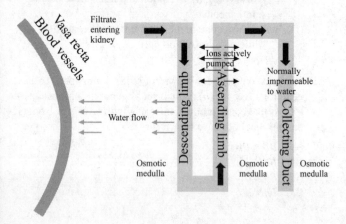

17. Which of the following statements correctly describes the state of things near the top of the descending limb?

(A) The fluid within the descending limb is hypotonic to fluid in the space surrounding the tubule.

(B) The blood within the vasa recta is hypotonic to the filtrate within the descending limb.

(C) The fluid in the area surrounding the tubule is hypertonic to the blood in the vasa recta.

(D) The water in the area surrounding the tubule has a higher water potential than the water in the descending tubule.

18. What type of transport is occurring when water flows out of the descending tubule?

(A) Simple diffusion
(B) Facilitated diffusion
(C) Active transport
(D) Secondary active transport

19. If an inhibitor of anti-diuretic hormone, such as caffeine, was ingested, what would be the result?

(A) Aquaporins would appear in the collecting duct.
(B) Aquaporins would increase in number in the collecting duct.
(C) It would block aquaporins like a competitive inhibitor.
(D) Aquaporins would not appear in the collecting duct.

20. The ability to reclaim water from the collecting duct is directly related to the osmotic pull of the medulla. Kangaroo rats are known to produce extremely concentrated urine. Compared to a human, the nephrons in kangaroo rats must have

(A) thick walls that are impermeable to water
(B) shorter Loops of Henle
(C) longer Loops of Henle
(D) shorter collecting ducts

Questions 21 and 22 refer to the following graph.

The graph below shows two growth curves for bacterial cultures, A and B.

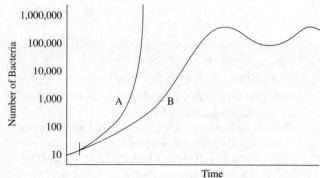

21. Which of the following represents the carrying capacity for culture B?

(A) 10
(B) 50
(C) 100,000
(D) 1,000,000

22. What could explain the differnce between culture A and culture B?

(A) Culture B started with more bacteria than culture A.
(B) Culture A was grown with a competitive inhibitor.
(C) Culture B was not measured as often.
(D) Culture A has not yet exhausted its space and resources.

GO ON TO THE NEXT PAGE.

23. In plants, the process of phototropism in which an organism responds to a light stimulus would be best exemplified by

    (A) a sunflower photographed pointing one direction in the morning and another direction at night.
    (B) a vine climbing and twisting around a trellis.
    (C) a flowering tree that blooms only in the springtime.
    (D) a flower that changes color to attract more pollinators.

24. Females with Turner's syndrome have a high incidence of hemophilia, a recessive, X-linked trait. Based on this information, it can be inferred that females with this condition

    (A) have an extra X chromosome
    (B) have an extra Y chromosome
    (C) have one less X chromosome than normal
    (D) have one less Y chromosome than normal

25. When a retrovirus inserted its DNA into the middle of a bacterial gene, it altered the normal reading frame by one base pair. This type of mutation is called

    (A) duplication
    (B) translocation
    (C) inversion
    (D) frameshift mutation

26. The principal inorganic compound found in living things is

    (A) carbon
    (B) oxygen
    (C) water
    (D) glucose

27. Metafemale syndrome, a disorder in which a female has an extra X chromosome, is the result of nondisjunction. This failure in oogenesis would first be apparent when

    (A) the homologous chromosomes are lined up in the middle
    (B) the sister chromatids are lined up in the middle
    (C) the nuclear envelope breaks down before meiosis
    (D) the homologous chromosomes are pulling apart

28. Suppose scientists discovered in the fossil record or molecular biology that tiger-like ancestors lived for millions of years before a mutation occurred. This mutation caused a striped phenotype that allowed the animals with this mutation to feed and reproduce at a much faster rate due to this new camouflage adaptation. Which theory of evolution would this discovery support?

    (A) Gradualism
    (B) Punctuated equilibrium
    (C) Genetic drift
    (D) Migration

29. *E. Coli* have rod-shaped cells that closely resemble cylinders. Which size *E. Coli* cell (measured in μm) would be most metabolically efficient?

    (A) radius = 0.5, height = 1.5
    (B) radius = 0.5, height = 2
    (C) radius = 1, height = 3
    (D) radius = 1, height = 4

Questions 30–32 refer to the following passage.

The following are important pieces of replication and transcription machinery:

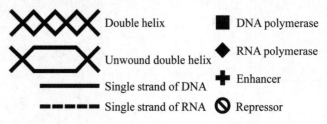

Double helix	DNA polymerase
Unwound double helix	RNA polymerase
Single strand of DNA	Enhancer
Single strand of RNA	Repressor

30. Which of the following figures would be present without helicase?

    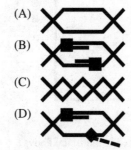

    (A)
    (B)
    (C)
    (D)

31. What might have occurred to produce the following situation?

    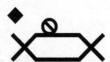

    (A) Stalled DNA replication
    (B) Initiation of transcription
    (C) Repression of transcription
    (D) Crossing-over in meiosis

GO ON TO THE NEXT PAGE.

32. Put these four situations in the correct order:

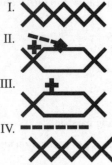

I.

II.

III.

IV.

(A) I, III, IV, II
(B) I, IV ,II, III
(C) IV, I, III, II
(D) I, III, II, IV

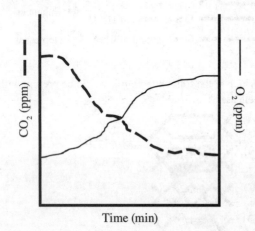

Time (min)

33. Which process is demonstrated in the figure above?

(A) Kreb's cycle
(B) Photosynthesis
(C) Glycolysis
(D) Fermentation

34. Avery, MacLeod, and McCarty performed a transformation experiment in which they added DNAse, protease, and RNAse to degrade DNA, proteins, and RNA respectively. They found no change in bacteria treated with protease and RNAse, but the transformation did not occur in those treated with DNAse. These results are most consistent with which of the following conclusions?

(A) The double helix is the structure of a DNA molecule.
(B) DNA is the hereditary material.
(C) DNA replication is semiconservative.
(D) DNA polymerase adds bases to only the 3' side of DNA.

35. Which of the following processes occur in the cytoplasm of an eukaryotic cell?

  I. DNA replication

 II. Transcription

III. Translation

(A) I only
(B) III only
(C) II and III only
(D) I, II, and III

36. Crossing-over during meiosis permits scientists to determine

(A) the chance for variation in zygotes
(B) the rate of mutations
(C) the distance between genes on a chromosome
(D) which traits are dominant or recessive

37. An animal cell that is permeable to water but not salts has an internal NaCl concentration of 10%. If placed in freshwater, the cell will

(A) plasmolyze
(B) swell and eventually lyse
(C) endocytose water into a large central vacuole
(D) shrivel

38. Three distinct bird species, flicker, woodpecker, and elf owl, all inhabit a large cactus, *Cereus giganteus*, in the desert of Arizona. Since competition among these birds rarely occurs, the most likely explanation for this phenomenon is that these birds

(A) have a short supply of resources
(B) have different ecological niches
(C) do not live together long
(D) are unable to breed

39. Lampreys attach to the skin of lake trout and absorb nutrients from its body. This relationship is an example of

(A) commensalism
(B) parasitism
(C) mutualism
(D) gravitropism

40. The nucleotide sequence of a template DNA molecule is 5'-C-A-T-3'. An mRNA molecule with a complementary codon is transcribed from the DNA. What would be the sequence of the anticodon that binds to this mRNA?

(A) 5'-G-T-A-3'
(B) 5'-G-U-A-3'
(C) 5'-C-A-U-3'
(D) 5'-U-A-C-3'

GO ON TO THE NEXT PAGE.

41. Viruses are considered an exception to the cell theory because they

    (A) require a host cell
    (B) do not contain a genome
    (C) do not contain a nuclei
    (D) cannot evolve via mechanisms that cells evolve from

Questions 42–44 refer to the following passage.

It is difficult to determine exactly how life began. Answer the following questions as if the following data had been collected billions of years ago.

	4 billion years ago	3.5 billion years ago	3.25 billion years ago	3 billion years ago
Atmospheric carbon dioxide	Present	Present	Present	Present
Atmospheric oxygen	Absent	Absent	Absent	Present
RNA	Absent	Present	Present	Present
Protein	Absent	Absent	Present	Present
DNA	Absent	Absent	Present	Present
Life	Absent	Present	Present	Present

42. Which best describes the origin of life according to the data?

    (A) Life required an environment with atmospheric oxygen and any type of nucleic acids.
    (B) Life required an environment with atmospheric carbon dioxide, but not atmospheric oxygen.
    (C) Life required an environment with self-replicating nucleic acids that can take on many shapes.
    (D) Life required an environment with nucleic acids and proteins, but not atmospheric oxygen.

43. When is the earliest that functional ribosomes could have been found?

    (A) Between 4 billion and 3.5 billion years ago
    (B) Between 3.5 billion and 3.25 billion years ago
    (C) Between 3.25 billion and 3 billion years ago
    (D) Between 3 billion years ago and the present

44. Photosynthesis likely began_____ billion years ago when the first _____ appeared.

    (A) 4.5; autotrophs
    (B) 3.2; autotrophs
    (C) 3.5; heterotrophs
    (D) 3.2; heterotroph

45. The sequence of amino acids in hemoglobin molecules of humans is more similar to that of chimpanzees than it is to the hemoglobin of dogs. This similarity suggests that

    (A) humans and dogs are more closely related than humans and chimpanzees
    (B) humans and chimpanzees are more closely related than humans and dogs
    (C) humans are related to chimpanzees but not to dogs
    (D) humans and chimpanzees are closely analogous

46. Two individuals, one with type B blood which can be I^Bi or I^BI^B and one with type AB blood with genotype I^AI^B, have a child. The probability that the child has type O blood which is caused by genotype ii, is

    (A) 0%
    (B) 25%
    (C) 50%
    (D) 100%

Questions 47 and 48 refer to the following bar graph, which shows the relative biomass of four different populations of a particular food pyramid.

Relative Biomass

Population A
Population B
Population C
Population D

47. The largest amount of energy is available to

    (A) population A
    (B) population B
    (C) population C
    (D) population D

48. Which of the following would be the most likely result if there was an increase in the number of organisms in population C?

    (A) The biomass of population D will remain the same.
    (B) The biomass of population B will decrease.
    (C) The biomass of population A will steadily decrease.
    (D) The food source available to population C would increase.

GO ON TO THE NEXT PAGE.

Questions 49–52 refer to the following illustration and information.

The cell cycle is a series of events in the life of a dividing eukaryotic cell. It consists of four stages: G_1, S, G_2, and M. The duration of the cell cycle varies from one species to another and from one cell type to another. The G_1 phase varies the most. For example, embryonic cells can pass through the G_1 phase so quickly that it hardly exists, whereas neurons are arrested in the cell cycle and do not divide.

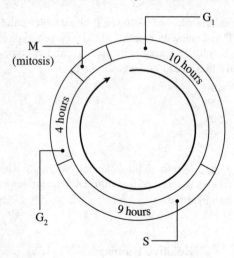

49. During which phase do chromosomes replicate?

(A) G_1
(B) S
(C) G_2
(D) M

50. In mammalian cells, the first sign of mitosis beginning is the

(A) appearance of chromosomes
(B) separation of chromatids
(C) disappearance of the cell membrane
(D) replication of chromosomes

51. If the cell cycle fails to progress, which of the following is NOT a possible explanation?

(A) There are inadequate phosphate groups available for the cyclin dependent kinase.
(B) A tumor suppressor protein has signaled for apoptosis.
(C) A cyclin is unable to release from its cyclin dependent kinase.
(D) An inhibitor of a cyclin gene has been highly expressed.

52. Since neurons are destined never to divide again, what conclusion can be made?

(A) These cells will go through cell division.
(B) These cells will be permanently arrested in the G_0 phase.
(C) These cells will be permanently arrested in the M phase.
(D) These cells will quickly enter the S phase.

GO ON TO THE NEXT PAGE.

Questions 53–56 refer to the following figures which show 5 species of insects that were discovered on a previoulsy unknown island and were named as shown in the table. Proteomic analysis was performed on a highly conserved protein in the insects and the number of amino acid differences was calculated and included in the table below. The scientists used this data to create the phylogenetic tree shown below with positions labeled I, II, III, and IV as well as O, P, Q, R, and S.

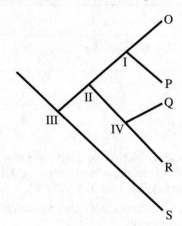

	Snippeiq	Fixxels	Snorfak	Gerdellen	Sqellert
Snippeiq	0	13	3	12	3
Fixxels		0	3	8	8
Snorflak			0	9	9
Gerdellen				0	4
Sqellert					0

53. Based on the data, which is the LEAST possible location for Gerdellen to be placed?

(A) Position O
(B) Position P
(C) Position R
(D) Position S

54. Which location contains the oldest common ancestor?

(A) Position I
(B) Position II
(C) Position III
(D) Position IV

55. Which species is the outgroup?

(A) Snippeiq
(B) Fixxels
(C) Snorflak
(D) Sqellert

56. Which 2 species would you expect to have the most shared derived characters?

(A) Snippeiq and Sqellert
(B) Gerdellen and Sqellert
(C) Snorflak and Gerdellen
(D) Snorflak and Fixxels

GO ON TO THE NEXT PAGE.

Questions 57 and 58 refer to the data below concerning the general animal body plan of four organisms.

Characteristic	Sea anemone	Hagfish	Eel	Salamander
Vertebral column		+	+	+
Jaws			+	+
Walking legs				+

Note: + indicates a feature present in an organism.

57. The two most closely related organisms are

(A) sea anemone and hagfish
(B) eel and salamander
(C) hagfish and eel
(D) sea anemone and salamander

58. The correct order of evolution for the traits above is

(A) jaws – vertebral column – walking legs
(B) walking legs – jaws – vertebral column
(C) jaws – walking legs – vertebral column
(D) vertebral column – jaws – walking legs

59. Pre- and post-zygotic barriers exist that prevent two different species from producing viable offspring. All of the following are pre-zygotic barriers EXCEPT

(A) anatomical differences preventing copulation
(B) different temporality of mating
(C) sterility of offspring
(D) incompatible mating songs

60. Birds and insects have both adapted wings to travel by flight. The wings of birds and insects are an example of

(A) divergent evolution
(B) convergent evolution
(C) speciation
(D) genetic drift

STOP

END OF SECTION I

IF YOU FINISH BEFORE TIME IS CALLED, YOU MAY CHECK YOUR WORK ON THIS SECTION. DO NOT GO ON TO SECTION II UNTIL YOU ARE TOLD TO DO SO.

THIS PAGE INTENTIONALLY LEFT BLANK

BIOLOGY
SECTION II
6 Questions
Writing Time—90 minutes

<u>Directions:</u> Questions 1 and 2 are long free-response questions that should require about 25 minutes each to answer and are worth 8–10 points each. Questions 3 through 6 are short free-response questions that should require about 10 minutes each to answer and are worth 4 points each.

Read each question carefully and completely. Write your response in the space provided following each question. Only material written in the space provided will be scored. Answers must be written out in paragraph form. Outlines, bulleted lists, or diagrams alone are not acceptable unless specifically requested.

1. The human retina has two types of photoreceptors, rods and cones, which respond to light stimuli. Stimulation of photoreceptors is essential for the organisms to respond to visual stimuli. When the rods and cones absorb a photon, a molecule called retinal alters their conformation and allows a sodium channel to close.

The response of rods—that is, the voltage—can be measured with a microelectrode. When a rod cell is bathed in vitreous humor, the fluid found inside the eyeball, it has a resting membrane potential of –40mV. As the rod is exposed to light, the membrane potential becomes more negative, or hyperpolarized.

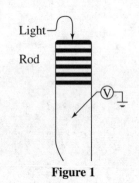

Figure 1

In Experiment 1, a rod was stimulated and its responses measured. Figure 2 shows the hyperpolarizations resulting from two light intensities (two separate stimuli are depicted on the same graph). Figure 3 is a plot of intensity versus hyperpolarization (depicted as a change in mV from –40mV).

In Experiment 2, the Na$^+$ was removed from the vitreous humor. The resting membrane potential changed to –65mV and was unaffected by a light stimulus.

(a) **Explain** why the ability to respond to a visual stimulus would be naturally selected. Would a rod receptor with a resting membrane potential of –65mV be selected for?

(b) **Explain** how the control in Experiment 1 would differ from the control in Experiment 2.

(c) **Describe** the relationship between a stimulus and hyperpolarization. Use evidence from the data.

(d) If the rods were bathed in fluid with a high sodium content, **explain** how the resting membrane potential would change. How does the sodium change the resting membrane potential?

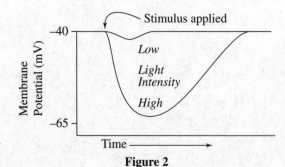

Figure 2

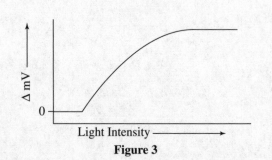

Figure 3

GO ON TO THE NEXT PAGE.

2. In a forest there are two types of wildflowers: red flowers and yellow flowers. The red flowers are the food of choice for a population of rodents. The yellow flowers are the food of choice for a type of mountain deer. Both populations of animals appear to be evenly distributed throughout the forest, but it often looks like there are more red flowers on one side of the forest than on the other side.

A researcher wanted to determine if the red flowers preferred to grow on the western side of the forest and the yellow preferred to grow on the eastern side of the forest. For 5 years, she counted the number of red flowers on each side of the forest. The results are in Table 1.

Year	Western Red Flowers	Eastern Red Flowers
2006	1,527	986
2007	1,324	1,011
2008	1,478	876
2009	1,048	788
2010	1,140	800
Average	1,304	892
Standard Error	208	103

Table 1

(a) **Explain** how biotic and abiotic factors affect where plants grow.
(b) **Construct** a graph of averages and standard error for the western and eastern flowers.
(c) **Perform** a chi-squared analysis using the average data for both flower types.
(d) Using your chi-squared analysis data, **explain** if you can reject the null hypothesis that flowers do not prefer the western side.

3. *Schizosaccharomyces pombe*, or fission yeast, has a cell cycle that resembles that of mammalian cells. During interphase, fission yeast grow to twice their normal size, and at the end of mitosis both daughter offspring are equal in size to the original parent cell. Genes that regulate the division of fission yeast are known as cell-division cycle, or *cdc*, genes.

The following experiments were conducted to determine the effect of cdc gene mutations on yeast cell division.

Experiment 1
In order to determine the effect of *cdc* mutations, wild-type cells and mutants were grown at 37°C in the presence of a radioactive drug that specifically binds to the spindle apparatus. Stages of the cell cycle were elucidated for both cell types incubated at this temperature. At 37°C, temperature-sensitive *cdc* mutants were unable to re-enter interphase after mitosis. The illustration below depicts the results for wild-type cells and *cdc* mutants.

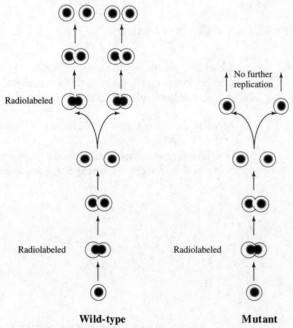

Wild-type   **Mutant**

Figure 1

Experiment 2
Cdc mutants were initially incubated at 25°C for 30 minutes and several rounds of mitosis occurred. Separate colonies were then exposed to temperatures near 37°C at different times during the cell cycle. For each colony, the cell cycle was arrested immediately after the last phase of mitosis.

(a) **Describe** how a mutation in DNA can lead to a protein that no longer functions correctly.
(b) **Explain** the impact of the radiolabeled dye on the cell cycle.
(c) **Predict** what would happen if the *cdc* mutants were initially incubated at 37°C and then switched to 25°C.
(d) **Justify** your prediction.

GO ON TO THE NEXT PAGE.

4. A prairie was sampled by entomological researchers to determine the species abundance and diversity. Three years into the study a fire swept through the prairie. Two years later quadrant sampling was completed again. Results were compiled in the table below.

Organism	Before Fire	2 years after fire
Hazel-hued bugs	3	10
Hairy bear moths	14	9
Telluride tiger beetles	6	15
Golden-winged butterflies	21	10

(a) **Describe** what happened to the biodiversity of the ecosystem before and after the fire.

(b) **Explain** your description in (A) using the Simpson's Diversity Index.

(c) **Predict** what would happen to the biodiversity if there was an invasive insect species introduced into this ecosystem.

(d) **Justify** your prediction.

5.

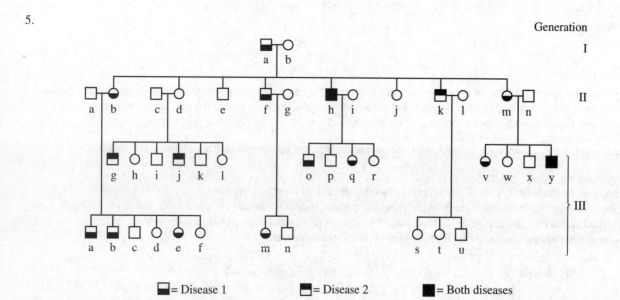

= Disease 1   = Disease 2   = Both diseases

The diagram above is a pedigree of three generations showing the occurrence of two genetically transmitted diseases. Phenotypes of affected individuals are shown. Circles represent females, and squares represent males. Assume that the diseases are rare, unlinked, and that individuals not blood related to I-a and I-b ("in-laws") do not have either disease-producing allele.

(a) **Explain** how it is possible for a person to have two rare unlinked genetic diseases at the same time.

(b) **Calculate** the ratios of affected males and affected females for Disease 2.

(c) **Identify** the likely inheritance patterns of Disease 1 and Disease 2.

(d) **Describe** how inheritance patterns are shaped by the chromosomes upon which the genes are located.

GO ON TO THE NEXT PAGE.

6. In this fictional scenario, an enzyme was discovered in reptile feces, gertimtonase, that was believed to aid in digestion and water retention. Researchers wanted to know how the enzyme's activity level changed with varying temperature. The scientists took fecal samples of Russian tortoises that were living in varying environmental temperatures. The data was summarized in the following graph.

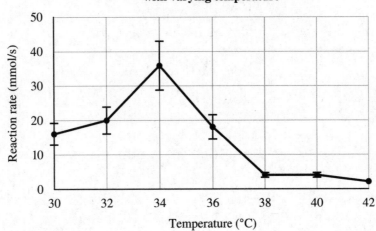

Russian tortoise gertimtonase reaction rate with varying temperature

(a) **Describe** how you can determine that tortoises raised at 34°C had the highest variability in enzyme reaction rates.

(b) Reptiles in the study needed to have a gertimtonase reaction rate exceeding 10 mmol/s for an observed improvement in water retention. **Describe** the range of temperatures that Russian tortoises could be raised in to achieve this benefit.

(c) Desert tortoises live in environments where the ground temperature can exceed 60°C. Using the data above, **predict and justify** how the graph would look for desert tortoise gertimtonase.

(d) **Explain** at a molecular level how the temperature change might affect gertimtonase.

STOP

END OF EXAM

Practice Test 2:
Answers and
Explanations

PRACTICE TEST 2 ANSWER KEY

1.	D	21.	C	41.	A
2.	C	22.	D	42.	C
3.	D	23.	A	43.	B
4.	B	24.	C	44.	B
5.	D	25.	D	45.	B
6.	D	26.	C	46.	A
7.	C	27.	D	47.	B
8.	C	28.	B	48.	B
9.	C	29.	A	49.	B
10.	A	30.	C	50.	A
11.	D	31.	C	51.	C
12.	B	32.	D	52.	B
13.	C	33.	B	53.	D
14.	A	34.	B	54.	C
15.	D	35.	B	55.	A
16.	A	36.	C	56.	D
17.	A	37.	B	57.	B
18.	B	38.	B	58.	D
19.	D	39.	B	59.	C
20.	C	40.	C	60.	B

PRACTICE TEST 2 EXPLANATIONS

Section I: Multiple Choice

1. **D**   Because of the cell wall, a plant cell will not lyse when placed in a hypotonic solution. Instead, it will just swell. Both plant and animal cells perform cellular respiration (A) and have transcription factors (B), and both have vacuoles (C).

2. **C**   If the contents of the cell separated from the cell wall, then water was moving out of the cell. This would cause the space between the cell wall and the cell membrane to expand, so eliminate (B). This also means that the concentration of solutes in the extracellular environment would be hypertonic with respect to the cell's interior. You can also eliminate (A) because the fluid in the cell was hypotonic to the sugar solution, fluid was moving out of the vacuole, which caused it to become smaller. Choice (C) is the correct answer because if the water was rushing out of the cell, then the water potential inside the vacuole would have been higher than outside the cell. High water potential means the water wants to move. Choice (C) says the water potential would be lower.

3. **D**   This question tests your ability to associate what happens when enzymes are denatured and what would happen in the synaptic cleft. Acetylcholinesterase is an enzyme that degrades acetylcholine in the synaptic cleft. If acetylcholinesterase is denatured, acetylcholine will still be released from the presynaptic membrane (the nerve before the cleft) and continue to diffuse across the synaptic cleft. Acetylcholine will still bind to the postsynaptic membrane (on the nerve after the cleft) because acetylcholine is not degraded. Therefore, you can eliminate (A), (B), and (C).

4. **B**   The ratio of purines to pyrimidines should be constant because purines always bind with pyrimidines, no matter which ones they may be.

5. **D**   Feedback inhibition occurs when something created by a process then inhibits that process. Choices (A) and (B) involve stimulations, so they can be eliminated. Choice (C) doesn't make sense because it says that "C" is an enzyme and the passage says that the enzymes are the numbers shown between the steps.

6. **D**   If substance F leads to the inhibition of enzyme 3, then substances D and E and enzymes 3, 4, and 5 will be affected. The activity of enzyme 5 will be decreased, not increased.

7. **C**   If enzyme 1 were inhibited, then everything that comes after that in the pathway would be reduced. Substance A would not be reduced, but both C and X would. Choice (C) is correct.

8. **C**   Based on the graph, fetal hemoglobin has a higher affinity for oxygen than maternal hemoglobin. Fetal hemoglobin does not give up oxygen more readily than maternal hemoglobin, so eliminate (A). You can also get rid of (B), as the dissociation curve of fetal hemoglobin is to the left of the maternal hemoglobin. Finally, eliminate (D) because fetal hemoglobin and maternal hemoglobin are different structurally, but you can't tell this from the graph.

9. **C**   The passage says that high CO_2 and low pH lead to reduced affinity. A right shift curve would represent a reduced affinity. Photosynthesis consumes CO_2, so this would not increase the CO_2. Mitosis and respiration are not related. Fermentation is done in cells that are doing anaerobic cell respiration, so CO_2 should be released, possibly along with lactic acid, which would lower pH.

10. **A**   Hemoglobin's affinity for O_2 decreases as the concentration of H^+ increases (or the pH decreases) and as the concentration of OH^- increases (or the pH increases). As the pH decreases, the affinity for oxygen will decrease, and as the pH increases, the affinity for oxygen will increase.

11. **D**   The CO_2-rich environment would decrease the affinity of hemoglobin for oxygen, so the curve should shift toward the right. The regular curve seems to hit 90% saturation around 45–50 pO_2, so the CO_2-rich curve would take more oxygen to get there.

12. **B**   Both prokaryotes and eukaryotes possess double-stranded DNA. Choices (A), (C), and (D) are correct differences between prokaryotes and eukaryotes.

13. **C**   *Bb* and *Bb* is the correct set of genotypes that represents a cross that could produce offspring with silver fur from parents that both have brown fur. Complete a Punnett square for this question. In order for the offspring to have silver fur, both parents must have the silver allele.

14. **A**   Retroviruses have an RNA genome (hereditary material, eliminating (B)) that must be reverse-transcribed into DNA (which is the opposite of the Central Dogma first step) and therefore (A) is correct. The codon sequence is used by all things that do translation (like viruses, eliminating (C)). Retroviruses use transcription as well after their genome has been reverse transcribed, so (D) is also incorrect.

15. **D**   All of the choices are examples of hydrolysis except the conversion of pyruvic acid to glucose. Hydrolysis is the breaking of a covalent bond by adding water. In all of the correct examples, complex compounds are broken down to simpler compounds. The conversion of pyruvic acid to glucose is an example of decarboxylation—a carboxyl group is removed as carbon dioxide and the 2-carbon fragments are oxidized.

16. **A**   Peroxisomes catalyze reactions that produce hydrogen peroxide, mitochondria contain enzymes involved in cellular respiration, and ribosomes are involved in protein synthesis. Eliminate (B) and (D) because lysosomes are the sites of degradation; they contain hydrolytic enzymes but do not produce hydrogen peroxide. Choice (C) is incorrect, as the Golgi apparatus sorts and packages substances that are destined to be secreted out of the cell.

17. **A**   The figure shows that water leaves the descending limb and flows by osmotic pressure into the surrounding space and then into the vasa recta. This means that each of those places is more hypertonic than the previous. The most hypotonic is the descending tubule. (This is why the water leaves.)

18. **B**   When water flows out of the tubule, it is moving by simple osmosis. However, because it is a polar molecule, it must pass through an aquaporin channel, which is facilitated diffusion.

19. **D**   ADH makes the collecting duct permeable to water. If it is inhibited, then it will be unable to do that, meaning choice (D) is correct. The question doesn't say that the inhibitor is inhibiting the aquaporins directly. Instead, the inhibitor says it is inhibiting the ADH.

20. **C**   To make concentrated urine, the medulla must be very osmotic because it must reclaim a lot of water from the collecting duct. To make a very osmotic medulla, the Loop of Henle must be very long. The collecting duct should be very permeable, not impermeable, because concentrated urine means that a lot of water has been reclaimed.

21. **C**   The carrying capacity is the maximum number of organisms of a given species that can be maintained in a given environment. Once a population reaches its carrying capacity, the number of

organisms will fluctuate around it.

22. **D**   The growth of curve A is exponential, meaning that the bacteria are replicated as fast as possible without any restrictions. It looks like the populations started at the same level, and an inhibitor would not cause high growth as shown in culture A. Not being measured as often would not cause a graph like the one shown for culture B.

23. **A**   Phototropism is the ability of a plant to move with regard to the sun. A climbing vine uses pressure to sense the trellis. A tree blooming once a year is using photoperiodism.

24. **C**   Females with Turner's syndrome lack an X chromosome. If females with this syndrome have a high rate of hemophilia, they must not have the second X to mask the expression of the disease.

25. **D**   This type of mutation is called frameshift mutation. The insertion of DNA leads to a change in the normal reading frame by one base pair. The other answer choices refer to chromosomal aberrations. Choice (A), duplication, occurs when an extra copy of a chromosome segment is introduced. Choice (B), translocation, occurs when a segment of a chromosome moves to another chromosome. Choice (C), inversion, occurs when a segment of a chromosome is inserted in the reverse orientation.

26. **C**   The principal inorganic compound found in living things is water. Water is a necessary component for life.

27. **D**   An extra X chromosome could be created by a problem separating the homologous chromosomes or sister chromatids. When they are lined up, it would not be apparent that they were not separating properly, but once they start to pull apart, one could see if a mistake had occurred.

28. **B**   This scenario describes natural selection, which eliminates (C) and (D). In this situation, the change in tigers is caused by a long period of stasis followed by a major event: this is punctuated equilibrium (B), and not gradualism (A).

29. **A**   Metabolic efficiency relates directly to the surface area-to-volume ratio. A higher ratio means higher efficiency. Since the ratio is $(2\pi rh + 2\pi r^2)/\pi r2h$ this simplifies to $2/r + 2/h$, so if the radius or height is higher, the ratio is smaller. Therefore the smallest radius and height (A) gives the highest ratio of 5.33 and the highest efficiency.

30. **C**   Helicase is the enzyme that unwinds the helix. The closed helix would exist without helicase.

31. **C**   The RNA polymerase is not bound, but there is a bound repressor. The transcription must be repressed.

32. **D**   These four phases must be transcription. Transcription begins with a closed helix, then an enhancer binds, and then RNA polymerase is seen making RNA. Finally, the closed helix and the completed mRNA strand exist.

33. **B**   Photosynthesis requires light, carbon dioxide, and water and produces oxygen and glucose.

34. **B**   Avery, MacLeod, and McCarty concluded that DNA is the hereditary material because it was able to transform non-pathogenic bacteria into pathogenic bacteria. Watson, Crick, and Franklin determined the double helix, and Meselson determined that replication is semi-conservative.

35. **B**   Translation, the synthesis of proteins from mRNA, occurs in the cytoplasm. DNA replication (I) occurs in the nucleus. Transcription (II), the synthesis of RNA from DNA, occurs in the nucleus.

36. **C**   Crossing-over permits scientists to determine chromosome mapping. Chromosome mapping is a detailed map of all the genes on a chromosome. The frequency of crossing-over between any two alleles is proportional to the distance between them. The farther apart the two linked alleles are on a chromosome, the more often the chromosome will break between them. Crossing-over does not tell us about the chance of variation in zygotes, the rate of mutations, or whether the traits are

dominant, recessive, or masked.

37. **B** In this scenario, the inside of the cell is hypertonic to the outside environment—this will cause water to move into the cell and can cause the cell to lyse.

38. **B** The most likely explanation for this phenomenon is that these birds have different ecological niches. An ecological niche is the position or function of an organism or population in its environment. Eliminate (A) because we do not know if there is a short supply of resources. Choice (C) can also be eliminated because we do not know how long the bird species live together. The breeding patterns of the bird species do not explain the lack of competition, which eliminates (D) as well.

39. **B** This relationship is an example of parasitism. Parasitism is a form of symbiosis in which one organism benefits and the other is harmed. Choice (A), commensalism, is a form of symbiosis in which one organism benefits and the other is unaffected. Choice (C), mutualism, is a form of symbiosis in which both organisms benefit. Choice (D), gravitropism, is the growth of a plant toward or away from gravity.

40. **C** If the nucleotide sequence of a DNA molecule is 5'-C-A-T-3', then the transcribed DNA strand (mRNA) would be 3'-G-U-A-5'. The nucleotide sequence of the tRNA codon would be 5'-C-A-U-3'.

41. **A** Viruses are considered an exception to the cell theory because they can survive only by invading a host. Choice (B) is incorrect, as all viruses have genomes and some have many genes. Choice (C) is also wrong because prokaryotes are cells and do not have nuclei. You can also eliminate (D), as viruses evolve through the same mechanisms that living things do.

42. **C** The origin of life refers to the period when the first life began. Life is shown to be present 3.5 billion years ago when no atmospheric oxygen was present yet, so (A) can be eliminated. Also, there was no life at 4 billion years despite having CO_2 and no O_2; therefore something else must be required, and (B) can be eliminated. There were no proteins when the first life arose so (D) is incorrect. Choice (C) states that self-replicating nucleic acids were necessary, which is why life could only occur when RNA appeared.

43. **B** Functional ribosomes would not likely be found before the proteins they make. Therefore, this would be between 3.5 and 3.25 billion years ago.

44. **B** The sign that photosynthesis began was when oxygen appeared. Oxygen was absent at 3.25 billion years and present at 3 billion years so photosynthesis must have begun between those times; eliminate (A) and (C). Autotrophs, not heterotrophs, perform photosynthesis. Heterotrophs rely on eating other things for their energy.

45. **B** The similarity suggests that humans and chimpanzees are more closely related than humans and dogs. Because these two organisms share similar amino acid sequences, humans must share more recent common ancestors with chimpanzees than with dogs.

46. **A** There is no way for these two parents to produce a type O child. The only genotype that will result in type O blood is two "O" alleles, one from each parent, since the allele for O blood is recessive to the alleles for A or B blood. The AB parent has alleles for only A and B blood, so it is impossible for these two individuals to produce a child with type O blood.

47. **B** The largest amount of energy is available to producers. Population B is most likely composed of producers because they have the largest biomass.

48. **B** An increase in the number of organisms in population C would most likely lead to a decrease in the biomass of B because population B is the food source for population C. Make a pyramid based on the biomasses given. If population C increases, population B will decrease. Eliminate (A) and

(C), as we cannot necessarily predict what will happen to the biomass of populations that are above population C. Choice (D) can also be eliminated because the food source available to population C would most likely decrease, not increase.

49. **B**   Chromosomes replicate during interphase, the S phase. Choices (A) and (C) are incorrect because during G_1 and G_2, the cell makes protein and performs other metabolic duties.

50. **A**   The first sign of prophase in mammalian cells is the appearance of chromosomes, which are usually invisible. They are condensed at the start of mitosis.

51. **C**   To induce cell cycle progression, an inactive CDK (cyclin-dependent kinase) binds a regulatory cyclin. Once together, the complex is activated, can affect many proteins in the cell, and causes the cell cycle to continue. To do this, the CDK transfers phosphate groups onto other molecules. If there are inadequate phosphate groups, CDK would not function properly and the cell cycle would not progress, which rules out (A). Tumor suppressor proteins inhibit cell cycle progression and can trigger apoptosis, which rules out (B). To inhibit cell cycle progression, CDKs and cyclins are kept separate. Together, they promote the cell cycle. If a cyclin is unable to release from its cyclin dependent kinase, the cell cycle would be promoted, not stopped, making (C) correct. Choice (D) could've been elimianted because an inhibitor of a cyclin gene would prevent expression of the cyclin gene. With lower levels of the cyclin protein, cell cycle progression would be inhibited.

52. **B**   Because neurons are not capable of dividing, it is reasonable to conclude that these cells will stay in the G_1 phase (specifically in an extension of G_1 called G_0). phase. This is a reading comprehension question. The passage states that cells that do not divide are arrested at the G_1 phase. Choice (A) is incorrect because these cells will not be committed to go through cell division. You can also eliminate (C) and (D), as the cells will not enter the M or S phase.

53. **D**   Since Snippeiq has the most differences with the other organisms it must be at position S. This means that Gerdellen cannot be there, which makes (D) the best answer. (Position Q or R is the best position for Gerdellen, though positions O and P could possibly be correct as well.)

54. **C**   Position III, (C), contains the oldest common ancestor because it is the node for all the current species designated by the lettered positions.

55. **A**   Snippeiq, (A), is the outgroup because it is the least related to the other species. The table shows that Snippeiq has the most differences from all of the other species.

56. **D**   Snorflak and Fixxels, (D), would have the most shared derived characters because they have the fewest amino acid differences. Snippeiq and Sqellert would be expected to have the least since they are so different, which rules out (A). Sqellert and Gerdellen, (B), were close with only 4 differences but still not as low as Snorflak and Fixxels. Snorflak and Gerdellen, (C), must be an intermediate with 9 differences observed in the table.

57. **B**   The two most closely related organisms are the two with the most shared derived characteristics.

58. **D**   Shared derived characteristics are newly evolved traits that are shared with every group on a phylogenic tree except for one. Vertebral columns are present in every group except for the sea anemone, so it must have evolved first. Walking legs are found only in the salamander, indicating that it most likely evolved most recently.

59. **C**   Pre-zygotic barriers to reproduction are those that prevent fertilization, so you can eliminate (A), (B), and (D). Choice (C) is an example of a post-zygotic barrier to reproduction.

60. **B**   Convergent evolution occurs when two organisms that are not closely related independently evolve similar traits, such as the wings of insects and birds. Divergent evolution occurs when two closely related individuals become different over time and can lead to speciation.

Section II: Free Response

Short student-style responses have been provided for each of the questions. These samples indicate an answer that would get full credit, so if you're checking your own response, make sure that the actual answers to each part of the question are similar to your own. The structure surrounding them is less important, although we've modeled it as a way to help organize your own thoughts and to make sure that you actually respond to the entire question.

Note that the rubrics used for scoring periodically change based on the College Board's analysis of the previous year's test takers. This is especially true as of the most recent Fall 2019 changes to the AP Biology exam! We've done our best to approximate their structure, based on our institutional knowledge of how past exams have been scored and on the information released by the test makers. However, the 2020 exam's free-response questions will be the first of their kind.

Our advice is to over-prepare. Find a comfortable structure that works for you, and really make sure that you're providing all of the details required for each question. Also, continue to check the College Board's website, as they may release additional information as the test approaches. For some additional help, especially if you're worried that you're not being objective in scoring your own work, ask a teacher or classmate to help you out. Good luck!

Question 1

a.   **Explain** why the ability to respond to a visual stimulus would be naturally selected. Would a rod receptor with a resting membrane potential of –65mV be selected for? (2 points)

If an organism could use visual cues, this could give it an advantage since it might be able to navigate the world more easily. It might access alternative ecological niches or visualize predators or other dangers in the world. Increasing survival will increase evolutionary fitness and having this trait will be selected for. A rod with a low resting membrane potential of –65mV would not be selected for because when the cell is hyperpolarized to such a large degree the passage says it does not respond to light stimuli anymore. This means that the visual cues will be absent and the trait won't be selected for.

b.   **Explain** how the control in Experiment 1 would differ from the control in Experiment 2. (3 points)

The control in Experiment 1 would be the rod cell in the vitreous humor that is not stimulated by light. Experiment 2 would have a similar control except that it would be in vitreous humor devoid of sodium and then without stimulation by light.

c.   **Describe** the relationship between a stimulus and hyperpolarization. Use evidence from the data. (3 points)

When the stimulus was a low light intensity, it caused a small hyperpolarization shown by the small downward bump in Figure 2. However, when the stimulus was a high light intensity, the membrane potential dipped very low, all the way to –65mV. The intensity of the light stimulus and the level of hyperpolarization have a direct relationship.

d.   If the rods were bathed in fluid with a high sodium content, **explain** how the resting membrane potential would change. How does the sodium change the resting membrane potential? (1 point)

If the sodium content were high in the fluid, it would make the resting membrane potential more positive/less negative, perhaps –20mV. This is the opposite of what happened when the sodium was removed from the vitreous humor. The cell would depolarize at rest because sodium is always flowing through the sodium channel and going into the cell and the positive sodium ions make it less negative. When a light stimulus occurs, the channel closes and the cell hyperpolarizes.

Question 2

a. **Explain** how biotic and abiotic factors affect where plants grow. (2 points)

Biotic factors are things that are alive, like plants and fungi and animals and bacteria. Some of these things compete with the plants, but other things like certain types of fungi and bacteria can help the plant acquire nutrients. Abiotic things are nonliving, like sunlight and soil, nutrients, water, and atmosphere composition. These things are also important, especially since plants need appropriate water and sunlight for photosynthesis.

b. **Construct** a graph of averages and standard error for the western and eastern flowers. (4 points)

        1 point—proper axes
        1 point—proper size of bars showing mean
        1 point—correct labels
        1 point—graphing error bars

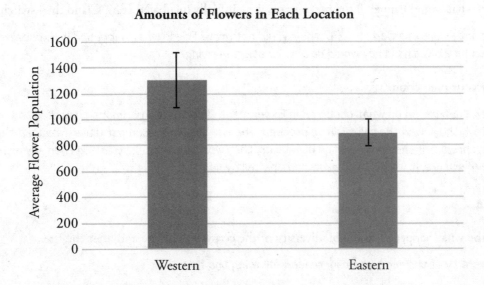

Amounts of Flowers in Each Location

c. **Perform** a chi-squared analysis using the average data for both flower types. (3 points)

Using the null hypothesis that the flowers do not prefer the Western side, we would have expected values of equal numbers living in each place. There were 2,196 flowers total in the average. So, we can expect that there would be 1,098 flowers in each location for the null hypothesis. For the western flowers observed – expected = 1,303 – 1,098 = 205 and for the eastern side observed – expected = 892 – 1,098 = –206. When we square each of them, it is 42,025 and 42,436. Divide each by the expected = 42,025/1,098 and 42,436/1,098 = 38.27 and 38.64. Those get added together = 76.91 for the chi-squared value.

d. Using your chi-squared analysis data, **explain** if you can reject the null hypothesis that flowers do not prefer the western side. (1 point)

Yes, the null hypothesis can be rejected. The flowers appear to grow significantly more on the Western side.

Question 3

a. **Describe** how a mutation in DNA can lead to a protein that no longer functions correctly.

Mutations in DNA are a mistake in the recipe used to make a protein. When the DNA gets expressed, it gets turned into an RNA during transcription. This causes an error in the mRNA. Then, the mRNA gets translated and the codons get read and amino acids get added to build the protein. If there was a mistake in the RNA, then the wrong amino acid would be added or maybe a Stop codon would be added. This can cause the protein to be the wrong shape when it folds up. Protein shape is important for protein function and if it is the wrong shape, it probably won't work correctly.

b. **Explain** the impact of the radiolabeled dye on the cell cycle.

The radiolabeled dye does not impact the cell cycle. It is only being used to help track the phases of the cell cycle and visualize them. We know the dye has no effect because it is used in the wild-type cells and it doesn't cause them to have an arrested cell cycle.

c. **Predict** what would happen if the *cdc* mutants were initially incubated at 37°C and then switched to 25°C.

If the mutants were started at 37°C, they would not grow. This is what happened in Experiment 1. If they were switched to 25°C, then they would be able to start growing.

d. **Justify** your prediction.

I predicted that based on what occurred in Experiment 2. In that experiment they started at the lower temperature which allowed the cell cycle to proceed. Then at the higher temperature it caused them to stop the cell cycle. I predict if they begin at the high temp, they won't have a cell cycle and then when they shift to the lower temperature, it will begin to grow and the cell cycle will continue.

Question 4

a. **Describe** what happened to the biodiversity of the ecosystem before and after the fire.

The biodiversity of the ecosystem increased following the fire.

b. **Explain** your description in (a) using the Simpson's Diversity Index.

Before the fire:

Insect	n	n/N	(n/N)²
Hazel-hued bugs	3	3/44 = 0.068	0.004624
Hairy bear moths	14	14/44 = 0.318	0.101124
Telluride tiger beetles	6	6/44 = 0.136	0.018496
Golden-winged butterflies	21	21/44 = 0.477	0.227529

D = 0.004624 + 0.101124 + 0.018496 + 0.227529 = 0.351773
Simpson's Diversity Index = 1 − D = 0.648227

After the fire:

Insect	n	n/N	(n/N)²
Hazel-hued bugs	10	10/44 = 0.227	0.05129
Hairy bear moths	9	9/44 = 0.205	0.042025
Telluride tiger beetles	15	15/44 = 0.341	0.098596
Golden-winged butterflies	10	10/44 = 0.227	0.05129

$D = 0.05129 + 0.042025 + 0.098596 + 0.05129 = 0.24301$

Simpson's Diversity Index = $1 - D = 0.756799$

The Simpson's Diversity Index increased, meaning that biodiversity of the ecosystem increased.

c. **Predict** what would happen to the biodiversity if there was an invasive insect species introduced into this ecosystem.

The addition of an invasive species is likely to decrease the biodiversity of the ecosystem.

d. **Justify** your prediction.

Since invasive species are unlikely to have a natural predator, they will likely increase in number and consume a large portion of the resources leading to other organisms being unable to consume those numbers and decreasing in number. Therefore a greater percentage of the organisms would be made up of the invasive species while fewer would be made up of the original species decreasing the biodiversity.

Question 5

a. **Explain** how it is possible for a person to have two rare unlinked genetic diseases at the same time.

Everybody has thousands of genes, and it is possible for a person to have mutations in more than one gene. This causes them to have a phenotype that is the combination of all their genes. If a person has the bad luck to have more than one mutated gene, then they could have more than one genetic disease at the same time.

b. **Calculate** the ratios of affected males and affected females for Disease 2.

There are 21 males and 4 are affected = 4/21

There are 20 females and 0 are affected = 0/21

c. **Identify** the likely inheritance patterns of Disease 1 and Disease 2.

Disease 1 is autosomal dominant

Disease 2 is X-linked recessive

d. **Describe** how inheritance patterns are shaped by the chromosomes upon which the genes are located.

If a gene is located on an X or a Y chromosome, then it will be inherited differently in males and female since males have XY and females have XX. This is sex-linked inheritance. In this pedigree with X-linked recessive, the affected males never have affected fathers. They get the mutated X from their mothers, but the mothers are never affected with the disease themselves. Things on autosomes get inherited the same in males and females because they each have the same number of autosomes.

Question 6

a. **Describe** how you can determine that tortoises raised at 34°C had the highest variability in enzyme reaction rates.

Russian tortoises raised at 34°C had the largest error bars and therefore the largest variability.

b. Reptiles in the study needed to have a gertimtonase reaction rate exceeding 10 mmol/s for an observed improvement in water retention. **Describe** the range of temperatures that Russian tortoises could be raised in to achieve this benefit.

Russian tortoises could be raised between 37°C and 30°C for gertimtonase to have the water retention effect.

c. Desert tortoises live in environments where the ground temperature can exceed 60°C. Using the data above, **predict and justify** how the graph would look for desert tortoise gertimtonase.

The graph's peak would shift to the higher temperatures. Because the tortoise has lived in the desert for many years, its enzymes you would expect to be optimized for a higher temperature. Therefore, we would expect to see the peak around 60°C with reaction rates dropping significantly at lower and higher temperatures.

d. **Explain** at a molecular level how the temperature change might affect gertimtonase.

Raising temperatures for Russian tortoises to those exceeding 37°C would likely denature the enzyme. Denaturation of the enzyme greatly limits its function which would lower the reaction rate.

Practice Test 3

AP® Biology Exam

SECTION I: Multiple-Choice Questions

DO NOT OPEN THIS BOOKLET UNTIL YOU ARE TOLD TO DO SO.

At a Glance

Total Time
1 hour and 30 minutes
Number of Questions
60
Percent of Total Score
50%
Writing Instrument
Pencil required

Instructions

Section I of this examination contains 60 multiple-choice questions.

Indicate all of your answers to the multiple-choice questions on the answer sheet. No credit will be given for anything written in this exam booklet, but you may use the booklet for notes or scratch work. After you have decided which of the suggested answers is best, completely fill in the corresponding oval on the answer sheet. Give only one answer to each question. If you change an answer, be sure that the previous mark is erased completely. Here is a sample question and answer.

Sample Question   Sample Answer

Chicago is a

(A)   state
(B)   city
(C)   country
(D)   continent

Use your time effectively, working as quickly as you can without losing accuracy. Do not spend too much time on any one question. Go on to other questions and come back to the ones you have not answered if you have time. It is not expected that everyone will know the answers to all the multiple-choice questions.

About Guessing

Many candidates wonder whether or not to guess the answers to questions about which they are not certain. Multiple-choice scores are based on the number of questions answered correctly. Points are not deducted for incorrect answers, and no points are awarded for unanswered questions. Because points are not deducted for incorrect answers, you are encouraged to answer all multiple-choice questions. On any questions you do not know the answer to, you should eliminate as many choices as you can, and then select the best answer among the remaining choices.

BIOLOGY
SECTION I
60 Questions
Time—90 minutes

<u>Directions:</u> Each of the questions or incomplete statements below is followed by four suggested answers or completions. Select the one that is best in each case and then fill in the corresponding oval on the answer sheet.

1. The resting membrane potential depends on which of the following?

   I. Active transport

   II. Selective permeability

   III. Differential distribution of ions across the axonal membrane

   (A) III only
   (B) I and II only
   (C) II and III only
   (D) I, II, and III

2. The Krebs cycle in humans releases

   (A) carbon dioxide
   (B) pyruvate
   (C) glucose
   (D) lactic acid

3. A heterotroph

   (A) obtains its energy from sunlight, harnessed by pigments
   (B) obtains its energy by catabolizing organic molecules
   (C) makes organic molecules from CO_2
   (D) obtains its energy by consuming exclusively autotrophs

4. Regarding meiosis and mitosis, one difference between the two forms of cellular reproduction is that in meiosis

   (A) there is one round of cell division, whereas in mitosis there are two rounds of cell division
   (B) separation of sister chromatids occurs during the second division, whereas in mitosis separation of sister chromatids occurs during the first division
   (C) chromosomes are replicated during interphase, whereas in mitosis chromosomes are replicated during the first phase of mitosis
   (D) spindle fibers form during interphase, whereas in mitosis the spindle fibers form during prophase

5. A feature of amino acids that is NOT found in carbohydrates is the presence of

   (A) carbon atoms
   (B) oxygen atoms
   (C) nitrogen atoms
   (D) hydrogen atoms

6. Which of the following is NOT a characteristic of prokaryotic cells?

   (A) Circular double-stranded DNA
   (B) Membrane-bound cellular organelles
   (C) Plasma membrane consisting of lipids and proteins
   (D) Ribosomes that synthesize polypeptides

7. Which of the following best explains why a population is described as the evolutionary unit?

   (A) Genetic changes can occur only at the population level.
   (B) The gene pool in a population remains fixed over time.
   (C) Natural selection affects individuals, not populations.
   (D) Individuals cannot evolve, but populations can.

8. The endocrine system maintains homeostasis using many feedback mechanisms. Which of the following is an example of positive feedback?

   (A) Infant suckling causes a mother's brain to release oxytocin, which in turn stimulates milk production.
   (B) An enzyme is allosterically inhibited by the product of the reaction it catalyzes.
   (C) When ATP is abundant, the rate of glycolysis decreases.
   (D) When blood sugar levels decrease to normal after a meal, insulin is no longer secreted.

GO ON TO THE NEXT PAGE.

9. A scientist carries out a cross between two guinea pigs, both of which have black coats. Black hair coat is dominant over white hair coat. Three quarters of the offspring have black coats, and one quarter have white coats. The genotypes of the parents were most likely

   (A) *bb   bb*
   (B) *Bb   Bb*
   (C) *Bb   bb*
   (D) *BB   Bb*

10. A large island is devastated by a volcanic eruption. Most of the horses die except for the heaviest males and heaviest females of the group. They survive, reproduce, and perpetuate the population. If weight is a highly heritable trait, which graph represents the change in population before and after the eruption?

    (A) A higher mean weight compared with their parents
    (B) A lower mean weight compared with their parents
    (C) The same mean weight as members of the original population
    (D) A higher mean weight compared with members of the original population

11. All of the following play a role in early embryogenesis EXCEPT

    (A) apoptosis
    (B) regulatory factors
    (C) operons
    (D) differentiation

12. During the period when life is believed to have begun, the atmosphere on primitive Earth contained abundant amounts of all the following gases EXCEPT

    (A) oxygen
    (B) hydrogen
    (C) ammonia
    (D) methane

Questions 13–14 refer to the following passage.

The digestive system in humans can be divided into two parts: the alimentary canal and the accessory organs. The canal comprised of the esophagus, stomach, and intestines is where the food actually passes during its transition into waste. The accessory organs are any organs that aid in the digestion by supplying the organs in the alimentary canal with digestive hormones and enzymes.

13. The small intestine is the main site of absorption. It can accomplish it so efficiently because of villi and microvilli that sculpt the membrane into hair-like projections. They likely aid in reabsorption by

    (A) increasing the surface area of the small intestine
    (B) decreasing the surface area of the small intestine
    (C) making the small intestine more hydrophilic
    (D) making the small intestine more hydrophobic

14. The pancreas is a major accessory organ in the digestive system. Which of the following would destroy the function of the digestive products produced by the pancreas?

    (A) A decrease in absorption rates within the alimentary canal
    (B) Removing the excess water from the food waste
    (C) Increased acidity due to the inability to neutralize stomach acid
    (D) An increase in peristalsis and subsequent diarrhea

15. In animal cells, which of the following represents the most likely pathway that a secreted protein takes as it is synthesized in a cell?

    (A) Plasma membrane–Golgi apparatus–ribosome–secretory vesicle–rough ER
    (B) Ribosome–Golgi apparatus–rough ER–secretory vesicle–plasma membrane
    (C) Plasma membrane–Golgi apparatus–ribosome–rough ER–secretory vesicle
    (D) Ribosome–rough ER–Golgi apparatus–secretory vesicle–plasma membrane

16. All of the following statements are correct regarding alleles EXCEPT

    (A) alleles are alternative forms of the same gene
    (B) alleles are found on corresponding loci of homologous chromosomes
    (C) a gene can have more than two alleles
    (D) an individual with two identical alleles is said to be heterozygous with respect to that gene

GO ON TO THE NEXT PAGE.

17. Once specific genes, such as the gene coding for ampicillin, have been incorporated into a plasmid, the plasmid may be used to carry out a transformation, which is

    (A) inserting it into a bacteriophage
    (B) treating it with a restriction enzyme
    (C) inserting it into a suitable bacterium
    (D) running a gel electrophoresis

18. Although mutations occur at a regular and predictable rate, which of the following statements is the LEAST likely reason the frequency of mutation often *appears* to be low?

    (A) Some mutations produce alleles that are recessive and may not be expressed.
    (B) Some undesirable phenotypic traits may be prevented from reproducing.
    (C) Some mutations cause such drastic phenotypic changes that they are soon removed from the gene pool.
    (D) The predictable rate of mutation results in ongoing variability in a gene pool.

19. A scientist wants to test the effect of temperature on seed germination. Which of the following should be part of the experimental design?

    (A) Use temperature as the dependent variable and alter the germination times
    (B) Use temperature as the independent variable and measure the rate of germination
    (C) Use temperature as the controlled variable and keep everything identical between groups
    (D) Use the variable natural outside temperature as a control group

20. A mustard plant seed undergoes a polyploidy event resulting in the new plant's pollen being unable to pollinate the plant that originally produced the seed despite them being less than a meter apart. This best exemplifies which of the following?

    (A) Allopatric speciation because the plants remain in close contact
    (B) Allopatric speciation because there is a significant geographic barrier
    (C) Sympatric speciation because it is the simplest form of speciation
    (D) Sympatric speciation because the plants are not separated

21. Which of the following is most correct concerning cell differentiation in vertebrates?

    (A) Cells in different tissues contain different sets of genes, leading to structural and functional differences.
    (B) Differences in the timing and expression levels of different genes lead to structural and functional differences.
    (C) Differences in the reading frame of mRNA lead to structural and functional differences.
    (D) Differences between tissues result from spontaneous morphogenesis.

Questions 22–23 refer to the following passage.

Pumping blood through the human heart must be carefully organized for maximal efficiency and to prevent backflow. In the figure below, the blood enters the heart through the vena cava (1), passes through the right atrium and right ventricle and then goes through the pulmonary artery toward the lungs. After the lungs, the blood returns through the pulmonary vein and then passes into the left atrium and the left ventricle before leaving the heart via the aorta.

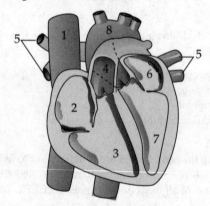

22. Which of the following chambers or vessels carry deoxygenated blood in the human heart?

    (A) 1 only
    (B) 2 and 3
    (C) 1, 2, 3, 4
    (D) 4 and 5

23. Blood is pumped via heart contractions triggered by action potentials spreading through the heart muscle. If there is a sudden increase in blood in chamber 3, which chamber of the heart received an increased number of action potentials?

    (A) Left atrium
    (B) Left ventricle
    (C) Right atrium
    (D) Right ventricle

GO ON TO THE NEXT PAGE.

24. Some strains of viruses can change normal mammalian cells into cancer cells in vitro. Which of the following is the best explanation for this impact on the mammalian cell?

    (A) A pilus is formed between the mammalian cell and the virus.
    (B) The viral genome incorporates into the mammalian cell's nuclear DNA.
    (C) The host's genome is converted into the viral DNA.
    (D) There is a viral release of spores into the mammalian cell.

25. All of the following correctly describe meiosis EXCEPT

    (A) meiosis produces four haploid gametes
    (B) homologous chromosomes join during synapsis
    (C) sister chromatids separate during meiosis I
    (D) crossing-over increases genetic variation in gametes

26. All of the following are examples of events that can prevent interspecies breeding EXCEPT

    (A) the potential mates experience geographic isolation
    (B) the potential mates experience behavioral isolation
    (C) the potential mates have different courtship rituals
    (D) the potential mates have different alleles

27. Which of the following is NOT a characteristic of asexual reproduction in animals?

    (A) Progeny cells have the same number of chromosomes as the parent cell.
    (B) Progeny cells are identical to the parent cell.
    (C) The parent cell produces diploid cells.
    (D) The progeny cells fuse to form a zygote.

28. Transpiration is a result of special properties of water. The special properties of water include all of the following EXCEPT

    (A) cohesion
    (B) adhesion
    (C) capillary action
    (D) hydrophobicity

Questions 29–32 refer to the following passage.

An experiment was performed to assess the growth of two species of plants when they were grown in different pHs, given different volumes of water, and watered at different times of day over 6 weeks. Two plants were grown of each species and the average heights (in cm) are shown in the table.

		Species A	Species B
pH	2	3.2	4.1
	4	37.6	20.6
	7	62.3	22.4
	10	48.4	13.5
	13	4.1	2.7
Volume (mL)	10	4.9	12.4
	20	19.2	38.9
	40	56.2	45.6
	80	65.1	21.5
	160	2.6	1.8
Time	12:00 A.M.	62.3	20.3
	7:00 A.M.	61.1	21.8
	12:00 P.M.	66.7	18.4
	7:00 P.M.	65.3	19.3

29. For which conditions do the species have different preferences?

    (A) pH
    (B) Volume
    (C) Volume and watering time
    (D) pH and volume and watering time

30. What are the preferred growth conditions for Species B?

    (A) pH 7, 40 mL, any time of day
    (B) pH 10, 40 mL, 7:00 A.M.
    (C) pH 7, 80 mL, any time of day
    (D) pH 10, 80 mL, 12:00 P.M.

31. Which pH and volume were likely used for the watering time experiment?

    (A) pH 4 and 40 mL
    (B) pH 7 and 40 mL
    (C) pH 4 and 80 mL
    (D) pH 7 and 80 mL

GO ON TO THE NEXT PAGE.

32. Which of the following would most improve the statistical significance of the results?

    (A) Let the plants grow for a longer period of time.
    (B) Add more conditions to test, such as amount of light and amount of soil.
    (C) Test the same plants with more pHs and more volumes and times of day.
    (D) Increase the number of plants in each group.

33. Photoperiodism in plants, in which plants respond to the stimulus of the day lengthening or shortening, can be best compared to which of the following phenomena in animals?

    (A) Viral infection
    (B) Increased appetite
    (C) Meiotic cell divison
    (D) Circadian rhythms

34. In most ecosystems, gross primary productivity, the total amount of chemical energy that producers create in a given time, is not entirely available for the consumers to utilize due to which explanation below?

    (A) Not all solar energy is in the correct spectrum for plants to absorb it.
    (B) Plants utilize some energy for the cellular respiration.
    (C) Heterotrophs do not absorb energy from autotrophs.
    (D) Very little biomass is available at the producer level.

35. Hawkmoths are insects that are similar in appearance and behavior to hummingbirds. Which of the following is LEAST valid?

    (A) These organisms are examples of convergent evolution.
    (B) These organisms were subjected to similar environmental conditions.
    (C) These organisms are genetically related to each other.
    (D) These organisms have analogous structures.

36. Which of the following describes a mutualistic relationship?

    (A) A tapeworm feeds off its host's nutrients, causing the host to lose large amounts of weight.
    (B) Certain plants grow on trees in order to gain access to sunlight, not affecting the tree.
    (C) Remora fish eat parasites off sharks. The sharks stay free of parasites, and the remora fish are protected from predators.
    (D) Meerkats sound alarm calls to warn other meerkats of predators.

37. The pancreas is an organ that makes insulin and glucagon in its beta and alpha cells, respectively. Insulin is released when blood glucose is high and glucagon is released when blood glucose is low. Anti-beta cell antibodies, which bind to their target and inhibit functionality, will cause which of the following to occur?

    (A) Glucagon secretion will stop, and blood glucose levels will not decrease.
    (B) Glucagon secretion will stop, and blood glucose levels will decrease.
    (C) Glucagon secretion will stop, and digestive enzymes will be secreted.
    (D) Insulin secretion will stop, and blood glucose levels will not decrease.

Questions 38–40 refer to the following passage.

The rainfall and biomass of several trophic levels in an ecosystem were measured over several years. The results are shown in the graph below.

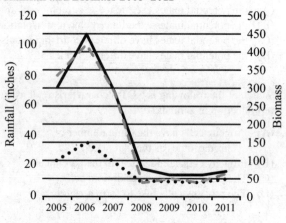

Rainfall and Biomass 2005–2011

— Producers
•••• Primary Consumers
– – Rainfall

38. Which of the following concepts is best demonstrated by this experiment?

    (A) Populations with higher genetic variation can withstand droughts better.
    (B) Meteorological impacts will affect the evolution of populations.
    (C) Environmental changes can affect all the levels of the ecosystem.
    (D) Unoccupied biological niches are dangerous because they attract invasive species.

GO ON TO THE NEXT PAGE.

39. If it rained 120 inches, what would you project the primary consumer biomass to be?

   (A) 150–200
   (B) 60
   (C) 45
   (D) 20

40. Which of the following graphs best depicts the projected biomass of secondary consumers if they were measured?

   (A) **Rainfall and Biomass 2005–2011**

   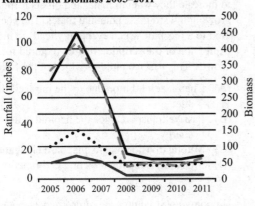

   Producers
   •••• Primary Consumers
   Secondary Consumers
   – – Rainfall

   (B) **Rainfall and Biomass 2005–2011**

   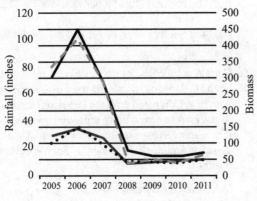

   Producers
   •••• Primary Consumers
   Secondary Consumers
   – – Rainfall

   (C) **Rainfall and Biomass 2005–2011**

   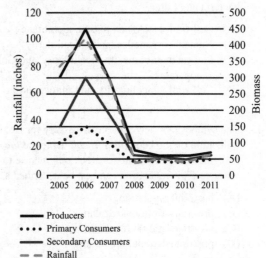

   Producers
   •••• Primary Consumers
   Secondary Consumers
   – – Rainfall

   (D) **Rainfall and Biomass 2005–2011**

   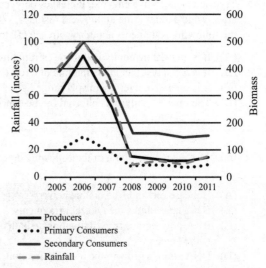

   Producers
   •••• Primary Consumers
   Secondary Consumers
   – – Rainfall

41. The calypso orchid, *Calypso bulbosa*, grows in close association with mycorrhizae fungi. The fungi penetrate the roots of the flower and take advantage of the plant's food resources. The fungi concentrate rare minerals, such as phosphates, in the roots and make them readily accessible to the orchid. This situation is an example of

   (A) parasitism
   (B) commensalism
   (C) mutualism
   (D) endosymbiosis

GO ON TO THE NEXT PAGE.

42. Which of the following are characteristics of both bacteria and fungi?

    (A) Cell wall, DNA, and plasma membrane
    (B) Nucleus, organelles, and unicellularity
    (C) Plasma membrane, multicellularity, and Golgi apparatus
    (D) Cell wall, unicellularity, and mitochondria

43. The synthesis of new proteins necessary for lactose utilization by the bacterium *E. coli* using the *lac* operon is regulated by the ability of RNA polymerase to bind and advance. This regulation can best be described as

    (A) bacterial regulation
    (B) pre-transcriptional regulation
    (C) pre-translational regulation
    (D) post-translational regulation

44. Trypsin is a digestive enzyme. It cleaves polypeptides after lysine and arginine amino acid residues. Which of the following statements about trypsin is NOT true?

    (A) It is an organic compound made of proteins.
    (B) It is a catalyst that alters the rate of a reaction.
    (C) It is operative over a wide pH range.
    (D) The rate of catalysis is affected by the concentration of substrate.

45. In DNA replication, which of the following does NOT occur?

    (A) Helicase unwinds the double helix.
    (B) DNA ligase links the Okazaki fragments.
    (C) RNA polymerase is used to elongate both chains of the helix.
    (D) DNA strands grow in the 5′ to 3′ direction.

46. Which of the following is true about genetic variation?

    (A) Mutation is the greatest source of genetic variation.
    (B) Ecosystems with high levels of genetic variation are not resistant to stress.
    (C) Mitosis provides genetic variation to most somatic cells.
    (D) Crossing-over prevents plants from undergoing speciation.

47. The energy given up by electrons as they move through the electron transport chain is used to

    (A) break down glucose
    (B) make glucose
    (C) produce ATP
    (D) make NADH

48. If a plant undergoing the light-dependent reactions of photosynthesis began to release $^{18}O_2$ instead of normal oxygen, one could most reasonably conclude that the plant had been supplied with

    (A) H_2O containing radioactive oxygen
    (B) CO_2 containing radioactive oxygen
    (C) $C_6H_{12}O_6$ containing radioactive oxygen
    (D) NO_2 containing radioactive oxygen

49. Yeast haploid cells secrete pheromones to other yeast to indicate they want to mate, and others respond by growing toward these potential mates. The pheromones bind to a receptor which eventually leads to increased expression of transcription factors required for this growth. This best exemplifies which of the following?

    (A) Chemical inhibition due to the intracellular binding site
    (B) Passive transport since the haploid cells require less energy
    (C) Mitotic division due to anaphase elongation
    (D) Signal transduction affecting cell function

50. Homologous structures are often cited as evidence for the process of natural selection. All of the following are examples of homologous structures EXCEPT

    (A) the forearms of a cat and the wings of a bat
    (B) the flippers of a whale and the arms of a man
    (C) the pectoral fins of a porpoise and the flippers of a seal
    (D) the forelegs of an insect and the forelimbs of a dog

51. Certain populations of finches have long been isolated on the Galapagos Islands off the western coast of South America. Compared with the larger stock population of mainland finches, these separate populations exhibit far greater variation over a wider range of species. The variation among these numerous finch species is the result of

    (A) convergent evolution
    (B) divergent evolution
    (C) disruptive selection
    (D) stabilizing selection

52. Which of the following contributes the MOST to genetic variability in a population?

    (A) Sporulation
    (B) Binary fission
    (C) Vegetative propagation
    (D) Mutation

GO ON TO THE NEXT PAGE.

Questions 53–55 refer to the following information and table.

A marine ecosystem was sampled in order to determine its food chain. The results of the study are shown below.

Type of Organism	Number of Organisms
Shark	2
Small crustaceans	400
Mackerel	20
Phytoplankton	1,000
Herring	100

53. Which of the following organisms in this population are secondary consumers?

    (A) Sharks
    (B) Phytoplankton
    (C) Herrings
    (D) Small crustaceans

54. Which of the following organisms has the largest biomass in this food chain?

    (A) Phytoplanktons
    (B) Mackerels
    (C) Herrings
    (D) Sharks

55. If the herring population is reduced by predation, which of the following would most likely be a secondary effect on the ecosystem?

    (A) The mackerels will be the largest predator in the ecosystem.
    (B) The small crustacean population will be greatly reduced.
    (C) The phytoplankton population will be reduced over the next year.
    (D) The small crustaceans will become extinct.

GO ON TO THE NEXT PAGE.

Questions 56–57 refer to the following information and diagram.

Scientists used embryology, morphology, paleontology, and molecular biology to create the phylogenetic tree below.

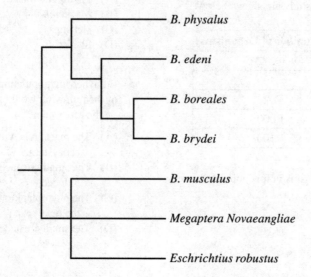

56. Which of the following is the least closely related according to the phylogenetic tree?

(A) *B. physalus* and *B. brydei*
(B) *B. boreales* and *B. brydei*
(C) *B. musculus* and *B. brydei*
(D) *B. musculus* and *Eschrichtius robustus*

57. Which of the sources of evolutionary evidence would be the most reliable of those listed?

(A) Embryology
(B) Morphology
(C) Paleontology
(D) Molecular biology

58. The brain of the frog is destroyed. A piece of acid-soaked paper is applied to the frog's skin. Every time the piece of paper is placed on its skin, one leg moves upward. Which of the following conclusions is best supported by the experiment?

(A) Reflex actions are not automatic.
(B) Some reflex actions can be inhibited.
(C) All behaviors in frogs are primarily reflex responses.
(D) This reflex action does not require the brain.

GO ON TO THE NEXT PAGE.

Questions 59–60 refer to the figure and chart below.

First Position (5' end)	Second Position	Third Position (3' end)			
		U	C	A	G
U	U	UUU UUC *Phenylalanine*		UUA UUG *Lucine*	
	C	UCU UCC UCA UCG *Serine*			
	A	UAU UAC *Tytrosine*		UAA UAG	
	G	UGU UGC *Cysteine*		UGA	UGG *Tryptophan*
C	U	CUU CUC CUA CUG *Leucine*			
	C	CCU CCC CCA CCG *Proline*			
	A	CAU CAC *Histidine*		CAA CAG *Glutamine*	
	G	CGU CGC CGA CGG *Arginine*			
A	U	AUU AUC AUA *Isoleucine*			AUG
	C	ACU ACC ACA ACG *Threonine*			
	A	AAU AAC *Asparagine*		AAA AAG *Lysine*	
	G	AGU AGC *Serine*		AGA AGG *Arginine*	
G	U	GUU GUC GUA GUG *Valine*			
	C	GCU GCC GCA GCG			
	A	GAU GAC *Aspartic Acid*		GAA GAG *Glutamic acid*	
	G	GGU GGC GGA GGG *Glycine*			

Table title: The Genetic Code: Codons of mRNA that Specify a Given Amino Acid

mRNA: A U G C C A C U A G C A C G U

Protein:

Met	Pro	Leu	Ala	Arg
Methionine	Proline	Leucine	Alanine	Arginine

Formation of a Protein

59. Which of the following DNA strands is the template strand that led to the amino acid sequence shown above?

(A) 3'-ATGCGACCAGCACGT-5'
(B) 3'-AUGCCACUAGCACGU-5'
(C) 3'-TACGGTGATCGTGCA-5'
(D) 3'-UACGGUGAUCGUGCA-5'

60. Immediately after the translation of methionine, a chemical is added which deletes all remaining uracil nucleotides in the mRNA. Which of the following represents the resulting amino acid sequence?

(A) Serine–histidine–serine–threonine
(B) Methionine–proline–glutamine–histidine
(C) Methionine–proline–leucine–alanine–arginine
(D) Methionine–proline–alanine–arginine–arginine

STOP

END OF SECTION I

IF YOU FINISH BEFORE TIME IS CALLED, YOU MAY CHECK YOUR WORK ON THIS SECTION. DO NOT GO ON TO SECTION II UNTIL YOU ARE TOLD TO DO SO.

BIOLOGY
SECTION II

6 Questions
Writing Time—90 minutes

<u>Directions:</u> Questions 1 and 2 are long free-response questions that should require about 25 minutes each to answer and are worth 8–10 points each. Questions 3 through 6 are short free-response questions that should require about 10 minutes each to answer and are worth 4 points each.

Read each question carefully and completely. Write your response in the space provided following each question. Only material written in the space provided will be scored. Answers must be written out in paragraph form. Outlines, bulleted lists, or diagrams alone are not acceptable unless specifically requested.

1. At what level does natural selection operate: the individual or the group? This is a central question in the field of sociobiology. In 1962, V. C. Wynne Edwards put forth his revolutionary group selection thesis, which states that animals avoid overexploitation of their habitats, especially with regard to food supply. In his theory, they accomplish this by altruistic restraint on the part of individuals who reduce their reproduction, or refrain altogether, to avoid overpopulation. Thus altruism is favored by natural selection.

   For example, small birds of the species *Parus major* typically produce nine or ten eggs per clutch, although they have been observed to produce as many as thirteen eggs per clutch. Data show that a clutch size larger than nine or ten actually produces fewer surviving offspring. See Figure 1; the vertical axis gives the percent occurrence of each brood size, and the numbers labelling the dots indicate the number of known survivors per nest.

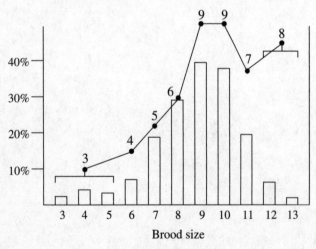

Figure 1

   Additional evidence for the group-selection theory is that there appears to be a relationship between reproductive success of individuals and the density of the population. When density is low, mortality is likewise low and

reproductive rate high. At high numbers resources are more scarce, and it is more difficult to stay alive and to reproduce, so mortality is high and reproductivity low. Figure 2 shows the number of surviving offspring per mating pair plotted against the number of breeding adults present (the graph covers several years).

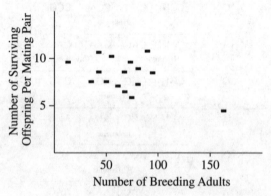

Figure 2

(a) **Describe** the carrying capacity in a population. Can a brood size have a carrying capacity? **Justify** your choice.

(b) This study involved observing birds in their natural habitat. **Identify** the independent variable in this brood size vs. percent survival study. **Describe** what makes an observational study different from a study inside a lab environment.

(c) **Describe** the brood size(s) in which 100% of the offspring survive. **Identify** which brood size has the highest percent of mortality.

(d) Over many more years the number of breeding adults was measured, and it was found that over a long time the number of breeding adults averaged around 100. **Describe** how this would change the information in Figure 1. **Justify** your answer.

GO ON TO THE NEXT PAGE.

2. Diabetes mellitus type 2 (T2DM) is a metabolic disorder characterized by insulin resistance on the part of the body's tissues. Onset is usually much later in life and highly associated with obesity. A recent study evaluated the relative dysfunction of mitochondria of individuals who were lean, obese, and diabetic (type 2). Muscle biopsies were taken both prior to and after a twenty-week exercise program. The mitochondrial mass was measured by cardiolipin, the citric acid cycle was measured by citrate synthase, and electron transport chain activity was measured by NADH oxidase levels. The results are shown below in the table.

Mean	Cardiolipin		Citrate synthase		NADH oxidase	
	Before	After	Before	After	Before	After
Lean ($n = 10$)	70	109	2.4	3.2	0.45	0.63
Obese ($n = 9$)	72	83	3.7	5.2	0.16	0.29
T2DM ($n = 11$)	58	86	3.3	5.1	0.15	0.28

Std Error	Before	After	Before	After	Before	After
Lean ($n = 10$)	6.1	5.2	0.2	0.4	0.1	0.12
Obese ($n = 9$)	7.9	4.8	0.3	0.5	0.05	0.1
T2DM ($n = 11$)	6.5	7.3	0.2	0.5	0.04	0.08

Table 1 Mean (upper rows) and standard error (lower rows) measurements for markers of ETC activity, in relative units normalized to creatine kinase activity.

(a) **Explain** how the citric acid cycle and the electron transport chain are related and how they are affected by exercise.

(b) **Construct** a graph of the data indicating citric acid cycle activity. Be sure to use error bars.

(c) **Identify** which group did not show an increase in cardiolipin after exercise was done. **Justify** your choice with data.

(d) If another measure of the citric acid cycle were added in a future experiment, **explain** what results you would expect.

3. Maternal inheritance is one pattern of inheritance which does not follow the rules of Mendelian genetics. It is an example of uniparental inheritance in which all progeny have the genotype and phenotype of the female parent.

Maternal inheritance can be demonstrated in the haploid fungus *Neurospora* by crossing the fungi in such a way that one parent contributes the bulk of the cytoplasm to the progeny. This cytoplasm-contributing parent is called the female parent, even though no true sexual reproduction occurs. The inheritance patterns of a mutant strain of *Neurospora* called poky have been studied using such crosses. Poky differs from the wild-type in that it is slow-growing and has abnormal quantities of cytochromes.

Investigators suspected that the poky mutation was carried in the mitochondria, instead of in the nuclear genome. The following experiments were designed to test this hypothesis.

Step 1: Mitochondria were extracted from poky *Neurospora* mutants.

Step 2: An ultrafine needle and syringe was used to inject these mitochondria into wild-type *Neurospora* cells.

Step 3: These recipient cells were cultured for several generations, and the phenotypes were examined.

Results: The poky phenotype was observed in some of the cultured fungi.

(a) In mitochondrial inheritance the offspring have the genotype and phenotype of the mother. Under Mendelian inheritance it is possible for the offspring to have a different genotype but the same phenotype as the mother. **Describe** how this is possible.

(b) **Explain** what important control(s) the scientists should do to account for the procedure in Step 2.

(c) If the nuclear DNA was injected into the wild-type *Neurospora* cells instead of the mitochondria, **explain** whether the resulting fungi would have the poky phenotype?

(d) **Justify** your prediction.

GO ON TO THE NEXT PAGE.

4. A diverse flock of finches that can eat many sizes of seeds lives on a small island in Polynesia, which recently experienced a drought. Trees producing robust, medium-sized seeds died out in large numbers but more resilient trees that produced both small and large-sized seeds are thriving. The finches can shelter in the trees or in burrows underground. The island is home to several tree species that do not produce edible seeds and small rodents that also eat the seeds.

(a) Finches are incapable of digging burrows yet approximately 25% of the finches found on the island live in underground dwellings. **Describe** the source of the finch dwellings.

(b) **Describe** the relationship between the finches and rodents on the island.

(c) Protein HITB8 is found to be present in high levels in medium-sized beak finches and less so in other finch types. **Predict** what will happen to the frequency of this protein product in the finch population after two generations if the drought continues.

(d) **Justify** your answer to (c).

5.

Chesapeake Bay Waterbird Food Web

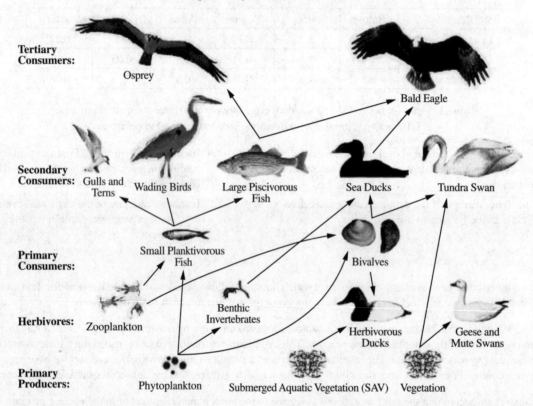

Tertiary Consumers: Osprey, Bald Eagle

Secondary Consumers: Gulls and Terns, Wading Birds, Large Piscivorous Fish, Sea Ducks, Tundra Swan

Primary Consumers: Small Planktivorous Fish, Bivalves

Herbivores: Zooplankton, Benthic Invertebrates, Herbivorous Ducks, Geese and Mute Swans

Primary Producers: Phytoplankton, Submerged Aquatic Vegetation (SAV), Vegetation

(a) **Describe** the importance of being the sole occupant of a niche in an ecosystem.

(b) **Describe** the impact of an invasive species that competes with phytoplankton and zooplankton.

(c) **Identify** the number of species with no natural predator on this food web.

(d) **Explain** why toxins are more of a concern for tertiary consumers than they are for primary consumers.

GO ON TO THE NEXT PAGE.

6. A teacher observes squirrels frequently crossing the road and wants to determine if squirrels prefer one side of the road to the other. At 4:00 P.M. for three non-consecutive days the teacher counts the number of squirrels on each side. The north side has predominantly pine trees, while the south side has mostly red maple trees.

	North Side	South Side
Monday	21	15
Wednesday	18	18
Friday	22	14
Average		

Table 1

(a) **Identify** the null hypothesis of this experiment.

(b) **Complete** the table with the average data and values necessary to test the hypothesis. (All white empty blanks should be filled.)

	O	E	O-E	$(O-E)^2$	$(O-E)^2/E$
North Side					
South Side					
				$\chi^2 =$	

(c) Use statistics to **determine** whether the null hypothesis should be rejected.

(d) **Explain** how the results of the experiment support the theory that animal behavior changes as their environment changes.

STOP

END OF EXAM

Practice Test 3:
Answers and
Explanations

PRACTICE TEST 3 ANSWER KEY

1.	D	21.	B	41.	C
2.	A	22.	C	42.	A
3.	B	23.	C	43.	B
4.	B	24.	B	44.	C
5.	C	25.	C	45.	C
6.	B	26.	D	46.	A
7.	D	27.	D	47.	C
8.	A	28.	D	48.	A
9.	B	29.	B	49.	D
10.	D	30.	A	50.	D
11.	C	31.	D	51.	B
12.	A	32.	D	52.	D
13.	A	33.	D	53.	C
14.	C	34.	B	54.	A
15.	D	35.	C	55.	C
16.	D	36.	C	56.	C
17.	C	37.	D	57.	D
18.	D	38.	C	58.	D
19.	B	39.	A	59.	C
20.	D	40.	A	60.	B

PRACTICE TEST 3 EXPLANATIONS

Section I: Multiple Choice

1. **D** The resting potential depends on active transport (the Na^+K^+-ATPase pump) and the selective permeability of the axon membrane to K^+ than to Na^+, which leads to a differential distribution of ions across the axonal membrane.

2. **A** The Krebs cycle releases carbon dioxide as the carbon molecules are broken down and electron carriers are generated. Glucose is at the start of glycolysis and pyruvate is at the end of glycolysis. Lactic acid is generated through fermentation.

3. **B** A heterotroph obtains its energy from organic molecules. An autotroph obtains energy from sunlight utilizing pigments such as chlorophyll and uses CO_2 and water to make organic molecules. Therefore, (A) and (C) can be eliminated. Heterotrophs can obtain their energy from ingesting autotrophs, but they can also consume other heterotrophs. So you can eliminate (D), leaving (B) as your answer.

4. **B** In meiosis, the sister chromatids separate during the second metaphase of meiosis (Meiosis II), whereas the sister chromatids separate during metaphase of mitosis. Choice (A) is incorrect because in meiosis there are two rounds of cell division, whereas in mitosis there is only one round of cell division. Chromosomes are replicated during interphase in both meiosis and mitosis, so (C) is incorrect. Choice (D) can also be eliminated because spindle fibers never form during interphase.

5. **C** Amino acids are organic molecules that contain carbon, hydrogen, oxygen, and nitrogen, so eliminate (A), (B), and (D). Don't forget to associate amino acids with nitrogen because of the amino group (NH_2).

6. **B** Unlike eukaryotes, prokaryotes (which include bacteria) do not contain membrane-bound organelles. Bacteria contain circular double-stranded DNA, ribosomes, and a cell wall, so (A) and (D) are incorrect. Also eliminate (C) because bacterial cell membranes are made up of a bilipid layer with proteins interspersed.

7. **D** Populations can be described as the evolutionary unit because changes in the genetic makeup of populations can be measured over time. Eliminate (A), as genetic changes occur only at the individual level. Only under Hardy-Weinberg equilibrium does the gene pool remain fixed over time in a population. However, this statement does not explain why the population is the evolving unit, so (B) is incorrect. Choice (C) is true but does not address the question.

8. **A** Positive feedback occurs when a stimulus causes an increased response. Choices (B), (C), and (D) are examples of negative feedback.

9. **B** In order to determine the genotype of the parents, use the ratio of the offspring given in the question and work backward. The ratio of black-haired to white-haired guinea pigs is 3:1. In order to get a white-haired offspring, each parent must have been able to contribute a *b* allele. However, since the parents were both black in color, they must have each been *Bb*.

10. **D** The mean weight of the offspring in the next generation will be heavier than the mean weight of the original population because all the lighter horses in the original population died off. The normal distribution for weight will therefore shift to the heavier end (to the right of the graph). You can therefore eliminate (C) because the mean weight should increase. The mean weight of the offspring could be heavier or lighter than their parents, so you can also eliminate (A) and (B).

11. **C** Apoptosis (programmed cell death), hox and homeotic genes (genes that control differentiation), and differentiation itself play a role in morphogenesis. Operons are sets of multiple genes regulated by a single regulatory unit in bacteria.

12. **A** The primitive atmosphere lacked oxygen (O_2). It contained methane (CH_4), ammonia (NH_3), and hydrogen (H_2).

13. **A** Villi and microvilli are fingerlike projections present in the small intestine which dramatically increase the surface area available for nutrient absorption.

14. **C** The pancreas produces digestive enzymes, and enzymes are sensitive to pH. The inability to neutralize stomach acid would disrupt the function of the enzymes. Food moving too quickly would decrease digestion/absorption rates but the function of the enzyme would be intact. Removal of water should not greatly affect enzyme function.

15. **D** Ribosomes are the site of protein synthesis. Therefore, the correct answer should start with ribosome. So eliminate (A) and (C). The polypeptide then moves through the rough ER to the Golgi apparatus, where it is modified and packaged into a vesicle. The vesicle then floats to the plasma membrane and is secreted. Choice (D) is your answer.

16. **D** Choice (D) is false because an individual with two identical alleles is said to be *homozygous,* not *heterozygous,* with respect to that gene. Alleles are different forms of the same gene found on corresponding positions of homologous chromosomes, so (A) and (B) are incorrect. More than two alleles can exist for a gene, but a person can have only two alleles for each trait.

17. **C** A transformation is the uptake of DNA by a bacterium. DNA uptake by a virus (A) would be a transduction, although this requires a virus to gain the DNA via infection of a bacterium.

18. **D** The least likely explanation for why mutations are low is that mutations produce variability in a gene pool. Any gene is bound to mutate. This produces a constant input of new genetic information into a gene pool. Choice (D) is therefore correct, because it doesn't give us any additional information about the rate of mutations. Some mutations are subtle and cause only a slight decrease in reproductive output, so eliminate (A). Some mutations are harmful and decrease the productive success of the individual, so (B) is incorrect. Some mutations are deleterious and lead to total reproductive failure. The zygote fails to develop. Therefore, (C) can also be eliminated.

19. **B** The independent variable is the condition the scientist sets up. The dependent variable is the thing that is the measured outcome of the experiment. The scientist should set different temperatures and then measure the germination rate. The outdoor temperature is interesting, but it is too variable to be a good control for the groups with a set temperature.

20. **D** Sympatric speciation occurs when there is no geographic barrier separating the two species (as is the case here). Allopatric speciation requires a geographic barrier.

21. **B** Every cell in an organism has the same set of genes. Differences in the timing and expression levels of these genes lead to structural and functional differences.

22. **C** Deoxygenated blood flows through all chambers before going to the lungs. This means the correct answer is chambers 1, 2, 3, and 4.

23. **C** The action potentials cause the heart to contract. If chamber 3 is full of more blood, then that would indicate that the preceding chamber, chamber 2 (right atrium), had suddenly contracted a larger amount.

24. **B** Normal cells can become cancerous when a virus invades the cell and takes over the replicative machinery. This would occur if the mammalian genome is altered, such as if the viral genome inserted itself and disrupted mammalian gene expression. A pilus forms between two bacteria, so (A) is wrong. Also eliminate (C) because the host's genome is not converted to the viral genome. Choice (D) is incorrect because spores are released by fungi, not viruses.

25. **C** Crossing-over and synapsis occur during meiosis, which produces haploid gametes. Separation of homologous chromosomes occurs during meiosis I, while separation of sister chromatids does not occur until meiosis II.

26. **D** If potential mates have different alleles, this doesn't keep them from mating. Having different alleles of a gene is just a normal part of genetic variation. Use common sense to eliminate the other answer choices. If the organisms don't meet, they won't reproduce; eliminate (A). Also eliminate (B) and (C): if potential mates do not share the same behaviors (such as courtship rituals), they may not mate.

27. **D** There is no union of gametes in mitosis. Choices (A) and (C) are incorrect: asexual reproduction involves the production of two new cells with the same number of chromosomes as the parent cell. If the parent cell is diploid, then the daughter cells will be diploid. The daughter cells are identical to the parent cell, so eliminate (B).

28. **D** Cohesion, adhesion, and capillary action are special properties of water that are necessary for transpiration to occur. Water is a hydrophilic, polar molecule.

29. **B** The two species differ in their preference for water volume. Species A prefers 80 mL, and Species B prefers 40 mL. The different watering times do not seem to play a role in growth as there is no obvious pattern and all of the times had similar growth. Both plants prefer pH 7.

30. **A** The top growth conditions for Species B are pH 7, 40 mL, and any time of day. In those conditions, the Species B plants grew the tallest.

31. **D** The plants in the watering time experiment seemed to be around 60 and 20 cm in height. This corresponds to a pH of 7 rather than 4 and a watering volume of 80 mL rather than 40 mL.

32. **D** Choices (A), (B), and (C) would all give interesting information about the experiment, but the only one that would make the results more statistically significant is to increase the number of plants in each group. Two plants is not enough to be sure about the results of the experiment.

33. **D** Photoperiodism is a response of plants to the changing length of day/night. It is a way that plants respond to a signal from the environment. Animal circadian rhythms are also a response to the environment (D). Viral infection (A), increased appetite (B), and meiotic division (C) are not responses to the environment.

34. **B** The gross primary productivity is the total amount of energy stored, but the plant has energy needs of its own that are fulfilled with cellular respiration. Choice (A) is true but not relevant to this question. Choice (C) is not true. Heterotrophs must get energy from autotrophs and (D) is not true because most biomass is stored at the producer level.

35. **C** Choices (A) and (B) are incorrect: these organisms exhibit the same behavior because they were subjected to the same environmental conditions and similar habitats. This is an example of convergent evolution. However, they are not genetically similar, so (C) is the answer. (One is an insect, and the other a bird.) They are analogous, so (D) is also incorrect. They exhibit the same function but are structurally different.

36. **C** A mutualistic relationship is a relationship among two organisms in which both benefit. Choice (A) describes parasitism, and (B) describes commensalism. Choice (D) is an example of altruistic behavior.

37. **D** Beta cells secrete insulin. Binding of antibodies to the beta cells in the pancreas will halt the production of insulin. Therefore, eliminate (A), (B), and (C). This will lead to an increase in blood glucose levels.

38. **C** The graph does show a drought, but it says nothing about genetic variation. It is true that the rainfall could affect evolution of the population, but the graph doesn't address evolution. It shows only the decrease in biomass. Choice (C) is the best answer since it addresses how the environmental effects ripple through the levels of the ecosystem. Invasive species do not need an unoccupied niche, and that is not shown in the graph.

39. **A** The biomass seems to correlate fairly well with the rainfall. Therefore, a rainfall higher than any recorded would give a biomass higher than any recorded. The axis on the right shows the biomass. With a rainfall of 120, the biomass should be greater than 150.

40. **A** The secondary consumers' biomass should always be less than that of the primary consumers, so the only option is (A).

41. **C** This is an example of mutualism. Both organisms benefit. Choice (A), parasitism, is a type of symbiotic relationship in which one organism benefits and the other is harmed. Choice (B), commensalism, occurs when one organism benefits and the other is unaffected. Choice (D), endosymbiosis, is the idea that some organelles originated as symbiotic prokaryotes that live inside larger cells.

42. **A** Both bacteria and fungi contain genetic material (DNA), a plasma membrane, and a cell wall. Unlike fungi, bacteria lack a definite nucleus. Therefore, eliminate (B). Bacteria are unicellular, whereas fungi are both unicellular and multicellular. Therefore, eliminate (C) and (D).

43. **B** This question tests your understanding of what stage of gene expression utilizes RNA polymerase. Since the presence of RNA polymerase is required for transcription, the regulation is pretranscriptional.

44. **C** Trypsin is an enzyme. Enzymes are proteins and organic catalysts that speed up reactions without altering them. They are not consumed in the process. Therefore, you can eliminate (A) and (B). The rate of reaction can be affected by the concentration of the substrate up to a point, so (D) can be eliminated.

45. **C** DNA polymerase, not RNA polymerase, is the enzyme that causes the DNA strands to elongate. DNA helicase unwinds the double helix, so (A) is true and therefore incorrect. Choice (B), which states that DNA ligase seals the discontinuous Okazaki fragments, is also true. Eliminate it. In the presence of DNA polymerase, DNA strands always grow in the 5′ to 3′ direction as complementary bases attach. Therefore, (D) is also incorrect.

46. **A** Ecosystems high in variation are more resilient than those with low biodiversity (so (B) is out). Meiosis provides genetic variation but mitosis should provide almost none, so eliminate (C). Since crossing-over is a source of genetic variation it can be a source of speciation (making (D) incorrect). Choice (A) is the only correct answer.

47. **C** Electrons passed down along the electron transport chain from one carrier to another lose energy and provide energy for making ATP. Glucose is decomposed during glycolysis, but this process is not associated with energy given up by electrons; eliminate (A). Glucose is made during photosynthesis, so eliminate (B). NADH is an energy-rich molecule, which accepts electrons during the Krebs cycle. Therefore, (D) is incorrect as well.

48. **A** The oxygen released during the light reaction comes from splitting water. (Review the reaction for photosynthesis.) Therefore, water must have originally contained the radioactive oxygen. Carbon dioxide is involved in the dark reaction and produces glucose, so eliminate (B). Glucose is the final product and would not be radioactive unless carbon dioxide was the radioactive material, so (C) is incorrect. Finally, eliminate (D), because nitrogen is not part of photosynthesis.

49. **D** A ligand binding to a receptor and causing a cellular response is signal transduction. In this case, the ligand is the pheromone and the response is increased expression of transcription factors. All other answers do not make sense in this context.

50. **D** Homologous structures are organisms with the same structure but different functions. The forelegs of an insect and the forelimbs of a dog are not structurally similar. (One is an invertebrate, and the other a vertebrate.) They do not share a common ancestor. However, both structures are used for movement. All of the other examples are vertebrates that are structurally similar.

51. **B** Speciation occurred in the Galapagos finches as a result of the different environments on the islands. This is an example of divergent evolution. The finches were geographically isolated. Choice (A), convergent evolution, is the evolution of similar structures in distantly related organisms. Choice (C), disruptive selection, is selection that favors both extremes at the expense of the intermediates in a population. Choice (D), stabilizing selection, is selection that favors the intermediates at the expense of the extreme phenotypes in a population.

52. **D** Mutations produce genetic variability. All of the other answer choices are forms of asexual reproduction.

53.   **C**   Secondary consumers feed on primary consumers. If you set up a pyramid of numbers, you'll see that the herrings belong to the third trophic level.

54.   **A**   The biomass is the total bulk of a particular living organism. The phytoplankton population has both the largest biomass and the most energy.

55.   **C**   A decrease in the herring population will lead to an increase in the number of crustaceans and a decrease in the phytoplankton population. Reorder the organisms according to their trophic levels and determine which populations will increase and decrease accordingly.

56.   **C**   Relatedness is shown on a phylogenetic tree by how far back you have to go to reach the common ancestor of the two species (shown by a branch node). Since the recent species are on the right, traveling to the left goes back in time. Because you have to go to the oldest common ancestor to connect *B. musculus* and *B. brydei*, they are the least closely related. Even though they are near each other on the tree, clades can rotate around a node so this does not matter.

57.   **D**   Because molecular biology demonstrates where the actual mutations occured (rather than just what phenotype is observed) it is the most reliable, because it allows us to detect changes that are not phenotypically observable.

58.   **D**   Because the brain is destroyed, it is not associated with the movement of the leg. Choice (A) is incorrect, as reflex actions are automatic. Choices (B) and (C) can also be eliminated; both of these statements are true but are not supported by the experiment.

59.   **C**   The DNA template strand is complementary to the mRNA strand. Using the mRNA strand, work backward to establish the sequence of the DNA strand. Don't forget that DNA strands do not contain uracil, so eliminate (B) and (D).

60.   **B**   Use the amino acid chart to determine the sequence after uracil is deleted. The deletion of uracil creates a frameshift.

Section II: Free Response

Short student-style responses have been provided for each of the questions. These samples indicate an answer that would get full credit, so if you're checking your own response, make sure that the actual answers to each part of the question are similar to your own. The structure surrounding them is less important, although we've modeled it as a way to help organize your own thoughts and to make sure that you actually respond to the entire question.

Note that the rubrics used for scoring periodically change based on the College Board's analysis of the previous year's test takers. This is especially true as of the most recent Fall 2019 changes to the AP Biology exam! We've done our best to approximate their structure, based on our institutional knowledge of how past exams have been scored and on the information released by the test makers. However, the 2020 exam's free-response questions will be the first of their kind.

Our advice is to over-prepare. Find a comfortable structure that works for you, and really make sure that you're providing all of the details required for each question. Also, continue to check the College Board's website, as they may release additional information as the test approaches. For some additional help, especially if you're worried that you're not being objective in scoring your own work, ask a teacher or classmate to help you out. Good luck!

Question 1

a. **Describe** the carrying capacity in a population. Can a brood size have a carrying capacity? **Justify** your choice. (2 points)

A carrying capacity is the maximum amount of individuals in a population that an environment can support with the biotic and abiotic factors available. There just isn't enough space, food, etc., for more individuals to survive well. A brood could probably have a carrying capacity because the mother/father might be limited in resources that the babies will need. For instance, as the number of babies increases, it becomes more difficult to protect them from predators, or to gather enough food. In this data, it looks like the carrying capacity is around 9.

b. This study involved observing birds in their natural habitat. **Identify** the independent variable in this brood size vs. percent survival study. **Describe** what makes an observational study different from a study inside a lab environment. (3 points)

The independent variable would be the number of offspring in a brood; the scientist would count how many survive from each brood size. In an observational study, it is difficult to control anything other than the independent variable. Maybe the nests being studied are in slightly different environments. Maybe the parents of different broods are better than others. There is a lot that might be different between groups in an observational study. In a lab study, variables are more easily controlled, but this is an artificial setup that doesn't quite mimic the natural environment.

c. **Describe** the brood size(s) in which 100% of the offspring survive. **Identify** which brood size has the highest percent of mortality. (2 points)

For a brood size of 3 it shows that 3 offspring survive, so 100%. For a brood size of 13, only 8 survive. 8/13 is the least surviving percentage so a brood size of 13 has the highest mortality percentage.

d. Over many more years the number of breeding adults was measured, and it was found that over a long time the number of breeding adults averaged around 100. **Describe** how this would change information in Figure 1. **Justify** your answer. (3 points)

It would not change Figure 1, which measures only the percentage of surviving offspring per brood size. If the number of surviving offspring were counted, then the information could be put on Figure 2, but just the number of mating adults is not enough for Figure 1 or Figure 2.

Question 2

a. **Explain** how the citric acid cycle and the electron transport chain are related and how they are affected by exercise. (2 points)

The citric acid cycle (Krebs cycle) is an important part of cell respiration. The reduced electron carriers created during the citric acid cycle bring their electrons to the electron transport chain, which is used to generate ATP. During exercise the muscle cells will need more energy. The process of cell respiration will speed up. When muscle cells have exhausted their stores of oxygen, then the electron transport chain will cease and fermentation will take over.

b. **Construct** a graph of the data indicating citric acid cycle activity. Be sure to use error bars. (4 points)

    1 point—proper axes
    1 point—proper labels/data organization
    1 point—proper size of bars
    1 point—graphing error bars

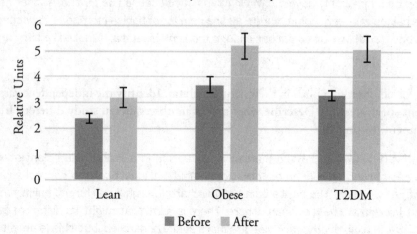

Citrate Synthase Activity

■ Before   ■ After

c. **Identify** which group did not show an increase in cardiolipin after exercise was done. **Justify** your choice with data. (2 points)

The obese group did not show an increase after exercise. Although the numbers 72 and 83 seem different, when the std error is considered, the values overlap. 72 + 7.9 = 79.9 and 83 – 4.8 = 78.2. The lean and T2DM participants each saw an increase in cardiolipin with exercise.

d. If another measure of the citric acid cycle were added in a future experiment, **explain** what results you would expect. (2 points)

If another measure were included, I would expect an increase for all groups after exercise.

Question 3

a. In mitochondrial inheritance the offspring have the genotype and phenotype of the mother. Under Mendelian inheritance it is possible for the offspring to have a different genotype but the same phenotype as the mother. **Describe** how this is possible.

It is possible for a mother and offspring to have a different genotype and the same phenotype under classic Mendelian inheritance because there are more than one genotype that have the same phenotype. For example, a homozygous dominant and a heterozygous genotype will both have the dominant phenotype.

b. **Explain** what important control(s) the scientists should do to account for the procedure in Step 2.

It would be important to make sure the injection process itself is not causing a problem. There should be a control group of wild-type cells that is injected with saline to account for the injection and make sure it is the actual mitochondria that is causing the mutant phenotype.

c. If the nuclear DNA was injected into the wild-type *Neurospora* cells instead of the mitochondria, **explain** whether the resulting fungi would have the poky phenotype.

They should NOT have the poky phenotype.

d. Justify your prediction.

The resulting fungi should not have the poky phenotype because the trait is not carried on a nuclear genome. It is part of the mitochondrial genome, and giving the nuclear genome is not going to provide the gene for the poky phenotype.

Question 4

a. Finches are incapable of digging burrows yet approximately 25% of the finches found on the island live in underground dwellings. **Describe** the source of the finch dwellings.

The source of the burrows must come from some other animal on the island. Small rodents could dig the burrows which the finches live in opportunistically either simultaneously or after the rodents have abandoned the burrows.

b. Describe the relationship between the finches and rodents on the island.

The finches and the rodents have a symbiotic relationship. The rodents may provide shelters for the birds and the birds spread seeds which will eventually develop into a food source for the rodents as well. This describes a mutualistic behavior, or the relationship could be commensalistic where the birds benefit but the rodents do not benefit from the birds.

c. Protein HITB8 is found to be present in high levels in medium-sized beak finches and less so in other finch types. **Predict** what will happen to the frequency of this protein product in the finch population after two generations if the drought continues.

The protein HITB8 will decrease in abundance.

d. Justify your answer to (c).

A drought will make surviving on the island more difficult due to the birds' decreased ability to find the food with their beak type. As such, the frequency of the gene encoding the HITB8 protein will be less plentiful and the protein will be expressed in lower numbers.

Question 5

a. Describe the importance of being the sole occupant of a niche in an ecosystem.

A niche is helpful because it is a set of resources that an organism is well-suited to utilize. Having no other competitors in the niche means that there is a set of resources (physical space, food, water, etc.) that the organism will have exclusive access to.

b. Describe the impact of an invasive species that competes with phytoplankton and zooplankton

If an invasive species competed with phytoplankton and zooplankton they would possibly disappear. Invasive species often have no natural enemies and they can steal resources from the other occupants in their niche that still do have to worry about predators. The consumers of the zooplankton and phytoplankton might also disappear when their food source disappears and an invasive species can cause trouble throughout the food web.

c. **Identify** the number of species with no natural predator on this food web.

There are 7 species that have no natural predator (they have no lines pointing away from them).

d. **Explain** why toxins are more of a concern for tertiary consumers than they are for primary consumers.

Toxins are a concern higher up the food web because they can often become concentrated. For example, it is bad for a toxin to be absorbed by a plant, but then a fish eats 10 toxin-affected plants and a bigger fish eats 10 of the little fish, which is like 100 toxin-affected plants. Higher up the food web toxins can concentrate and the animals can begin to be harmed by the level of toxin even though the primary consumers were not.

Question 6

a. **Identify** the null hypothesis of this experiment

Squirrels do not prefer one side of the road over another.

b. **Complete** the table with the average data and values necessary to test the hypothesis. (All white empty blanks should be filled.)

	O	E	O-E	$(O-E)^2$	$(O-E)^2/E$
North Side	20.33	18	2.33	5.4289	0.3016
South Side	15.66	18	–2.33	5.4289	0.3016
				$\chi^2 =$	0.603211

c. Use statistics to **determine** whether the null hypothesis should be rejected.

Since our degree of freedom is 1 and we are using $p = 0.05$, our critical value is 3.841. Our χ^2 value is less than our critical value so we fail to reject our null hypothesis.

d. **Explain** how the results of the experiment support the theory that animal behavior changes as their environment changes.

According to our data, the squirrels do not prefer one side of the road over the other. As the road was constructed, what was likely a continuous forest became separated. As both pine trees and maples provide food and shelter resources, the squirrels have adapted to the change in their habitat by migrating to both sides of the road frequently.

(Any answer properly linking the experimental data to changes in the environment and the squirrel behavior earns this point.)

Practice Test 4

AP® Biology Exam

SECTION I: Multiple-Choice Questions

DO NOT OPEN THIS BOOKLET UNTIL YOU ARE TOLD TO DO SO.

At a Glance

Total Time
1 hour and 30 minutes
Number of Questions
60
Percent of Total Score
50%
Writing Instrument
Pencil required

Instructions

Section I of this examination contains 60 multiple-choice questions.

Indicate all of your answers to the multiple-choice questions on the answer sheet. No credit will be given for anything written in this exam booklet, but you may use the booklet for notes or scratch work. After you have decided which of the suggested answers is best, completely fill in the corresponding oval on the answer sheet. Give only one answer to each question. If you change an answer, be sure that the previous mark is erased completely. Here is a sample question and answer.

Sample Question   Sample Answer

Chicago is a

(A)   state
(B)   city
(C)   country
(D)   continent

Use your time effectively, working as quickly as you can without losing accuracy. Do not spend too much time on any one question. Go on to other questions and come back to the ones you have not answered if you have time. It is not expected that everyone will know the answers to all the multiple-choice questions.

About Guessing

Many candidates wonder whether or not to guess the answers to questions about which they are not certain. Multiple-choice scores are based on the number of questions answered correctly. Points are not deducted for incorrect answers, and no points are awarded for unanswered questions. Because points are not deducted for incorrect answers, you are encouraged to answer all multiple-choice questions. On any questions you do not know the answer to, you should eliminate as many choices as you can, and then select the best answer among the remaining choices.

BIOLOGY
SECTION I
60 Questions
Time—90 minutes

Directions: Each of the questions or incomplete statements below is followed by four suggested answers or completions. Select the one that is best in each case and then fill in the corresponding oval on the answer sheet.

1. Klinefelter's syndrome is a genetic condition in males that results in the creation of an extra X chromosome during the early stages of embryo development, which must be inactivated. Which of the following karyotypes represents a person with Klinefelter's?

(A)

(B)

(C)

(D)

2. The restriction enzyme BAMHI predictably cuts the following sequence at the locations designated by the stars.

5' G*GATCC   3'

3' CCTAG*G   5'

Which of the following DNA helix sequences would be cut into 3 pieces by BAMHI?

(A) 5' GGATCC   3'
     3' CCTAGG   5'
(B) 5' GGATCCGGATCCGGATCC   3'
     3' CCTAGGCCTAGGCCTAGG   5'
(C) 3' GGATCC   5'
     5' CCTAGG   3'
(D) 5' AAGGATCCGGATCCAA   3'
     3' TTCCTAGGCCTAGGTT   5'

Questions 3–5 refer to the following passage.

In a species of peas, green (*G*) peas are known to be classically dominant over yellow (*g*) peas. A true breeding green plant and a true breeding yellow plant are crossed. The resulting F_1 generation is evaluated, and its phenotype distribution is shown in Table 1. Two green pea producing members of the F_1 generation are then crossed, and the resulting F_2 generation phenotype distribution is shown in Table 2.

Table 1. F_1 Generation

Type of Peas Produced	# of Plants
Green	1,432
Yellow	1

Table 2. F_2 Generation

Type of Peas Produced	# of Plants
Green	1,196
Yellow	374

3. You are given a green pea plant with an unknown genotype. You want to figure out the genotype by crossing it with another plant and looking at the offspring. Which of the following plants would be the most helpful in determining the genotype of the original plant?

(A) Any green pea plant
(B) A homozygous green plant
(C) A yellow pea plant
(D) None of the above

4. Which of the following best explains the single yellow pea producing plant in the F_1 generation?

(A) Crossing-over occurred in the parental pea plant to provide genetic variation.
(B) The law of independent assortment: one homologous pair separated independently of the others during gamete production in the parental pea plant.
(C) Nondisjunction in one of the parental pea plants caused an imbalance in chromosome distribution.
(D) A DNA polymerase error occurred during DNA replication just prior to meiosis.

5. If a yellow plant from the F_2 generation is crossed with a green plant from the F_1 generation, what would be the green-to-yellow phenotype ratio?

(A) 1:1
(B) 3:1
(C) 1:0
(D) 1:3

GO ON TO THE NEXT PAGE.

Question 6 refers to the following experiment.

A group of *Daphnia*, small crustaceans known as water fleas, was placed in one of three culture jars of different sizes to determine their reproductive rate. There were 100 females in the jar. The graph below shows the average number of offspring produced per female each day in each jar of pond water.

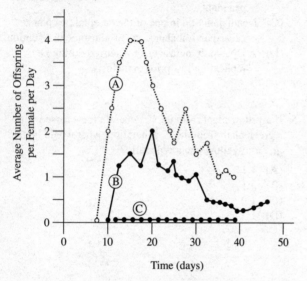

Key: (A) Water fleas in a 1-liter jar of pond water
(B) Water fleas in a 0.5-liter jar of pond water
(C) Water fleas in a 0.25-liter jar of pond water

6. The data in the figure above would best support which conclusion?

(A) If you decreased the number of females, the container would have to remain constant.
(B) The number of offspring produced scales proportionally with the container's size.
(C) The number of offspring produced increases with time.
(D) *Daphnia* prefer high-density conditions to have the most efficient reproductive rate.

7. Compact bone contains rings of osteons, each of which contains a central canal housing the blood vessels, which can be accessed only by those osteocytes adjacent to it. Gap junctions, which are tunnels between neighboring cells, then allow the cells that are not adjacent to the central canal to exchange nutrients and wastes with the bloodstream. Which statement is true about the exchange of materials with the bloodstream?

(A) Hydrophobic carbon dioxide waste must be passed via gap junctions from high concentration to low concentration.
(B) Hydrophilic Ca^{2+} is passed via gap junctions from low concentration to high concentration.
(C) Hydrophobic carbon dioxide waste does not require the gap junctions to travel from high concentrations to low concentrations.
(D) Hydrophilic Ca^{2+} does not require gap junctions to travel from high concentration to low concentrations.

Questions 8–10 refer to the following passage.

The somatic cells in a newly identified sexually reproducing species are found to be octoploidy (8n), and each cell contains 32 chromosomes, slightly fewer than the 46 chromosomes in a somatic human cell. The cell cycle of this species is similar to ours, and gametes are made during meiosis. Two gametes will come together at fertilization to create octoploidy offspring.

8. How many chromosome segments are present in a somatic cell at the completion of mitosis?

(A) 64
(B) 32
(C) 16
(D) 8

9. How many unique, non-homologous chromosomes are present in this species?

(A) 4
(B) 8
(C) 16
(D) 32

10. How many chromosome segments will be in each gamete if you count each homologous member individually?

(A) 32
(B) 16
(C) 8
(D) 4

GO ON TO THE NEXT PAGE.

Questions 11–14 refer to the following passage.

A population of goldfish in a large, isolated pond were studied in 1958 and again in 2008. The fishes' pigment level varied from pale white-orange fish to dark brown-orange fish. The color of each of the fish was recorded in the figure below.

1958 Goldfish Phenotype Distribution

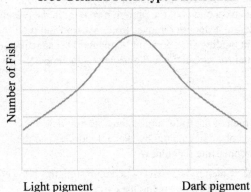

2008 Goldfish Phenotype Distribution

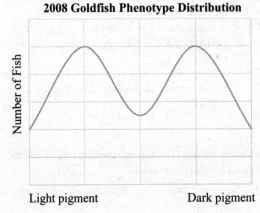

11. Which of the following best describes the fish population in the year 1958?

   (A) All medium-orange in color
   (B) Mostly medium-orange fish with some nearly white and some nearly brown
   (C) Mostly white fish and brown fish with a few orange fish
   (D) Equal numbers of orange fish, white fish, and brown fish

12. Which of the following theories is supported by the evidence?

   (A) The pigment trait in fish demonstrates incomplete dominance.
   (B) The pigment trait in fish demonstrates classical dominance.
   (C) The pigment trait in fish demonstrates codominance.
   (D) None of the above

13. If a dark-colored, poisonous fish and a bird that can see only light-colored fish are added to the ecosystem in 2008, what will the graph likely look like in 50 years?

   (A)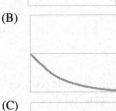

   (B)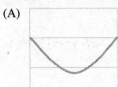

   (C)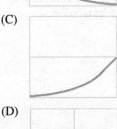

   (D)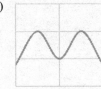

14. Which addition to the pond did NOT likely contribute to the change between 1958 and 2008?

   (A) A poisonous fish with a medium orange pigment
   (B) Runoff from fields that makes the water dark and murky
   (C) A light-orange water grass that grows in the pond
   (D) Predatory birds that can easily see medium-orange pigment

GO ON TO THE NEXT PAGE.

Questions 15–16 refer to the following figure.

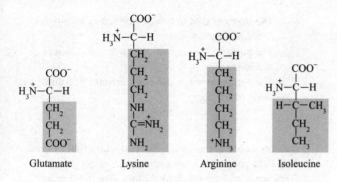

15. Which of the following amino acids would likely be found in the transmembrane domain?

(A) Lysine
(B) Arginine
(C) Isoleucine
(D) Glutamate

16. An enzyme's active site contains arginine residues. Which residues will likely be found on the corresponding substrate region?

(A) Arginine
(B) Lysine
(C) Glutamate
(D) Isoleucine

Questions 17–19 refer to the following passage.

The extracellular environment of the human body is typically abundant in sodium and calcium. In skeletal muscle cells, the sarcoplasmic reticulum is a specialized organelle that actively sequesters calcium from the cytosol. It stockpiles the calcium until a motor neuron triggers its release through an action potential, which opens voltage-gated calcium channels in the sarcoplasmic reticulum membrane. The calcium is necessary for the calcium-mediated functions of the protein troponin. In the presence of calcium, troponin removes tropomyosin from the myosin binding sites on actin filaments. This attachment is essential for sarcomere and muscle contraction.

17. Which of the following statements best describes the uptake of calcium by skeletal muscles and the sarcoplasmic reticulum?

(A) Calcium is passively taken up from the extracellular environment and the cytosol.
(B) Calcium is actively taken up from the extracellular environment and the cytosol.
(C) Calcium is passively taken up from the extracellular environment and actively taken up from the cytosol.
(D) Calcium is actively taken up from the extracellular environment and passively taken up from the cytosol.

18. Which statement best summarizes the role of calcium in skeletal muscle contraction?

(A) It prevents muscle contractions during action potentials.
(B) It changes the hypertonic nature of the cytosol to allow action potentials.
(C) It amplifies the physical contraction of the sarcomere.
(D) It connects the electrical neuronal signal to the actual physical contraction.

19. If a muscle cell fails to contract, which of the following could be a reason?

(A) The cell is lacking tropomyosin.
(B) Too much calcium is in the sarcoplasmic reticulum.
(C) The cell is lacking troponin.
(D) Too many action potentials are reaching the cell.

GO ON TO THE NEXT PAGE.

20. Dehydration synthesis is a key part of the creation of many macromolecules. It is best described as

    (A) loss of a water molecule in order to make something else
    (B) water rushing out of a cell during the process of osmosis
    (C) life moving out of the ocean and becoming complex
    (D) kidneys filtering hydrophilic compounds during urine formation

21. When humans experience excessive heat loss, the nervous system sends random signals for skeletal muscles to contract resulting in shivering. This then raises the body temperature due to the production of heat. This physiological mechanism is most similar to which other process?

    (A) When a B-cell receptor binds to an infectious agent, signaling occurs to stimulate a larger immune response.
    (B) When blood sugar levels are sensed to be high in the body, insulin is released and blood sugar in the bloodstream drops.
    (C) When the uterus is stretched during contractions, this signals to the brain that oxytocin should be released that causes stronger contractions.
    (D) As fruit begins to ripen it releases ethylene that causes other fruit to ripen as as well.

22. A molecule of ADP is dephosphorylated once and then phosphorylated twice. What molecule will result?

    (A) AMP
    (B) ADP
    (C) ATP
    (D) AUP

Questions 23–25 refer to the following figure.

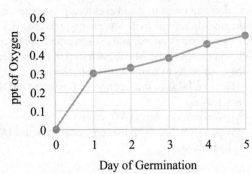

Oxygen Consumed vs. Day of Germination of Peas at 37°C

23. In the above figure, _____ is the dependent variable and _____ is the independent variable.

    (A) ppt of oxygen consumed; day of germination
    (B) ppt of oxygen consumed; 37°C
    (C) day of germination; ppt of oxygen consumed
    (D) 37°C; day of germination

24. In the above figure, which process is likely occurring?

    (A) Photosynthesis
    (B) Cellular respiration
    (C) Fermentation
    (D) All of the above

25. If the trend continues, what will be the oxygen consumed on day 7?

    (A) 0.55
    (B) 0.6
    (C) 0.75
    (D) 0.8

GO ON TO THE NEXT PAGE.

Questions 26–28 refer to the following passage.

The unit of contraction within skeletal muscle cells is called the sarcomere. A sarcomere contracts when the filamentous protein myosin stretches into a high-energy conformation and binds to the filamentous protein actin. When the myosin returns to its low-energy, relaxed conformation, actin is pulled, and the sarcomere contracts. The following steps relate ATP to each step of this process.

1—Myosin binds to actin (ADP is attached)

2—Myosin returns to low-energy conformation (ADP is released)

3—Myosin releases actin (ATP binds)

4—Myosin stretches to high-energy conformation (ATP is hydrolyzed)

26. What is bound to myosin when it is in its high-energy conformation?

   I. Actin

  II. ATP

 III. ADP

(A) II only
(B) III only
(C) I and II
(D) I and III

27. If the cell runs out of ATP, what would be the state of the sarcomere?

(A) Myosin is bound to actin in the high-energy conformation.
(B) Myosin is alone in the high-energy conformation.
(C) Myosin is bound to actin in the low-energy conformation.
(D) Myosin is alone in the low-energy conformation.

28. A calcium ion is required for the binding of myosin to actin. If a calcium chelator, such as EDTA, is added to a muscle cell, which of the following graphs shows how it will affect muscle contraction?

(A)
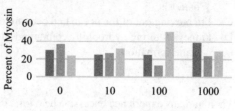

(B)
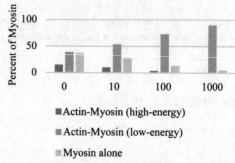

(C)
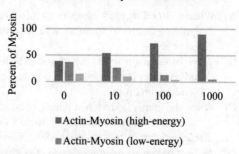

(D)
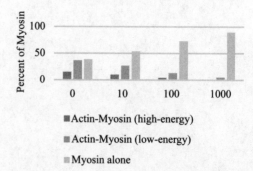

GO ON TO THE NEXT PAGE.

29. Which is NOT a true statement about acid rain?

    I. It increases the pH in the water and can harm aquatic plants.

    II. High levels of H+ can harm fish hatchlings.

    III. It can change the composition of the soil.

    (A) I only
    (B) I and II
    (C) II only
    (D) I and III

30. The Krebs cycle produces which of the following electron carriers?

    (A) NADPH
    (B) NADH
    (C) FADH
    (D) NAD+

31. Plant cells are well-known for having a structural dependence upon their large central vacuole. As this is a water-dependent structure, plants have developed many strategies for maintaining a state of hydration. One of these is a thick, waxy skin called a cuticle, which prevents water escaping from the plant surface. If a plant is misted with an enzyme designed to eat away the waxy cuticle, all of the following would be predicted outcomes EXCEPT

    (A) the plant would not stand up as tall
    (B) the plant would dry out more quickly
    (C) the plant would grow more roots
    (D) the plant would transport sugars more quickly

Questions 32–33 refer to the following passage.

Radiometric dating is a scientific technique based on predictable radioactive decay. The age of a rock or other substance that contains trace amounts of radioactive isotopes can be estimated by measuring how much of the original radioactive isotope is present and how much of the decayed version is present. Because the rate of decay occurs in an even, predictable manner, the original creation date of the rock can be estimated.

32. Two fossils found right next to each other are determined by radiometric dating to have similar levels of decayed isotope in their surrounding rock. Which of the following conclusions can be made?

    (A) The two life-forms had the same molecular DNA sequence.
    (B) The two life-forms were the same trophic level in an ecosystem.
    (C) The two life-forms were part of the same community.
    (D) The two life-forms are common ancestors to modern-day mammals.

33. Which of the following assumptions does radiometric dating NOT make?

    (A) The rock has not been in the presence of a strong magnetic field.
    (B) The rock formed at the same time that the radioactive isotope began decaying.
    (C) Neither the original isotope nor the decay product has escaped from the rock.
    (D) The rate of decay is predictable and has not greatly changed over time.

34. The membrane potential of cells is determined by the sodium-potassium pump, which

    (A) creates an intracellular space that is more negative than the extracellular space and possesses more potassium
    (B) creates an intracellular space that is more positive than the extracellular space and possesses more potassium
    (C) creates an intracellular space that is more negative than the extracellular space and possesses less potassium
    (D) creates an intracellular space that is more positive than the extracellular space and possesses less potassium

GO ON TO THE NEXT PAGE.

Questions 35–37 refer to the following passage.

Diabetes mellitus is a disease characterized by an inability of the cells to properly produce (type I) or respond (type II) to insulin, a hormone produced by the pancreas in response to high levels of blood glucose. Without insulin, glucose accumulates in the blood. In situations of low blood glucose, another pancreatic enzyme, glucagon, is released, which triggers the process of gluconeogenesis shown on the right side of the pathway below. The stimulators, activators, or inhibitors of each step are shown with + or – signs.

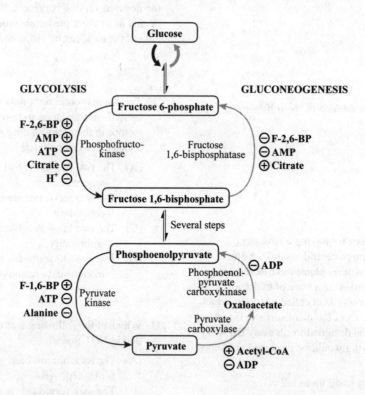

35. Which of the following conditions would lead to increased production of fructose 1,6-bisphosphate?

(A) High ATP and high citrate
(B) High AMP and high citrate
(C) High AMP and high F-2,6-BP
(D) High ATP and high F-2,6-BP

36. Patients with type I diabetes often require insulin injections. Which of the following situations would most require an insulin injection?

(A) After eating a stalk of celery
(B) After eating a cookie
(C) After skipping breakfast
(D) After drinking a lot of water

37. Which of the following situations likely stimulates gluconeogenesis?

(A) High levels of insulin
(B) High levels of F-2,6-BP
(C) High levels of glucagon
(D) High levels of ADP

GO ON TO THE NEXT PAGE.

<u>Questions 38–39</u> refer to the following passage.

Embryogenesis is a carefully timed and well-organized process. As a single-celled zygote divides and grows into hundreds and thousands of cells, a process called differentiation occurs wherein certain areas of the embryo become specialized to become different types of tissue. As differentiation continues, the level of specificity increases, and the cell potency decreases until highly specialized unique tissues and organs develop. The figure below shows 12 stages of development of human embryos.

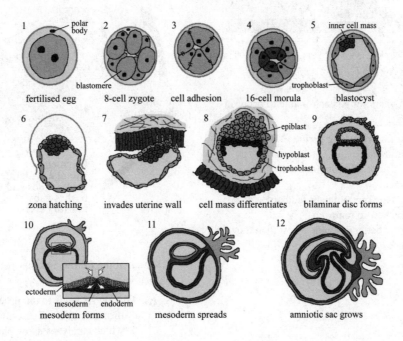

38. The inner cell mass is what eventually forms the embryo. During development, the embryo differentiates into various types of cell layers. Which of the following is NOT one of them?

    (A) Mesoderm
    (B) Hypoblast
    (C) Blastomere
    (D) Endoderm

39. A totipotent embryonic cell has the most cell potency. Which of the following is most likely to be totipotent?

    (A) 8-cell zygote
    (B) Inner cell mass
    (C) Mesoderm
    (D) Digestive tract

GO ON TO THE NEXT PAGE.

40. The following diagram demonstrates the ecological succession that occurs in an environment over time as it is colonized by different species.

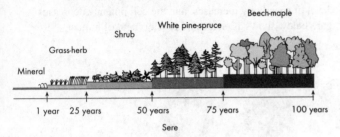

Why does it take 75 years for a beech-maple to occur in the figure above?

(A) Beech seeds have a very long period of dormancy prior to germination.

(B) It takes an average of 75 years for conifer trees to become extinct.

(C) Agriculture was the predominant industry, and hardwood trees were removed.

(D) Maple trees grow better in a pine forest than they do in a grassland.

41. During labor, pressure on the cervix and oxytocin form a positive feedback loop as shown below.

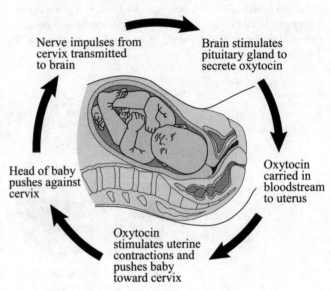

Which of the following other pathways also demonstrate positive feedback?

(A) Glycolysis leads to the production of ATP. ATP, in turn, turns off the enzyme phosphofructokinase, which catalyzes a key phosphorylation step in glycolysis.

(B) The anterior pituitary gland in the brain releases adrenocorticotropic hormone (ACTH). ACTH then causes the adrenal cortex to release glucocorticoids. Glucocorticoids then prevent the pituitary from releasing more ACTH.

(C) Lutenizing hormone triggers ovulation and the formation of the corpus luteum, which is a hormone-producing structure formed during ovulation. The corpus luteum secretes progesterone, which inhibits LH. The drop in LH causes the degradation of the corpus luteum.

(D) When a tissue is injured, it releases chemicals that activate platelets. Activated platelets themselves then release chemicals that activate more platelets. These activated platelets then release chemicals to activate more platelets.

GO ON TO THE NEXT PAGE.

Questions 42 and 43 refer to the following information.

A scientist studies the storage and distribution of oxygen in humans and Weddell seals to examine the physiological adaptations that permit seals to descend to great depths and stay submerged for extended periods. The figure below depicts the oxygen storage in both organisms.

Human (70 kilograms)

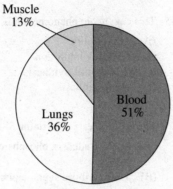

Total oxygen store: 1.95 liters

Weddell seal (450 kilograms)

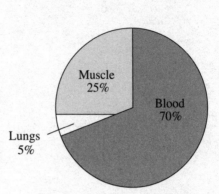

Total oxygen store: 25.9 liters

42. Compared with humans, approximately how many liters of oxygen does the Weddell seal store per kilogram of body weight?

(A) The same amount of oxygen
(B) Twice the amount of oxygen
(C) Three times the amount of oxygen
(D) Five times the amount of oxygen

43. During a dive, a Weddell seal's blood flow to the abdominal organs is shut off, and oxygen-rich blood is diverted to the eyes, brain, and spinal cord. Which of the following is the most likely reason for this adaptation?

(A) To increase the number of red blood cells in the nervous system
(B) To increase the amount of oxygen reaching the skeletomuscular system
(C) To increase the amount of oxygen reaching the central nervous system
(D) To increase the oxygen concentration in the lungs

GO ON TO THE NEXT PAGE.

Questions 44–46 refer to the following synthetic pathway of nRNA pyrimidine, cytidine 5′ triphosphate, CTP. This pathway begins with the condensation of two small molecules by the enzyme aspartate transcarbamylase (ATCase).

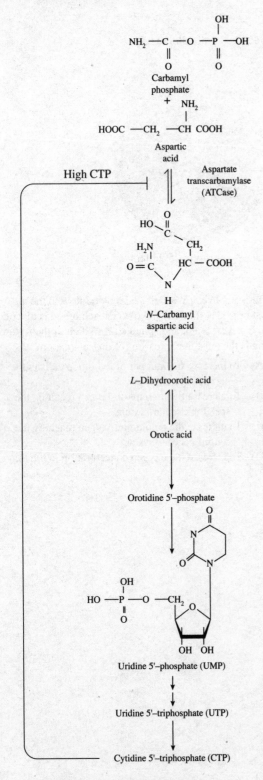

Regulation of CTP biosynthesis

44. Which of the following is true when the level of CTP is low in a cell?

(A) CTP is converted to ATCase.
(B) The metabolic traffic down the pathway increases.
(C) ATCase is inhibited, which slows down CTP synthesis.
(D) The final product of the pathway is reduced.

45. This enzymatic phenomenon is an example of

(A) transcription
(B) feedback inhibition
(C) dehydration synthesis
(D) photosynthesis

46. The biosynthesis of cytidine 5′-triphosphate requires

(A) a ribose sugar, a phosphate group, and a nitrogen base
(B) a deoxyribose sugar, a phosphate group, and a nitrogen base
(C) a ribose sugar, phosphate groups, and a nitrogen base
(D) a deoxyribose sugar, phosphate groups, and a nitrogen base

GO ON TO THE NEXT PAGE.

47. The following graph demonstrates 3 different strategies for survival.

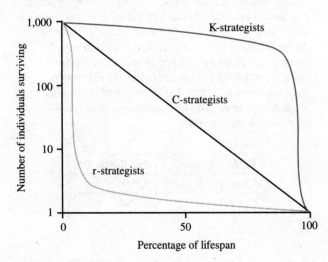

Which of the following are true statements according to the figure?

I. r-strategists are unlikely to die young.

II. The death rate for a C-strategist is constant.

III. Humans are an example of a K-stategist.

(A) I only
(B) II only
(C) II and III
(D) I, II, and III

Questions 48–49 refer to the following figure.

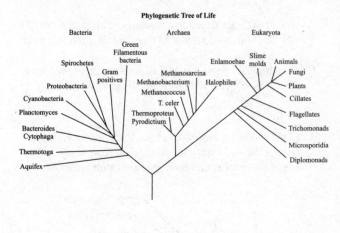

48. Which of the following are the most closely related?

(A) Aquifex and diplomonads
(B) Animals and fungi
(C) Halophiles and entamoebae
(D) There is not enough information given.

49. If a planctomyces is dividing by budding, which of the following will occur?

(A) A mitotic spindle will pull chromosome segments to opposite ends of the cell.
(B) Enzymes will unwind the helix and copy the entire bacterial genome.
(C) The nuclear envelope will break down and then reform after DNA replication.
(D) The mitochondria will replicate and be divided between the two cells.

GO ON TO THE NEXT PAGE.

50. Which of the following would make the Calvin Benson cycle unnecessary?

(A) If the light-dependent reactions made sugar and ATP
(B) If plants could make ATP in their electron transport chain
(C) If plants could use ATP to power cellular processes
(D) If NADPH could be created by photosystem I

51. In the following cladogram, a common ancestor (*) and species derived from it are illustrated. How many species have four or more common ancestors with Iguanodon?

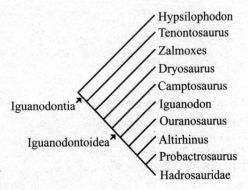

(A) 1
(B) 2
(C) 4
(D) 6

52. Which of the following does NOT affect higher order protein structure?

(A) Ester linkages between amino acids
(B) Amino acid sequence order
(C) Hydrogen bonding between R groups on the same polypeptide
(D) Hydrophobic interactions between side groups on different polypeptides

Questions 53–54 refer to the following passage.

The following study was carried out to examine sexual selection of beetles. Chemicals believed to be pheromones were isolated from males of certain species of butterfly. An environment was created in which a pheromone-containing droplet was applied to one side of the box and a control drop was applied to the other side. Female butterflies were introduced to the center of the box and given the opportunity to go to either side. The results are shown in the table below.

Preferred Side	# of Butterflies
Male chemical side	2,760
Control side	2,240

53. What is the best testable null hypothesis for the above experiment?

(A) Female butterflies prefer the scent of chemicals produced by males of their own species over the control chemicals.
(B) Female butterflies prefer the scent of the control chemicals over chemicals produced by males of their own species.
(C) Female butterflies cannot sense the chemicals produced by males of their own species.
(D) Female butterflies have no preference for either of the chemicals.

54. A chi-squared analysis would be performed in this experiment to make which of the following conclusions?

(A) To calculate that there is 1 degree of freedom
(B) To determine if the null hypothesis can be rejected
(C) To determine the standard deviation between two samples
(D) To prove that the working hypothesis is correct

GO ON TO THE NEXT PAGE.

Questions 55–56 refer to the following passage.

Two DNA sequences are shown below.

Sequence 1

5' GATTCCTACATCAG 3'
3' CTAAGGATGTAGTC 5'

Sequence 2

5' CGGCGAGACGCGGC 3'
3' GCCGCTCTGCGCCG 5'

55. If the mRNA sequence transcribed from one of the above sequences is 5' GAUUCCUACAUCAG 3', what is the sequence of the coding strand of DNA?

(A) 3' CTAAGGATGTAGTC 5'
(B) 5' GACTACATCCTTAG 3'
(C) 5' GATTCCTACATCAG 3'
(D) 3' CTGATGTAGGAATC 5'

56. Which of the following best describes the relationship between sequence 1 and sequence 2?

(A) Sequence 1 is less likely to be a coding sequence.
(B) Sequence 2 is more likely to degrade over time.
(C) Sequence 1 has more hydrogen bonds between base pairs.
(D) Sequence 2 has a higher melting temperature.

57. In a closed population, squirrel teeth can be either long or short. The long teeth allele is classically dominant. If there are 320 squirrels with long teeth and 40 squirrels with short teeth, what is the frequency of heterozygotes? Round to 2 decimal places.

(A) 0.11
(B) 0.33
(C) 0.44
(D) 0.67

58. Blood pressure is determined by the volume of the blood and the peripheral resistance of the blood vessels. Blood volume is dependent upon hydration level and osmotic pressure within the blood. Peripheral resistance refers to the volume of blood vessels, which is dependent upon constriction and dilation of arteries as blood flow is maximized and minimized in response to the body's needs.

Which of the following would BOTH help to raise blood pressure?

(A) Constriction of blood vessels and decreasing salt intake
(B) Dilation of blood vessels and decreasing salt intake
(C) Constriction of blood vessels and increasing salt intake
(D) Dilation of blood vessels and increasing salt intake

GO ON TO THE NEXT PAGE.

Questions 59–60 refer to the following passage.

Hydrogen peroxide (H_2O_2) is broken down into oxygen and water by the enzyme horseradish peroxidase (HRP). This reaction can be measured with an oxygen-sensitive color indicator, and the color change can be measured on a spectrophotometer. The rate of oxygen formation is measured with a constant HRP concentration and 5 different concentrations of H_2O_2 at room temperature. A follow-up experiment adds a known inhibitor of HRP (sodium azide). The results of both studies are shown in Figure 1.

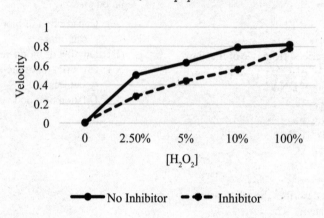

59 What is the substrate in this reaction?

(A) H_2O_2
(B) HRP
(C) Water
(D) Oxygen

60. Which of the following hypotheses about why the no-inhibitor graph plateaus after 10% H_2O_2 is supported?

(A) Water begins to act as an inhibitor.
(B) All the HRP is fully engaged.
(C) There is no more H_2O_2 available.
(D) The temperature is no longer optimal.

STOP

END OF SECTION I

IF YOU FINISH BEFORE TIME IS CALLED, YOU MAY CHECK YOUR WORK ON THIS SECTION. DO NOT GO ON TO SECTION II UNTIL YOU ARE TOLD TO DO SO.

THIS PAGE INTENTIONALLY LEFT BLANK

GO ON TO THE NEXT PAGE.

BIOLOGY
SECTION II
6 Questions
Writing Time—90 minutes

<u>Directions:</u> Questions 1 and 2 are long free-response questions that should require about 25 minutes each to answer and are worth 8–10 points each. Questions 3 through 6 are short free-response questions that should require about 10 minutes each to answer and are worth 4 points each.

Read each question carefully and completely. Write your response in the space provided following each question. Only material written in the space provided will be scored. Answers must be written out in paragraph form. Outlines, bulleted lists, or diagrams alone are not acceptable unless specifically requested.

1. A group of researchers wished to gain information about a type of bacteria that was known to actively uptake glucose across its cell membrane by use of a sodium-glucose cotransport mechanism whereby sodium and glucose enter the cell together. The researchers conducted an experiment in which bacterial cells with a relatively low intracellular sodium concentration were placed in glucose-rich media that had a relatively high sodium ion concentration. At regular intervals, the medium was analyzed for glucose and sodium concentrations.

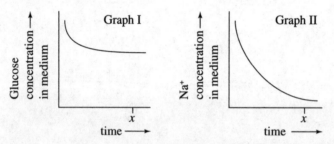

Figure 1. Glucose and Na+ concentrations in medium (no ATP in medium)

(a) **Describe** why a cell membrane is semipermeable. Include examples of molecules that can/cannot pass through.

(b) Bacterial cells contain a Na^+/K^+ ATPase pump that helps them maintain an osmotic balance with their surroundings. In the experimental cells, glycolysis was inhibited to prevent sodium from being pumped out of the cell. **Explain** how the inhibition of glycolysis likely achieves this.

(c) **Analyze** the data and **explain** how the levels of glucose and sodium inside the cell change throughout the experiment.

(d) **Predict** what would happen if the level of sodium inside the cell was not relatively low at the onset of the experiment.

GO ON TO THE NEXT PAGE.

2. Basking sharks, which can grow up to 10 m in length, have been recorded jumping out of the water as high and as fast as great white sharks. Marine biologists are unsure why they do this but have pointed to this phenomenon as evidence of how much we still have to learn about marine life. The sharks are speculated to jump out of the water only off the shores of Scotland, where they have been observed previously. A team of scientists wanted to determine if basking sharks in other areas of northern Europe can jump to similar heights. The following data were obtained by that team.

Location	Scotland	Ireland	Isle of Man	Cornwall
Jump Height Mean (m)	1.3175	1.1	0.2	1.5625
Confidence Intervals	0.058	0.125	0.091	0.024

(a) Basking sharks filter feed on plankton, tiny ocean organisms. **Describe** an adaptation and **explain** how that adaptation could increase a shark's fitness.

(b) **Construct** a graph plotting the mean jump height and confidence intervals represented by error bars.

(c) **Identify** which of the locations showed statistical differences in jump height compared to the Scotland sharks. **Explain** how you know this.

(d) Basking sharks are endotherms like most other sharks. **Predict** how the environment may differ between the Cornwall sampling site and the Isle of Man site. **Justify** your prediction.

GO ON TO THE NEXT PAGE.

3. Photosynthesis is the process plants use to derive energy from sunlight and is associated with a cell's chloroplasts. The energy is used to produce carbohydrates from carbon dioxide and water. Photosynthesis involves light and dark phases. Figure 1 represents two initial steps associated with the light phase.

The light phase supplies the dark phase with NADPH and a high-energy substrate.

A researcher attempted to produce a photosynthetic system outside the living organism according to the following protocols:

- Chloroplasts were extracted from green leaves and ruptured, and their membranes were thereby exposed, then a solution of hexachloroplatinate ions carrying a charge of –2 was added.
- The structure of the composite was analyzed, and the amount of oxygen produced by the system was measured.

The researcher concluded that the ions were bound to the membrane's Photosystem 1 site by the attraction of opposite charges. The resulting composite is shown in Figure 2. It was found that the hexachloroplatinate-membrane composite was photosynthetically active.

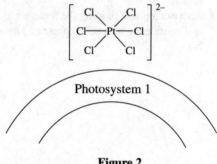

Figure 2

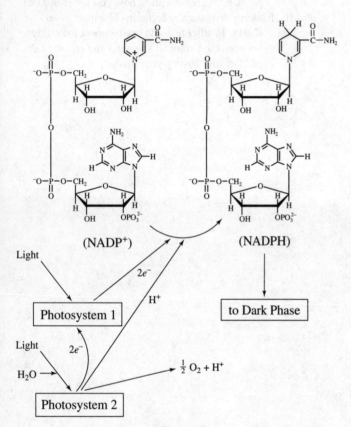

Figure 1

(a) **Describe** how photosynthesis is different from the energy-producing reactions of aerobic respiration.
(b) **Identify** the hypothesis the researcher was testing.
(c) **Predict** how the dark phases of photosynthesis were affected during this experiment.
(d) **Justify** your prediction.

GO ON TO THE NEXT PAGE.

4. A species of sunflower was studied in the 1980s by Professor Telly of Calicat Research Labs. Native to California, the *C. harriehazelet* sunflower blooms in late June and the blooms often draw photographers to the coast for the gorgeous fields of yellow flowers following the Sun. Professor Telly was interested in the preferences in soil moisture for the sunflower since California has often suffered extreme drought conditions spurring wildfires. The sunflower population at nine different moisture levels is shown in Figure 1.

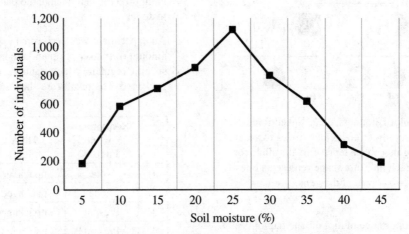

Figure 1. Sunflower populations vs. moisture levels 1987

After 40 years, Professor Telly decided to revisit the sunflower blooms and measure the soil moisture content at different sites again. It was discovered that sunflower species were primarily found at 2 tiers of moisture levels: 50–60% and 5–10%.

(a) Soil moisture and temperature are important for plant survival. **Explain** how plants react to environmental stimuli to maintain homeostasis.

(b) **Explain** how harsh environmental condition can change the phenotypes of a population over many generations.

(c) **Predict** what might have led to the change in water preference for the sunflower population.

(d) **Justify** your prediction.

GO ON TO THE NEXT PAGE.

5.

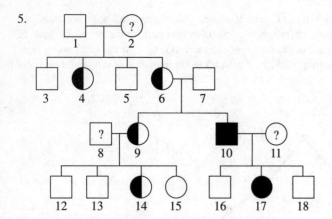

Family Tree for Color Blindness, an X-linked Recessive Disease: squares represent males and circles represent females. Shading in a shape represents the allele for color blindness. Each half of a shape represents one of the two copies of the gene.

(a) **Explain** the importance of diploid and haploid cells in sexual reproduction.

(b) **Explain** what the genotype of an individual with no shading would be. If you use abbreviations, be sure to indicate what they represent.

(c) Of the individuals with a "?", **identify** which of them is heterozygous.

(d) It is possible for a mother to pass on either of her two copies of a gene. **Explain** why this is helpful to prepare populations for selective pressures.

6. Regulation of genes prior to transcription is essential. The covalent modification of methylation is one such way that gene control occurs in eukaryotes, and it has been implicated in several diseases. DNA methylation turns off eukaryotic gene expression by physically blocking transcriptional proteins and by recruiting chromatin remodeling proteins that change the packing of DNA around histones.

An experiment was performed to assess the relative amount of transcription and the impact of a sequence-specific regulatory factor and the amount that DNA is packaged. The results are shown in Table 1.

	Sequence-Specific Factor	DNA	Relative Amount of Transcription
1.	None	unpackaged	0.74
2.	None	packaged	0.07
3.	Present	unpackaged	1.0
4.	Present	packaged	0.59

Table 1

(a) **Identify** the positive control in the data table.

(b) **Calculate** the percentage decrease in transcription with packaged DNA and no transcription factor.

(c) **Determine** which independent variable plays more of a role in regulating transcription.

(d) **Explain** how an epigenetic change can lead to a phenotype different from what is expected from the genotype.

STOP

END OF EXAM

Practice Test 4:
Answers and
Explanations

PRACTICE TEST 4 ANSWER KEY

1.	A	21.	B	41.	D
2.	D	22.	C	42.	B
3.	C	23.	A	43.	C
4.	D	24.	B	44.	B
5.	A	25.	B	45.	B
6.	B	26.	D	46.	C
7.	C	27.	C	47.	C
8.	B	28.	D	48.	B
9.	A	29.	A	49.	B
10.	B	30.	B	50.	A
11.	B	31.	D	51.	D
12.	D	32.	C	52.	A
13.	C	33.	A	53.	D
14.	A	34.	A	54.	B
15.	C	35.	C	55.	C
16.	C	36.	B	56.	D
17.	C	37.	C	57.	C
18.	D	38.	C	58.	C
19.	C	39.	A	59.	A
20.	A	40.	D	60.	B

PRACTICE TEST 4 EXPLANATIONS

Section I: Multiple Choice

1. **A**   Since Klinefelter's is a disease in males, this means the person will have a Y chromosome. Because it involves the creation of a Barr body, this also means that the person will have an extra X chromosome. The answers show karyotypes, which display a person's chromosomes. The 23rd pair represents the sex chromosomes. Choice (A) is the only one with a Y and two X chromosomes.

2. **D**   If BAMHI cuts between the guanines in the sequences 5′ GGATCC 3′, then each answer choice should be checked for that sequence. Choice (A) has one cut site, resulting in two chunks. Choice (B) has three cut sites, resulting in four chunks. Choice (D) has two sites, resulting in three chunks. Choice (C) has zero cut sites because the cut sequence is reversed from the correct 5′ → 3′ direction.

3. **C**   Green pea plants can result from either a homozygous dominant (*GG*) or a heterozygous (*Gg*) phenotype, so your mystery pea plant could be either. You need to figure out which it is. If you cross it with any green plant (A), you won't know what genotype you are crossing it with. If you cross it with a homozygous dominant plant (B), you will always get green plants no matter what your original mystery plant was. If you cross it with a yellow plant, then you will know it is a homozygous recessive. If your original plant is a homozygous dominant, you will get only green plants. If your original plant is a heterozygote, then you should get yellow plants 50% of the time.

4. **D**   The passage says the parental plants crossed are true breeding, so each is homozygous. That cross should result in only green plants in the F_1 generation. Crossing-over (A) will not give variation because the parental plants are homozygous. Their two copies are identical and crossing-over would just swap identical genes. The law of independent assortment (B) explains mixing of maternal and paternal versions of the different chromosomes. The two homozygous dominant pea color alleles are on the two copies of the same chromosome. A true breeding plant would also be unlikely to have gene alleles on other chromosomes that impact the color. Choice (D) is correct because mutation can occur at any time and result in changes to phenotype.

5. **A**   A yellow plant must be homozygous recessive (*gg*). All plants in the F_1 generation are heterozygotes (*Gg*) because the parents are a homozygous dominant (*GG*) and a homozygous recessive (*gg*). As shown in the Punnett square below, the ratio of green:yellow would be 1:1.

	G	*g*
g	*Gg* (green)	*gg* (yellow)
g	*Gg* (green)	*gg* (yellow)

6. **B** The passage does not say that the container would stay the same; in fact, that was the independent variable. Decreasing the number of females would not mean that the container could not be changed for a new experiment. The graph shows that both samples have a reproduction rate decrease after 10–20 days, so not (C). The highest density condition (same number of *Daphnia* but lower volume) shows no reproductive activity after trauma (so not (D)). Choice (B) is the correct answer because we see that as the container size increased, the number of offspring increased.

7. **C** Hydrophobic things, such as carbon dioxide, don't have gap junctions since they are not restricted by lipid bilayer membranes. They will simply diffuse from high concentrations to low concentrations. Hydrophilic things need the gap junctions, but they will never diffuse through them up their concentration gradient.

8. **B** A regular somatic cell after mitosis is just a regular somatic cell. The passage says 32, compared with 46 in humans. This means that there should be 32 unreplicated chromosomes. After replication, there would be 64 segments since each chromosome would have a sister chromatid.

9. **A** The passage says that humans have 46 chromosomes total. Humans are diploid and have two copies of each of the non-homologous chromosomes. Thus, humans have 23 unique, non-homologous chromosomes. If the species has 32 chromosomes and they are octoploidy, then they have 8 copies of each. So, they have only 4 unique chromosomes.

10. **B** The passage says that two gametes come together to make an octoploidy (8n) species. Therefore, each gamete should be tetraploidy (4n). If they have four non-homologous chromosomes, then $4 \times 4 = 16$.

11. **B** The peak in 1958 is in the middle, corresponding to medium-orange fish. However, there are also some fish on each extreme (light or dark).

12. **D** These graphs of phenotype tell us nothing about the mode of inheritance.

13. **C** If a dark-colored, poisonous fish existed, predators would soon learn to avoid dark-colored fish. The dark-colored fish would have an advantage. In the current situation, the dark-colored and the light-colored have an advantage. This would give the dark fish more advantage. Plus, the arrival of a bird that eats light-colored fish would really make the dark fish have the advantage.

14. **A** From 1958 to 2008, the graph shifted away from medium-orange and moved toward the extremes (light and dark). Choice (A) would make fewer orange fish be eaten, which would promote more medium-orange fish. Choices (B) and (C) would cause fewer light and dark fish to be eaten, respectively. This would promote the light and dark fish. Choice (D) would cause the medium-orange fish to be eaten more often and thus promote the light and dark extremes shown in 2008.

15. **C** Transmembrane domains are in the hydrophobic space and must not be polar. Of the amino acids shown above the question, glutamate is negatively charged and lysine and arginine are positively charged. Isoleucine (C) is the only non-charged choice.

16. **C** Enzyme active sites are the places where the substrate binds. If the active site contains positively charged residues, the substrate should have a negative charge because the active sites and substrate would attract each other.

17. **C** If calcium is prevalent outside the cell, then it will passively flow into the cell (as long as a calcium channel is open) because it wants to move down its concentration gradient. The passage states that the sarcoplasmic reticulum actively takes up calcium.

18. **D** Calcium links the action potential to contraction of the sarcomere (D). It stimulates contraction; it does not prevent it (A). It is not the effect on hypertonicity that causes the action potentials (B); action potentials are calcium specific. Calcium just relays the action potential message; it does not amplify the signal (C).

19. **C** Without troponin, the calcium could not move tropomyosin off the myosin binding sites, and then there would be no contraction. If there were no tropomyosin, then the contraction could occur at any time, regardless of calcium. An excess of calcium in the sarcoplasmic reticulum (B) or action potentials (D) would also lead to contraction, not a lack thereof.

20. **A** Dehydration is loss of water. Synthesis is building something. This is a common type of reaction during the polymerization of macromolecules.

21. **B** Human shivering is the result of negative feedback since a stimulus caused something to occur that reversed/stopped the original situation; the body temperature went up when there was a stimulus of a low body temperature. Choice (B) is another example of negative feedback as blood sugar is lowered when the level gets too high. All other choices are examples of positive feedback as they lead to a stimulus getting continued and amplified.

22. **C** ADP has two phosphates. If it is dephosphorylated, then it would be AMP, which has one phosphate. If it gets two added, then it would be ADP and then ATP.

23. **A** The x-axis is typically the independent variable, which is something set by the experimenter. The dependent variable is the thing that is measured as data. So germination is the independent variable, and ppt of oxygen is the dependent variable. The temperature is a constant in this instance as every point on the graph is meant to show germination at that given temperature. Often, temperature will be the independent variable in an experiment.

24. **B** The graph shows that oxygen consumed starts at 0 and then increases. In photosynthesis, oxygen is released. In fermentation, there is no oxygen present. This must be cellular respiration, which requires oxygen.

25. **B** The trend over the 4 points that seem to fall on a line averages 0.1 every 2 days. This would increase it to 0.6 on day 7.

26. **D** In its high-energy conformation, myosin is bound to ADP. ATP is hydrolyzed as it enters the high-energy state, so it should only have ADP bound once it gets there. It will also bind actin.

27. **C** Without a fresh ATP to bind, myosin will be bound to actin in the low-energy conformation and be unable to release it.

28. **D** Without calcium, myosin cannot bind to actin. With more EDTA, there should be more myosin alone, as shown in (D). The states bound to actin will disappear.

29. **A** Acid rain has high H^+ and thus a low pH. Therefore, it would NOT increase the water pH (I). However, these high H^+ levels can harm fish (II) and change soil composition (III).

30. **B** The Krebs cycle reduces NAD^+ to NADH. Therefore, NADH is produced. It also produces the electron carrier $FADH_2$.

31. **D** The waxy cuticle helps prevent water loss. Water is important in the plant for maintaining a turgid structure. The plant would not likely stand as tall (A), it would dry out rapidly (B), and it would grow more roots to seek out more water (C). This would not likely cause increased sugar transport (D).

32. **C** If the fossils had the same levels of decayed isotope, then they were likely living at the same time and if they were found near each other, they were likely in the same place. Therefore, they would be in the same community. They are not necessarily common ancestors to mammals (D). Maybe they are plant fossils. They might be predator and prey in different trophic levels (B). They have the same DNA only if they are identical twins or asexual reproducers (A).

33. **A** Radiometric dating assumes that the isotope started decaying when the rock was formed (B). It also assumes the full amount of the isotope or the decay product is there, so scientists can compare the levels of each (C). The measurement of decay is assumed to be predictable, which is why it can be used like a stopwatch (D). The presence of a magnetic field is not relevant, and anything on Earth is always in the presence of a strong magnetic field.

34. **A** The sodium-potassium pump brings two potassium into the cell and brings three sodium out of the cell. This makes the inside more negative. The pump also moves things against the concentration gradient, which means there is a high level of potassium in the cell and the pump makes it even more plentiful.

35. **C** ATP inhibits its production since it inhibits phosphofructokinase. This eliminates choices (A) and (D). Citrate stimulates turning fructose 1,6-bisphosphate into fructose 6-phosphate, which eliminates (B). Choice (C) includes two things that stimulate phosphofructokinase and lead to production of fructose 1,6-bisphosphate.

36. **B** Insulin is needed when blood glucose gets too high. Eating a cookie would increase blood sugar more than eating celery, fasting, or drinking water.

37. **C** Gluconeogenesis is the formation of glucose. If the body had high insulin, then it would already have lots of glucose, so (A) is eliminated, but if the body had high glucagon, it would need glucose (C). High ADP would inhibit gluconeogenesis as shown in the figure (D). High F-2,6-BP would also inhibit gluconeogenesis (B).

38. **C** The blastomere is shown prior to the inner cell mass; therefore, it cannot be a differentiation of the inner cell mass. The hypoblast comes from the inner cell mass, and the endoderm and mesoderm also come from it.

39. **A** Potency is the potential to become many things. The most potent cells are the least differentiated or the least specialized. The more specialized they become, the more they lose their potency. The 8-cell zygote is the least specialized and has the most potency.

40. **D** Few species can grow in barren areas, but once they are colonized with plants, then more species can grow and live there. It takes a long time for the beech and maple trees to grow because first species need to change the environment, allowing other species to colonize. The environment was not hospitable until it was a pine forest, which took a long time to reach as well.

41. **D** Positive feedback occurs when a process creates an end product whose production stimulates the process to create even more of the end product. In Choice (D), the platelets lead to more platelet creation. Choices (A), (B), and (C) illustrate negative feedback.

42. **B** The Weddell seal stores twice as much oxygen as humans. Calculate the liters per kilograms weight for both the seal and man using the information at the bottom of the chart. The Weddell seal stores 0.058 liters/kilograms (25.9 liters/450 kilograms) compared to 0.028 liters/kilograms (1.95 liters/70 kilograms) in humans.

43. **C** The most plausible answer is that blood is redirected toward the central nervous system, which permits the seal to navigate for long durations. Choice (A) is incorrect; the seal does not need to increase the number of red blood cells in the nervous system. Choice (B) can also be eliminated, as the seal does not need to increase the amount of oxygen to the skeletal system. Eliminate (D) because the diversion of blood does not increase the concentration of oxygen in the lungs.

44. **B** When the level of CTP is low in a cell, the metabolic traffic down the pathway increases. This pathway is controlled by feedback inhibition. The final product of the pathway inhibits the activity of the first enzyme. When the supply of CTP is low, the pathway will continue to produce CTP.

45. **B** This enzymatic phenomenon is an example of feedback inhibition. Feedback inhibition is the metabolic regulation in which high levels of an enzymatic pathway's final product inhibit the activity of its rate-limiting enzyme. Transcription, (A), is the production of RNA from DNA. Dehydration synthesis, (C), is the formation of a covalent bond by the removal of water. In photosynthesis, (D), radiant energy is converted to chemical energy.

46. **C** The biosynthesis of cytidine 5'-triphosphate requires a nitrogenous base, three phosphates, and a sugar ribose. Pyrimidines are a class of nitrogenous bases with a single ring structure. The sugar they contain is ribose, which is shown in the pathway diagram.

47. **C** Most r-strategists die early in their lifetime (A), but most K-strategists live long lives and then die when old. Humans are an example of K-strategists. The C-strategist survivorship line declines evenly, meaning that they are just as likely to die when old as they are when young (C).

48. **B** Animals and fungi might seem different, but in the grand scheme of life, they are not so distant at all. They branched away from each other most recently on the phylogenetic tree.

49. **B** Planctomyces are on the bacterial branch of the tree. Bacteria are prokaryotes and do not have segments of chromosomes because they have one circular chromosome. They do not need a mitotic spindle during division (A). They also do not have a nucleus (C) or mitochondria (D).

50. **A** The Calvin cycle converts the ATP and NADPH into sugar. If the light-dependent reactions made sugar and ATP, then the Calvin cycle would be unnecessary. Plants already do make ATP from the gradient of their electron transport chain (B). They also use ATP in their cellular processes (C). NADPH is made by photosystem I, and it still needs the Calvin cycle.

51. **D**   Iguanodon had one common ancestor with Hypsilophodon before Hypsilophodon diverged. Iguanodon had two with Tenontosaurus because one came from where Tenontosaurus diverged and they both share the common ancestor with Hypsilophodon. By that logic, Iguanodon has three common ancestors with Zalmoxes, four with Dryosaurus, and five with Camptosaurus. It will also share those five common ancestors with the other species on the tree, because everything above Camptosaurus still shares those original five. The higher levels also share the ancestor that existed when Iguanodon diverged. Either way, six species share at least four common ancestors with Iguanodon. Remember, don't count Iguanodon itself.

52. **A**   Higher order protein structure depends on the original amino acid sequence (B), the interactions between the side chains of the amino acids (C), and the interactions between other polypeptides (D). Ester linkages are not the linkages between proteins.

53. **D**   The null hypothesis would be that butterflies show no preference for either of the sides and the presence of the pheromone has no effect. The working hypothesis is that the females prefer the pheromones. If females cannot sense the chemicals, that would explain a possible reason the null hypothesis turns out to be true, but that is not the null hypothesis.

54. **B**   Chi-squared analysis can only prove something is unlikely; it cannot prove something is correct. There is 1 degree of freedom, but this is a fact and is not the conclusion of a chi-squared analysis. Standard deviation is a different statistical test.

55. **C**   The coding strand should always be the same as the mRNA except for the thymine in DNA and the uracil in RNA. mRNA is the complement of the template strand of DNA, not the coding strand.

56. **D**   Although the previous question showed that part of sequence 1 is a coding sequence, this does not mean it is more likely to be one than sequence 2. The two sequences differ by the composition of their sequences. Sequence 2 has more guanines and cytosines. The G-C bond has 3 hydrogen bonds between it rather than the 2 hydrogen bonds that the A-C bond has. This gives us a higher melting temperature because it is more difficult to break the strands apart.

57. **C**   There are 40/360 (0.11) homozygous recessives = the frequency of homozygous recessive = q^2, and then $q = 0.33$ and if $p + q = 1$, then $p = 0.67$. The frequency of heterozygotes is $2pq = 2 (0.33)(0.67) = 0.44$.

58. **C**   Blood pressure is increased when the space gets smaller (constriction of blood vessels) or when the volume gets bigger. When the blood gets too salty, it takes on water to dilute the salt and the volume increases. So, constricting vessels and increasing salt intake will increase blood pressure.

59. **A**   The substrate is H_2O_2 because that is what the HRP enzyme acts upon. Water and oxygen are the products.

60. **B**   The substrate increases from 10% to 100%, so we know the substrate is still available. However, it is the enzyme that remained constant. The temperature remains constant. There is also no evidence that the water product has become an inhibitor.

Section II: Free Response

Short student-style responses have been provided for each of the questions. These samples indicate an answer that would get full credit, so if you're checking your own response, make sure that the actual answers to each part of the question are similar to your own. The structure surrounding them is less important, although we've modeled it as a way to help organize your own thoughts and to make sure that you actually respond to the entire question.

Note that the rubrics used for scoring periodically change based on the College Board's analysis of the previous year's test takers. This is especially true as of the most recent Fall 2019 changes to the AP Biology exam! We've done our best to approximate their structure, based on our institutional knowledge of how past exams have been scored and on the information released by the test makers. However, the 2020 exam's free-response questions will be the first of their kind.

Our advice is to over-prepare. Find a comfortable structure that works for you, and really make sure that you're providing all of the details required for each question. Also, continue to check the College Board's website, as they may release additional information as the test approaches. For some additional help, especially if you're worried that you're not being objective in scoring your own work, ask a teacher or classmate to help you out. Good luck!

Question 1

a. **Describe** why a cell membrane is semipermeable. Include examples of molecules that can/cannot pass through. (2 points)

   Cells are semipermeable because certain things can pass through them and some cannot. The membrane is made of a phospholipid bilayer. This means hydrophobic things like oxygen can pass through and hydrophilic things like water and ions cannot pass through without assistance, such as a channel.

b. Bacterial cells contain a Na^+/K^+ ATPase pump that helps them maintain an osmotic balance with their surroundings. In the experimental cells, glycolysis was inhibited to prevent sodium from being pumped out of the cell. **Explain** how inhibition of glycolysis likely achieves this. (3 points)

   The sodium-potassium pump is an example of active transport. It pumps sodium ions out of the cell. This experiment was measuring the amount of sodium in the media and the pump might alter the results. To keep it turned off, the cell needs to lack the energy to run the pump. By preventing glycolysis, the cell is kept deficient in ATP and the pump will remain off. Glycolysis is part of cell respiration and it begins with glucose and turns it into pyruvate.

c. **Analyze** the data and **explain** how the levels of glucose and sodium inside the cell change throughout the experiment. (2 points)

   The glucose appears to disappear from the media and so it must be entering the cell. At first it enters the cell quickly and then slows down. The sodium enters the cell quickly, and keeps entering at the quick rate for longer than the glucose. The media is almost depleted of sodium so the sodium must be very high inside the cell.

d. **Predict** what would happen if the level of sodium inside the cell was not relatively low at the onset of the experiment. (2 points)

   If the sodium was not low, then there would not be a pull for the sodium to enter the cell. Sodium is entering due to its concentration gradient. If the sodium was high in the cell, then the sodium would probably remain in the media.

Question 2

a. Basking sharks filter feed on plankton, tiny ocean organisms. **Describe** (1 point) an adaptation and **explain** (1 point) how that adaptation could increase a shark's fitness.

A behavioral adaptation that basking sharks have is slow movement with their mouth open. By moving this way it would allow the shark to gather as much food as possible while using as little energy as possible. This would enable the shark to maximize growth, longevity, and mating opportunities which would increase its fitness.

(Any shark adaptation like fins, mouth, or tail that is linked to fitness would earn two points.)

b. **Construct** a graph plotting the mean jump height and confidence intervals represented by error bars.

    1 point—proper axes
    1 point—choice of a bar chart or histogram
    1 point—proper size of bars showing mean
    1 point—graphing confidence intervals as error bars

Figure 1. Breach heights observed at Northern Europe Locations

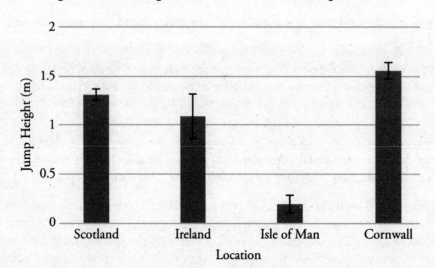

c. **Identify** (1 point) which of the locations showed statistical differences in jump height compared to the Scotland sharks. **Explain** (1 point) how you know this.

Isle of Man and Cornwall are statistically different in jump height compared to Scotland. I can see this due to the non-overlapping error bars.

d. Basking sharks are endotherms like most other sharks. **Predict** (1 point) how the environment may differ between the Cornwall sampling site and the Isle of Man site. **Justify** (1 point) your prediction.

As basking sharks are endothermic, this means that the temperature of their environment determines their metabolism. Therefore higher temperatures would mean increased activity. Since Cornwall sharks show an increase in jump height compared to Isle of Man we could predict that the Cornwall site had a higher temperature compared to the Isle of Man site.

AP Biology Prep

Question 3

a. Describe how photosynthesis is different from the energy-producing reactions of aerobic respiration.

Photosynthesis takes sunlight and water and carbon dioxide and produces sugar and oxygen. Cell respiration requires sugar and oxygen to make ATP and carbon dioxide and water. Photosynthesis gathers energy from sunlight. Cell respiration makes energy as ATP. The end products of photosynthesis get used as reactants in aerobic respiration.

b. Identify the hypothesis the researcher was testing.

The researcher's hypothesis was that a hexachloroplatinate composite made from chloroplasts and hexachloroplatinate ions would be photosynthetic.

c. Predict how the dark phases of photosynthesis were affected during this experiment.

The dark phases would be active.

d. Justify your prediction.

They would be active because the light phase supplies the NADPH energy for the dark phases and if the light phases were active as measured by oxygen production, then the NADPH must have been made too so as long as there is carbon dioxide, then the dark phases should be active.

Question 4

a. Soil moisture and temperature are important for plant survival. **Explain** how plants react to environmental stimuli to maintain homeostasis.

Plants can maintain homeostasis by changing their behavior in response to environmental stimuli. Sunflowers can turn to follow the sun as it moves across the sky. This allows them to get the maximum amount of sunlight throughout the day. It can also help them regulate temperature. Plants can also regulate their water by changing their behavior. Plants have large vacuoles that they store water in. If they are not getting much rain, then they can take out the water from their vacuoles to use it. Plants also have roots that they can grow and stretch in different directions. They also have pores on their surface called stomata that can open and shut to let water out or keep it in.

b. Explain how harsh environmental condition can change the phenotypes of a population over many generations.

A harsh condition could cause some individuals to die or to be less likely to reproduce. If they do not reproduce, then they don't contribute their genes and traits to the next generation. Over time, if this happens many times, the genes of the individuals that are less "fit" will become less and less and the phenotypes of the population will change.

c. **Predict** what might have led to the change in water preference for the sunflower population.

Something must have happened that caused the middle amount of water to be bad for the plants. Plants that liked to live in the middle range of moisture were not reproducing as well. This might have been because there was a fungus or an insect that also liked that moisture level and so the plants that lived in that level of moisture all died. The ones that could live in the extremes of dry or moist soil survived to make up the next generation.

d. **Justify** your prediction.

The population used to have a bell-curve with most of the individuals liking medium moisture and then a few liking the extremes. After 40 years it changed and the extremes were favored. This is called disruptive selection. Something happened that made the extreme amounts of soil moisture better to live at than the medium level of moisture.

Question 5

a. **Explain** the importance of diploid and haploid cells in sexual reproduction.

Diploid cells have two copies of every gene. Haploid cells have only one copy of every gene. In sexual reproduction two haploid gametes must come together and create diploid offspring. This is better than asexual reproduction because it creates genetic diversity.

b. **Explain** what the genotype of an individual with no shading would be. If you use abbreviations, be sure to indicate what they represent.

An individual without any shading would have two copies of the normal non-colorblind allele.

c. Of the individuals with a "?", **identify** which of them is heterozygous.

Individual 11 must be heterozygous. She must have passed on a colorblind copy of an X-chromosome since her daughter is color-blind and must have received a color-blind allele from each of her parents. She is not homozygous recessive since she has sons that received an X from her and they are not affected so the X must have been normal. So she must be a heterozygote.

Individual 2 must also be heterozygous since some of the daughters have a copy, but they could not have received one from the father since he is unaffected, so they must have received the copy from the mother. Again, she can't be a homozygous recessive since she gave her sons a good normal X.

d. It is possible for a mother to pass on either of her two copies of a gene. **Explain** why this is helpful to prepare populations for selective pressures.

It is helpful because it creates genetic diversity among the offspring, and more diverse populations withstand selective pressures more because there are more chances of having an individual with a phenotype that can survive the selective pressure.

Question 6

a. **Identify** the positive control in the data table.

The positive control is the one that expects full transcription, so the 3rd sample with the factor present and the unpackaged DNA is the positive control. The others are all relative to that full amount of transcription.

b. **Calculate** the percentage decrease in transcription with packaged DNA and no transcription factor.

The decrease would be 93%.

c. **Determine** which independent variable plays more of a role in regulating transcription.

The packaging status of DNA plays a larger role in regulating transcription. The samples with packaged DNA were always worse, even if the factor was present.

d. **Explain** how an epigenetic change can lead to a phenotype different from what is expected from the genotype.

An epigenetic change can essentially turn off a gene. So, it is as though that allele does not even exist, even though it is there. Based on the genotype a certain allele might be expected, but if the transcription machinery cannot access it then it will never be expressed.

The Princeton Review®

Completely darken bubbles with a No. 2 pencil. If you make a mistake, be sure to erase mark completely. Erase all stray marks.

1. YOUR NAME:
(Print) Last   First   M.I.

SIGNATURE: _____   DATE: ___ / ___ / ___

HOME ADDRESS: _____
(Print)   Number and Street
_____
City   State   Zip Code

PHONE NO. : _____
(Print)

IMPORTANT: Please fill in these boxes exactly as shown on the back cover of your test book.

2. TEST FORM

3. TEST CODE

4. REGISTRATION NUMBER

0	A	0	0	0	0	0	0	0	0	0
1	B	1	1	1	1	1	1	1	1	1
2	C	2	2	2	2	2	2	2	2	2
3	D	3	3	3	3	3	3	3	3	3
4	E	4	4	4	4	4	4	4	4	4
5	F	5	5	5	5	5	5	5	5	5
6	G	6	6	6	6	6	6	6	6	6
7		7	7	7	7	7	7	7	7	7
8		8	8	8	8	8	8	8	8	8
9		9	9	9	9	9	9	9	9	9

6. DATE OF BIRTH

Month	Day		Year	
○ JAN				
○ FEB				
○ MAR	0	0	0	0
○ APR	1	1	1	1
○ MAY	2	2	2	2
○ JUN	3	3	3	3
○ JUL		4	4	4
○ AUG		5	5	5
○ SEP		6	6	6
○ OCT		7	7	7
○ NOV		8	8	8
○ DEC		9	9	9

7. SEX
○ MALE
○ FEMALE

The Princeton Review®

5. YOUR NAME

First 4 letters of last name				FIRST INIT	MID INIT
A	A	A	A	A	A
B	B	B	B	B	B
C	C	C	C	C	C
D	D	D	D	D	D
E	E	E	E	E	E
F	F	F	F	F	F
G	G	G	G	G	G
H	H	H	H	H	H
I	I	I	I	I	I
J	J	J	J	J	J
K	K	K	K	K	K
L	L	L	L	L	L
M	M	M	M	M	M
N	N	N	N	N	N
O	O	O	O	O	O
P	P	P	P	P	P
Q	Q	Q	Q	Q	Q
R	R	R	R	R	R
S	S	S	S	S	S
T	T	T	T	T	T
U	U	U	U	U	U
V	V	V	V	V	V
W	W	W	W	W	W
X	X	X	X	X	X
Y	Y	Y	Y	Y	Y
Z	Z	Z	Z	Z	Z

1 A B C D
2 A B C D
3 A B C D
4 A B C D
5 A B C D
6 A B C D
7 A B C D
8 A B C D
9 A B C D
10 A B C D
11 A B C D
12 A B C D
13 A B C D
14 A B C D
15 A B C D
16 A B C D

17 A B C D
18 A B C D
19 A B C D
20 A B C D
21 A B C D
22 A B C D
23 A B C D
24 A B C D
25 A B C D
26 A B C D
27 A B C D
28 A B C D
29 A B C D
30 A B C D
31 A B C D
32 A B C D

33 A B C D
34 A B C D
35 A B C D
36 A B C D
37 A B C D
38 A B C D
39 A B C D
40 A B C D
41 A B C D
42 A B C D
43 A B C D
44 A B C D
45 A B C D
46 A B C D
47 A B C D
48 A B C D

49 A B C D
50 A B C D
51 A B C D
52 A B C D
53 A B C D
54 A B C D
55 A B C D
56 A B C D
57 A B C D
58 A B C D
59 A B C D
60 A B C D

Completely darken bubbles with a No. 2 pencil. If you make a mistake, be sure to erase mark completely. Erase all stray marks.

1. YOUR NAME:
(Print) Last First M.I.

SIGNATURE: _____ DATE: ____ / ____ / ____

HOME ADDRESS: _____
(Print) Number and Street

_____
City State Zip Code

PHONE NO. : _____
(Print)

IMPORTANT: Please fill in these boxes exactly as shown on the back cover of your test book.

2. TEST FORM

3. TEST CODE

4. REGISTRATION NUMBER

5. YOUR NAME

First 4 letters of last name | FIRST INIT | MID INIT

6. DATE OF BIRTH

Month	Day	Year
JAN		
FEB		
MAR		
APR		
MAY		
JUN		
JUL		
AUG		
SEP		
OCT		
NOV		
DEC		

7. SEX
MALE
FEMALE

The Princeton Review®

1 Ⓐ Ⓑ Ⓒ Ⓓ
2 Ⓐ Ⓑ Ⓒ Ⓓ
3 Ⓐ Ⓑ Ⓒ Ⓓ
4 Ⓐ Ⓑ Ⓒ Ⓓ
5 Ⓐ Ⓑ Ⓒ Ⓓ
6 Ⓐ Ⓑ Ⓒ Ⓓ
7 Ⓐ Ⓑ Ⓒ Ⓓ
8 Ⓐ Ⓑ Ⓒ Ⓓ
9 Ⓐ Ⓑ Ⓒ Ⓓ
10 Ⓐ Ⓑ Ⓒ Ⓓ
11 Ⓐ Ⓑ Ⓒ Ⓓ
12 Ⓐ Ⓑ Ⓒ Ⓓ
13 Ⓐ Ⓑ Ⓒ Ⓓ
14 Ⓐ Ⓑ Ⓒ Ⓓ
15 Ⓐ Ⓑ Ⓒ Ⓓ
16 Ⓐ Ⓑ Ⓒ Ⓓ

17 Ⓐ Ⓑ Ⓒ Ⓓ
18 Ⓐ Ⓑ Ⓒ Ⓓ
19 Ⓐ Ⓑ Ⓒ Ⓓ
20 Ⓐ Ⓑ Ⓒ Ⓓ
21 Ⓐ Ⓑ Ⓒ Ⓓ
22 Ⓐ Ⓑ Ⓒ Ⓓ
23 Ⓐ Ⓑ Ⓒ Ⓓ
24 Ⓐ Ⓑ Ⓒ Ⓓ
25 Ⓐ Ⓑ Ⓒ Ⓓ
26 Ⓐ Ⓑ Ⓒ Ⓓ
27 Ⓐ Ⓑ Ⓒ Ⓓ
28 Ⓐ Ⓑ Ⓒ Ⓓ
29 Ⓐ Ⓑ Ⓒ Ⓓ
30 Ⓐ Ⓑ Ⓒ Ⓓ
31 Ⓐ Ⓑ Ⓒ Ⓓ
32 Ⓐ Ⓑ Ⓒ Ⓓ

33 Ⓐ Ⓑ Ⓒ Ⓓ
34 Ⓐ Ⓑ Ⓒ Ⓓ
35 Ⓐ Ⓑ Ⓒ Ⓓ
36 Ⓐ Ⓑ Ⓒ Ⓓ
37 Ⓐ Ⓑ Ⓒ Ⓓ
38 Ⓐ Ⓑ Ⓒ Ⓓ
39 Ⓐ Ⓑ Ⓒ Ⓓ
40 Ⓐ Ⓑ Ⓒ Ⓓ
41 Ⓐ Ⓑ Ⓒ Ⓓ
42 Ⓐ Ⓑ Ⓒ Ⓓ
43 Ⓐ Ⓑ Ⓒ Ⓓ
44 Ⓐ Ⓑ Ⓒ Ⓓ
45 Ⓐ Ⓑ Ⓒ Ⓓ
46 Ⓐ Ⓑ Ⓒ Ⓓ
47 Ⓐ Ⓑ Ⓒ Ⓓ
48 Ⓐ Ⓑ Ⓒ Ⓓ

49 Ⓐ Ⓑ Ⓒ Ⓓ
50 Ⓐ Ⓑ Ⓒ Ⓓ
51 Ⓐ Ⓑ Ⓒ Ⓓ
52 Ⓐ Ⓑ Ⓒ Ⓓ
53 Ⓐ Ⓑ Ⓒ Ⓓ
54 Ⓐ Ⓑ Ⓒ Ⓓ
55 Ⓐ Ⓑ Ⓒ Ⓓ
56 Ⓐ Ⓑ Ⓒ Ⓓ
57 Ⓐ Ⓑ Ⓒ Ⓓ
58 Ⓐ Ⓑ Ⓒ Ⓓ
59 Ⓐ Ⓑ Ⓒ Ⓓ
60 Ⓐ Ⓑ Ⓒ Ⓓ

Completely darken bubbles with a No. 2 pencil. If you make a mistake, be sure to erase mark completely. Erase all stray marks.

1. YOUR NAME:
(Print)   Last   First   M.I.

SIGNATURE: _____ DATE: ___/___/___

HOME ADDRESS: _____
(Print)   Number and Street

_____
City   State   Zip Code

PHONE NO. : _____
(Print)

IMPORTANT: Please fill in these boxes exactly as shown on the back cover of your test book.

5. YOUR NAME

First 4 letters of last name					FIRST INIT	MID INIT
Ⓐ	Ⓐ	Ⓐ	Ⓐ		Ⓐ	Ⓐ
Ⓑ	Ⓑ	Ⓑ	Ⓑ		Ⓑ	Ⓑ
Ⓒ	Ⓒ	Ⓒ	Ⓒ		Ⓒ	Ⓒ
Ⓓ	Ⓓ	Ⓓ	Ⓓ		Ⓓ	Ⓓ
Ⓔ	Ⓔ	Ⓔ	Ⓔ		Ⓔ	Ⓔ
Ⓕ	Ⓕ	Ⓕ	Ⓕ		Ⓕ	Ⓕ
Ⓖ	Ⓖ	Ⓖ	Ⓖ		Ⓖ	Ⓖ
Ⓗ	Ⓗ	Ⓗ	Ⓗ		Ⓗ	Ⓗ
Ⓘ	Ⓘ	Ⓘ	Ⓘ		Ⓘ	Ⓘ
Ⓙ	Ⓙ	Ⓙ	Ⓙ		Ⓙ	Ⓙ
Ⓚ	Ⓚ	Ⓚ	Ⓚ		Ⓚ	Ⓚ
Ⓛ	Ⓛ	Ⓛ	Ⓛ		Ⓛ	Ⓛ
Ⓜ	Ⓜ	Ⓜ	Ⓜ		Ⓜ	Ⓜ
Ⓝ	Ⓝ	Ⓝ	Ⓝ		Ⓝ	Ⓝ
Ⓞ	Ⓞ	Ⓞ	Ⓞ		Ⓞ	Ⓞ
Ⓟ	Ⓟ	Ⓟ	Ⓟ		Ⓟ	Ⓟ
Ⓠ	Ⓠ	Ⓠ	Ⓠ		Ⓠ	Ⓠ
Ⓡ	Ⓡ	Ⓡ	Ⓡ		Ⓡ	Ⓡ
Ⓢ	Ⓢ	Ⓢ	Ⓢ		Ⓢ	Ⓢ
Ⓣ	Ⓣ	Ⓣ	Ⓣ		Ⓣ	Ⓣ
Ⓤ	Ⓤ	Ⓤ	Ⓤ		Ⓤ	Ⓤ
Ⓥ	Ⓥ	Ⓥ	Ⓥ		Ⓥ	Ⓥ
Ⓦ	Ⓦ	Ⓦ	Ⓦ		Ⓦ	Ⓦ
Ⓧ	Ⓧ	Ⓧ	Ⓧ		Ⓧ	Ⓧ
Ⓨ	Ⓨ	Ⓨ	Ⓨ		Ⓨ	Ⓨ
Ⓩ	Ⓩ	Ⓩ	Ⓩ		Ⓩ	Ⓩ

2. TEST FORM
_____

3. TEST CODE

	A	0	0	0	0	0	0	0	0	0
0	A	0	0	0	0	0	0	0	0	0
1	B	1	1	1	1	1	1	1	1	1
2	C	2	2	2	2	2	2	2	2	2
3	D	3	3	3	3	3	3	3	3	3
4	E	4	4	4	4	4	4	4	4	4
5	F	5	5	5	5	5	5	5	5	5
6	G	6	6	6	6	6	6	6	6	6
7		7	7	7	7	7	7	7	7	7
8		8	8	8	8	8	8	8	8	8
9		9	9	9	9	9	9	9	9	9

4. REGISTRATION NUMBER

6. DATE OF BIRTH

Month	Day		Year	
◯ JAN				
◯ FEB				
◯ MAR	0	0	0	0
◯ APR	1	1	1	1
◯ MAY	2	2	2	2
◯ JUN	3	3	3	3
◯ JUL		4	4	4
◯ AUG		5	5	5
◯ SEP		6	6	6
◯ OCT		7	7	7
◯ NOV		8	8	8
◯ DEC		9	9	9

7. SEX
◯ MALE
◯ FEMALE

The Princeton Review®

1 Ⓐ Ⓑ Ⓒ Ⓓ
2 Ⓐ Ⓑ Ⓒ Ⓓ
3 Ⓐ Ⓑ Ⓒ Ⓓ
4 Ⓐ Ⓑ Ⓒ Ⓓ
5 Ⓐ Ⓑ Ⓒ Ⓓ
6 Ⓐ Ⓑ Ⓒ Ⓓ
7 Ⓐ Ⓑ Ⓒ Ⓓ
8 Ⓐ Ⓑ Ⓒ Ⓓ
9 Ⓐ Ⓑ Ⓒ Ⓓ
10 Ⓐ Ⓑ Ⓒ Ⓓ
11 Ⓐ Ⓑ Ⓒ Ⓓ
12 Ⓐ Ⓑ Ⓒ Ⓓ
13 Ⓐ Ⓑ Ⓒ Ⓓ
14 Ⓐ Ⓑ Ⓒ Ⓓ
15 Ⓐ Ⓑ Ⓒ Ⓓ
16 Ⓐ Ⓑ Ⓒ Ⓓ

17 Ⓐ Ⓑ Ⓒ Ⓓ
18 Ⓐ Ⓑ Ⓒ Ⓓ
19 Ⓐ Ⓑ Ⓒ Ⓓ
20 Ⓐ Ⓑ Ⓒ Ⓓ
21 Ⓐ Ⓑ Ⓒ Ⓓ
22 Ⓐ Ⓑ Ⓒ Ⓓ
23 Ⓐ Ⓑ Ⓒ Ⓓ
24 Ⓐ Ⓑ Ⓒ Ⓓ
25 Ⓐ Ⓑ Ⓒ Ⓓ
26 Ⓐ Ⓑ Ⓒ Ⓓ
27 Ⓐ Ⓑ Ⓒ Ⓓ
28 Ⓐ Ⓑ Ⓒ Ⓓ
29 Ⓐ Ⓑ Ⓒ Ⓓ
30 Ⓐ Ⓑ Ⓒ Ⓓ
31 Ⓐ Ⓑ Ⓒ Ⓓ
32 Ⓐ Ⓑ Ⓒ Ⓓ

33 Ⓐ Ⓑ Ⓒ Ⓓ
34 Ⓐ Ⓑ Ⓒ Ⓓ
35 Ⓐ Ⓑ Ⓒ Ⓓ
36 Ⓐ Ⓑ Ⓒ Ⓓ
37 Ⓐ Ⓑ Ⓒ Ⓓ
38 Ⓐ Ⓑ Ⓒ Ⓓ
39 Ⓐ Ⓑ Ⓒ Ⓓ
40 Ⓐ Ⓑ Ⓒ Ⓓ
41 Ⓐ Ⓑ Ⓒ Ⓓ
42 Ⓐ Ⓑ Ⓒ Ⓓ
43 Ⓐ Ⓑ Ⓒ Ⓓ
44 Ⓐ Ⓑ Ⓒ Ⓓ
45 Ⓐ Ⓑ Ⓒ Ⓓ
46 Ⓐ Ⓑ Ⓒ Ⓓ
47 Ⓐ Ⓑ Ⓒ Ⓓ
48 Ⓐ Ⓑ Ⓒ Ⓓ

49 Ⓐ Ⓑ Ⓒ Ⓓ
50 Ⓐ Ⓑ Ⓒ Ⓓ
51 Ⓐ Ⓑ Ⓒ Ⓓ
52 Ⓐ Ⓑ Ⓒ Ⓓ
53 Ⓐ Ⓑ Ⓒ Ⓓ
54 Ⓐ Ⓑ Ⓒ Ⓓ
55 Ⓐ Ⓑ Ⓒ Ⓓ
56 Ⓐ Ⓑ Ⓒ Ⓓ
57 Ⓐ Ⓑ Ⓒ Ⓓ
58 Ⓐ Ⓑ Ⓒ Ⓓ
59 Ⓐ Ⓑ Ⓒ Ⓓ
60 Ⓐ Ⓑ Ⓒ Ⓓ

NOTES

NOTES

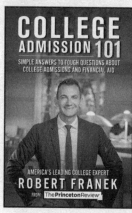